T0092028

Digital Twin Driven Service

Digital Twin Driven Service

Edited by

Fei Tao

School of Automation Science and Electrical Engineering,
Beihang University, Beijing, China

Qinglin Qi

School of Automation Science and Electrical Engineering,
Beihang University, Beijing, China;

Department of Industrial and Systems Engineering,
The Hong Kong Polytechnic University, Hong Kong, China

A.Y.C. Nee

Department of Mechanical Engineering,
National University of Singapore, Singapore

Academic Press is an imprint of Elsevier
125 London Wall, London EC2Y 5AS, United Kingdom
525 B Street, Suite 1650, San Diego, CA 92101, United States
50 Hampshire Street, 5th Floor, Cambridge, MA 02139, United States
The Boulevard, Langford Lane, Kidlington, Oxford OX5 1GB, United Kingdom

Library of Congress Cataloging-in-Publication Data
A catalog record for this book is available from the Library of Congress

British Library Cataloguing-in-Publication Data
A catalogue record for this book is available from the British Library

ISBN: 978-0-323-91300-3

For information on all Academic Press publications visit our website at https://www.elsevier.com/books-and-journals

Publisher: Matthew Deans
Acquisitions Editor: Brian Guerin
Editorial Project Manager: Franchezca A. Cabural
Production Project Manager: Prem Kumar Kaliamoorthi
Cover Designer: Christian Bilbow

Typeset by TNQ Technologies

Contents

Contributors

Jie Fan
School of Mechanical Automation, Wuhan University of Science and Technology, Wuhan, Hubei, China

Hongwei Fan
School of Mechanical Engineering, Xi'an University of Science and Technology, Xi'an, Shaanxi, China; Shaanxi Key Laboratory of Mine Electromechanical Equipment Intelligent Monitoring, Xi'an, Shaanxi, China

Ke He
School of Mechanical Engineering, Shanghai Jiao Tong University, Shanghai, China

Kendrik Yan Hong Lim
School of Mechanical and Aerospace Engineering, Nanyang Technological University, Singapore, Singapore; Advanced Remanufacturing and Technology Centre (ARTC), Agency for Science, Technology and Research (A*STAR), Singapore, Singapore

Ruiqi Hu
School of Mechanical Engineering, Shanghai Jiao Tong University, Shanghai, China

Mingxuan Hu
School of Mechanical Engineering, Shanghai Jiao Tong University, Shanghai, China

Zuguang Huang
Chinese Machine Quality Supervision and Inspection Center, Beijing, China

Zhenrui Ji
School of Information Engineering, Wuhan University of Technology, Wuhan, Hubei, China

Shiqian Ke
School of Mechanical Automation, Wuhan University of Science and Technology, Wuhan, Hubei, China

Hao Li
College of Mechanical and Electrical Engineering, Zhengzhou University of Light Industry, Zhengzhou, Henan, China

Yilin Li
School of Safety and Ocean Engineering, China University of Petroleum, Beijing, China

Dar Win Liew
School of Mechanical and Aerospace Engineering, Nanyang Technological University, Singapore, Singapore

Xinyuan Lv
School of Mechanical Engineering, Xi'an University of Science and Technology, Xi'an, Shaanxi, China

Yufei Ma
School of Mechanical Engineering, Shanghai Jiao Tong University, Shanghai, China

Yongli Ma
School of Information Engineering, Wuhan University of Technology, Wuhan, Hubei, China

Yanping Ma
School of Information Engineering, Wuhan University of Technology, Wuhan, Hubei, China

A.Y.C. Nee
Department of Mechanical Engineering, National University of Singapore, Singapore, Singapore

Qinglin Qi
School of Automation Science and Electrical Engineering, Beihang University, Beijing, China; Department of Industrial and Systems Engineering, The Hong Kong Polytechnic University, Hong Kong, China

Qianzhe Qiao
School of Safety and Ocean Engineering, China University of Petroleum, Beijing, China

Huibin Sun
School of Mechanical Engineering, Northwestern Polytechnical University, Xi'an, Shaanxi, China

Diyin Tang
School of Automation Science and Electrical Engineering, Beihang University, Beijing, China

Fei Tao
School of Automation Science and Electrical Engineering, Beihang University, Beijing, China

Jinjiang Wang
School of Safety and Ocean Engineering, China University of Petroleum, Beijing, China

Haoqi Wang
College of Mechanical and Electrical Engineering, Zhengzhou University of Light Industry, Zhengzhou, Henan, China

Yan Wang
School of Mechanical Engineering, Xi'an University of Science and Technology, Xi'an, Shaanxi, China; Shaanxi Key Laboratory of Mine Electromechanical Equipment Intelligent Monitoring, Xi'an, Shaanxi, China

Feng Xiang
School of Mechanical Automation, Wuhan University of Science and Technology, Wuhan, Hubei, China

Wenjun Xu
School of Information Engineering, Wuhan University of Technology, Wuhan, Hubei, China

Yuanpu Yao
School of Mechanical Engineering, Northwestern Polytechnical University, Xi'an, Shaanxi, China

Jinsong Yu
School of Automation Science and Electrical Engineering, Beihang University, Beijing, China

Xuhui Zhang
School of Mechanical Engineering, Xi'an University of Science and Technology, Xi'an, Shaanxi, China; Shaanxi Key Laboratory of Mine Electromechanical Equipment Intelligent Monitoring, Xi'an, Shaanxi, China

Zhinan Zhang
State Key Laboratory of Mechanical System and Vibration, Shanghai Jiao Tong University, Shanghai, China; School of Mechanical Engineering, Shanghai Jiao Tong University, Shanghai, China

Pai Zheng
Department of Industrial and Systems Engineering, The Hong Kong Polytechnic University, Hung Hom, Hong Kong, China

Zude Zhou
School of Information Engineering, Wuhan University of Technology, Wuhan, Hubei, China

Ying Zuo
Research Institute for Frontier Science, Beihang University, Beijing, China

Preface

The advance, maturity, and application innovation of new-generation information technologies (e.g., Internet of Things, cloud computing, big data analytics, etc.) greatly enhance the overall service capacities. Moreover, based on more refined and dynamic models as well as richer and more sourced data drives, digital twins have played an important role in product lifecycle process, supply chain optimization, predictive operation and maintenance, optimization of processes and control, etc. The combination of services and digital twin would radically revolutionize industry. From service to digital twin service, services could optimize the entire business processes and operation procedure to achieve a new higher level of productivity. Digital twin services enable companies to provide more valuable services to customers and form innovations in application services. Thereby, this book aims to introduce digital twin (DT), as a new approach to support services innovation.

In this book the authors propose to employ digital twin to reinforce the service theories, methods, and tools on a wide spectrum of service applications such as condition monitoring and forecasting, fault diagnosis, remaining useful life prediction, maintenance, test, prognostics and health management, energy-efficient assessment, optimization, decision-making, hands-on training, etc. The applicability of digital twin-driven service lies in its unique capability of high-fidelity modeling, integrative virtual and real data analytics, cyber-physical interconnection and interaction, etc. The authors hope that this book will contribute to the pervasive research and application of digital twin service in the future.

This book has 10 chapters, which are classified into two parts. Part 1 includes Chapters 1—4, which present relevant theories and methodologies of digital twin-driven service. Chapter 1, From Service to Digital Twin Service, analyzed the development and connotation of service, product service, manufacturing service, and then combined digital twin and service to propose digital twin service and discussed the from four perspectives. Chapter 2, Digital Twin-Driven Service Collaboration, applied digital twin to the process of manufacturing service collaboration to realize the optimal allocation of resources and efficient collaboration through the interaction and integration of the physical world and the information world. In Chapter 3, Digital Twin-driven Production Line Custom Design Service, presented the framework of digital twin-driven custom design service system, and introduced the connotation, basic components, key technologies, theories and the development methods of digital twin-driven production line custom design in detail. Chapter 4, Digital twin-enhanced product family design and optimization service, proposed a generic digital twin-enhanced approach to support the in-context virtual prototyping and reconfiguration/redesign of the complex product family with lifecycle considerations.

Part 2 includes Chapters 5–10, which focus on practical applications of digital twin-driven service. Chapter 5, Digital Twin-Driven Fault Diagnosis Service of Rotating Machinery, presents a digital twin-driven fault diagnosis method and demonstrates its application for rotating machinery fault diagnosis. Chapter 6, Digital Twin-Driven Energy-Efficient Assessment Service, presents a digital twin-driven energy-efficient assessment method for quantifying sustainable manufacturing capabilities, provides solution for encapsulating services and sharing assessment functions, and develops a digital twin-driven assessment service platform and the application for industrial robots. Chapter 7, Digital Twin-Driven Cutting Tool Service, introduces the digital twin-driven cutting tool service for cutting tool wear condition monitoring, tool condition forecast, remaining useful life prediction, cutting tool selection decision-making, etc., and develops a prototype to illustrate and validate it. Chapter 8, Digital Twin-Driven Prognostics and Health Management, showcases the applications of digital twin-driven prognostics and health management in practical engineering problems through three typical cases of the aerospace on-orbit control moment gyro, lithium-ion battery, and infrared imaging system. Chapter 9, Production Process Management for Intelligent Coal Mining Based on Digital Twin, constructs the digital twin application framework and technical system of coal mining face, exemplified by the typical application of digital twin-driven PHM for Shearer, digital twin-driven virtual-physical interaction, digital twin-driven MR-aided maintenance and hands-on training. Finally, Chapter 10, Digital Twin-Enhanced Tribo-test Service, proposes a new digital twin-enhanced tribo-test service framework and develops a remote service webpage, exemplified by the triboelectric test system.

The authors would like to acknowledge the invaluable support and suggestions from many collaborators, based on their research in digital twin service.

Many thanks for the strong support and cooperative work from Profs. Feng Xiang, Hao Li, Pai Zheng, Jinjiang Wang, Zuguang Huang, Wenjun Xu, Zude Zhou, Huibin Sun, Jinsong Yu, Diyin Tang, Xuhui Zhang, Zhinan Zhang, as well as Jie Fan, Shiqian Ke, Ying Zuo, Haoqi Wang, Kendrik Yan Hong Lim, Dar Win Liew, Yilin Li, Qianzhe Qiao, Zhenrui Ji, Yongli Ma, Yanping Ma, Yuanpu Yao, Xinyuan Lv, Yan Wang, Hongwei Fan, Yufei Ma, Mingxuan Hu, Ke He, and Ruiqi Hu to accomplish this book together.

In particular, the authors would like to express their gratitude for the invaluable contributions from the members in Digital Twin Research Group at Beihang University: He Zhang, Weiran Liu, Xin Ma, Jiangfeng Cheng, Meng Zhang, etc. They actively engaged in the research program on digital twin at Beihang University together with Prof. Fei Tao from 2015.

Thanks to all the participants who attended the Conference on Digital Twin and Smart Manufacturing Service from 2017 to 2021, who kindly helped promote the research and application of digital twin.

Some contents were previously published in the *Journal of Manufacturing Systems, International Journal of Production Research, Journal of Cleaner Production, IEEE Transactions on Systems, Man, and Cybernetics: Systems, IEEE Transactions on Reliability, IEEE Sensors Journal, Robotics and Computer-integrated Manufacturing, Microelectronics Reliability, International Journal of Advanced Manufacturing Technology, Computer-Integrated Manufacturing Systems (in Chinese), China Mechanical Engineering (in Chinese), Manufacturing Technology & Machine Tool (in Chinese), Journal of Shanghai Jiaotong University (in Chinese), Industry and Mine Automation (in Chinese), and Procedia CIRP, as well as IEEE 15th International Conference on Networking, Sensing and Control (ICNSC), IFIP International Conference on Advances in Production Management Systems, Asian Simulation Conference, etc.* Thanks go to all the anonymous reviewers who had provided many constructive comments and suggestions.

Some contents of this book were financially supported by the following research projects in China: Key Project of International (Regional) Cooperation and Exchange Programs of National Natural Science Foundation of China (No. 52120105008), The National Key Research and Development Program of China (No. 2020YFB1708400), National Natural Science Foundation of China (No. 52005024), and The Hong Kong Polytechnic University (No. G-YZ3N).

The authors are most grateful to Mr. Brian Guerin, the Senior Acquisitions Editor from Elsevier, who took the initiative to encourage them to publish this book. The authors are equally grateful to the anonymous reviewers who had provided constructive feedback on the book proposal. The strong support from Ms. Chezca Cabural as well as other colleagues from Elsevier is greatly appreciated.

Fei Tao, Qinglin Qi and A.Y.C. Nee
September 26, 2021

CHAPTER 1

From service to digital twin service

Qinglin Qi[1,2], Fei Tao[1] and A.Y.C. Nee[3]
[1]School of Automation Science and Electrical Engineering, Beihang University, Beijing, China; [2]Department of Industrial and Systems Engineering, The Hong Kong Polytechnic University, Hong Kong, China; [3]Department of Mechanical Engineering, National University of Singapore, Singapore, Singapore

1.1 Introduction

Currently, digitalization has become a consensus, especially when digital twin is revolutionizing the industry [1]. The advances in informatization of society, the rapid improvement of mobile networks, the network access of numerous mobile terminals and smart sensing devices, as well as the richness of application scenarios, have led to more people, things, scenes, etc. being digitized [2]. Digitalization provides new tools and new perspectives to understand things, which has promoted the arrival of the era of services. The advances, maturity, and application innovation of new generation of information technologies (New IT, such as Internet of things [IoT], cloud computing, big data analytics, and artificial intelligence [AI], etc.) have been applied to various aspects, such as software and hardware services, information acquisition, smart decision-making, etc., thus greatly enhancing the overall service capacity [3].

Taking the manufacturing industry as an example, pure product manufacturing or assembly is at the bottom of the "smile curve" [4], which added value is getting lower due to the advancement of science and technology. In the era of digital innovation, the transaction method, form, and performance of products are redefined. The value creation process is shifted from traditional company-centered product systems to customer-centric product and service systems [5]. In the face of much pressure such as emerging markets, intense global competition, and increasing labor costs, manufacturing companies have generally begun to adopt the logic of service [6]. By outsourcing non-core business, manufacturing companies can focus on their core business and collaborate with each other. By providing service-oriented production and productive services to each other, manufacturing companies provide customers with products and applications in an efficient, flexible, agile, and low-cost production method [7]. As an

Digital Twin Driven Service
ISBN 978-0-323-91300-3
https://doi.org/10.1016/B978-0-323-91300-3.00006-1

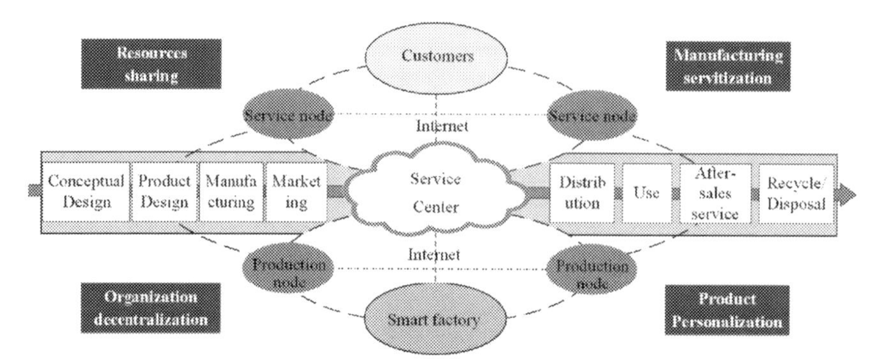

Figure 1.1 Servitization of manufacturing industry around the product lifecycle [9].

important source of value, service has become a dominant trend in the manufacturing industry. As shown in Fig. 1.1, more manufacturing companies are integrating value-added services into all links of the product lifecycle. The role of services in the manufacturing value chain is growing. The boundary between manufacturing and service industries is increasingly blurred and gradually merging with each other [8]. The status of service factors in the production and operation activities of manufacturing enterprises is constantly rising.

As shown in Fig. 1.2, the transformation of manufacturing enterprises to service-oriented manufacturing has the following stages. In the very beginning, pure product sales were still the main activity and the source of the greatest profit [10]. Services around products are only limited to after-sales services, such as fault handling, comprehensive maintenance, etc. Subsequently, manufacturing enterprises actively explored the potential needs of customers to provide them with value-added services along with the value chain extension. The value-added services include storage, logistics, installation, training, maintenance, etc. As the manufacturing of products becomes more complex, it is difficult for an enterprise to complete all the production activities independently [11]. And manufacturing enterprise also needs a variety of support activities such as research and development (R&D), finance, procurement, etc. Therefore, some companies outsource non-core activities and focus only on core links to reduce costs and improve competitiveness [12]. Servicalization of functional activities has evolved into service businesses that provide customers in markets, such as information services, financial services, specific manufacturing services, etc. Moreover, due to differences in product types, processes, and equipment resources, manufacturing companies may face situations of

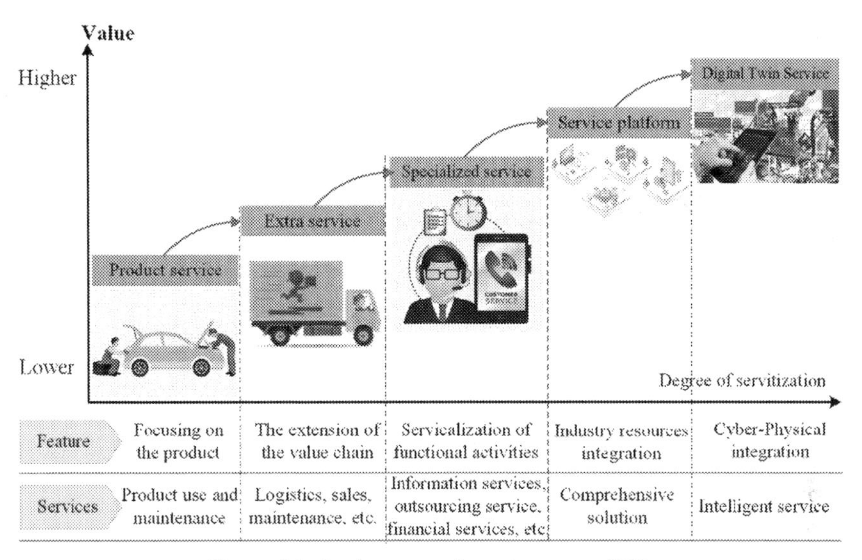

Figure 1.2 Evolvement of service scope [15].

insufficient capacities or idle resources. With the Internet and the IoTs, the service platform provides an effective way to improve the utilization efficiency of diversified and distributed manufacturing resources [13]. Service platform generally provides comprehensive solutions based on the specific needs of target customers. The solutions usually include a combination of certain core products and multiple services. The products and services are from many providers. Manufacturing companies with others jointly provide services to customers through establishing collaboration based on service scenarios [14].

At present, digital twin is revolutionizing industry [1]. Based on more refined and dynamic models as well as richer and more sourced data drives, digital twins have played an important role in operations monitoring, simulation, optimization, predictive maintenance, etc. [15] The cores of digital twin, data and model, are playing an increasingly important position in the manufacturing industry [16]. Digital twin captures data from the physical world through various sensing technologies and then analyzes and applies these data to help companies optimize production processes and improve operational efficiency. The digital twin model supports full life-cycle visualization of physical objects, tracking changes, and performing analysis to optimize production performance. Besides, driven by data and models, companies can continuously perceive customer needs and create new service models with digital twin. Therefore, digital twin not only can cope with fierce competition and obtain higher income, but also integrate complex manufacturing processes with services to achieve closed-loop optimization of product design, manufacturing, and services.

In view of the concept of Everything-as-a-Service (XaaS), services could fully unleash the potential of digital twin. Through services, digital twin as a service can be shared and used in a convenient "pay-as-you-go" manner, especially virtual models which are not easy to be created rapidly [17]. Moreover, digital twin-driven services have been widely used, for example, digital twin-driven monitoring, simulation, evaluation, prediction, optimization, control, etc. The combination of smart manufacturing services and digital twin would radically change product design, manufacturing, usage, maintenance, repair and operations (MRO), and other processes.

From the above analysis, the age of digital twin service has arrived. In this chapter, digital twin and service are combined, the following respectively introduce in detail the connotation of service, product service, manufacturing service, and digital twin service.

1.2 The development and connotation of service

The essence of human society is the division of labor and cooperation [19]. With the evolution of human beings and the continuous development of human society, the modes of division of labor and cooperation are gradually diversified, and the degree of division of labor and cooperation is also gradually deepened. Services arise from the division of labor and collaboration. As shown in Fig. 1.3, from the analysis of the development stages of human history, the development of services can be divided into the following four stages:

(1) Handicraft age: Since the emergence of division of labor in human society, service has been created. Prior to the first industrial revolution, human society had been in the manual manufacturing stage for a long time [18]. Artifacts were predominantly designed and manufactured by artisans, whose activities were of low complexity [18,20]. Therefore, the resources and specialization level of traditional service are limited, with the characteristics of low marketization degree and monopoly. For example, artisans produce commodities to serve others and provide maintenance services. Merchants serve the exchange of goods between producers and consumers. The learned man provides educational services. The government provides basic public administration services, etc. These services are very primitive.

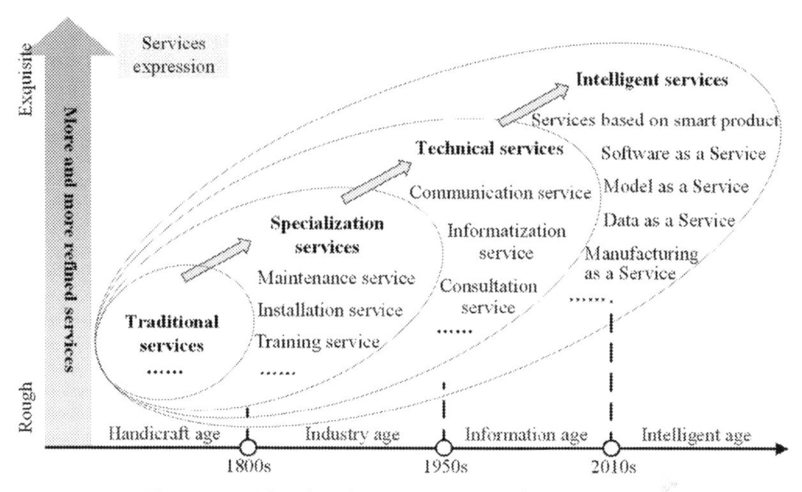

Figure 1.3 The development stages of services [18].

(2) Industry age: After the first industrial revolution, machines were employed as production tools instead of human and animal labor technology in the factories, leading to a significant increase in the scale of manufacturing [21]. The complexity of the production process and the diversification of the production system led to the refinement and specialization of the division of labor, so that the content of services continues to increase. Although the marketization degree of service, as the added-value of product, is still relatively low, the level of specialization has been significantly improved. For example, the enterprise has a special maintenance department to provide professional maintenance services for equipment or product. To ensure that the machines, equipment, or product purchased by users can be installed smoothly, debugged correctly, and run reliably, enterprises send technical personnel to the site to guide or directly install and debug the product, which are installation and commissioning services. In addition, the emergence of new modes of communication such as radio, telephone, and telegraph and new means of transportation such as automobiles, ships, and trams in the second industrial age [22] promoted the development of global transportation services, communications services, and social public services.

(3) Information era: In the information age (or the digital age), information technologies were widely applied. With the continuous evolution of the industrialization degree, social productivity development mainly relied on information technology and modern management [23]. Services have also entered a stage where information, technology, and knowledge are relatively intensive [24]. For example, software companies provide consultation services and informatization services to other enterprises, such as enterprise resource planning (ERP), manufacturing execution system (MES), and computer-aided process planning (CAPP), etc. of manufacturing enterprises. In addition, to realize the needs of information transmission between enterprises, departments, or individuals, the communication enterprise provided communication service with high quality and low price. During this period, services have the ability to operate on a large scale, to be deeply specialized, and to be highly integrated.

(4) Intelligent age: With the vigorous development and in-depth application of New IT, such as mobile communication, cloud computing, IoT, big data, etc., the value of all kinds of data can be deeply mined and brought into play [3]. For example, the advances in "microchip, sensor and IT technologies" contributed to the advent of smart products, which can track and communicate their operating conditions and thus allow to "feed" their product models with data about their status [20]. New IT has also been applied to hardware and software services, livelihood services, intelligent decision-making, and other aspects, thus promoting the arrival of the era of intelligent services. For example, the platform is used to improve the sharing level and utilization efficiency of information and resources, so as to reduce costs and improve the scope and level of services [25]. Software as a service uses the Internet to provide applications to users, thus greatly reducing the time and money spent on tedious tasks such as installing, managing, and upgrading software [26]. Through data collection and analysis, as well as model-driven evaluation, prediction, planning, and optimization, intelligent services enable relevant departments to make decisions more scientific and reasonable to achieve intelligent and precise decision-making. Moreover, manufacturing enterprises provide the customer with comprehensive services in the form of "manufacturing as a service" and "service based on smart product" to achieve high quality, efficient, low consumption, clean, and flexible production [7]. Intelligent services help companies optimize production processes, improve operational efficiency, and drive business growth.

From the above analysis, it is evident that services cover all aspects of production and marketing (e.g., manufacturing, logistics, installation, maintenance, finance, consulting, etc.), personal services (e.g., education, healthcare, catering, culture, tourism, retail, etc.), government public management services, as well as communication and information services, etc. The concept of service has been defined in many fields. While it is difficult to define commonly within or across fields on what services are, there is reasonable triangulation on services [27]. The triangular structure consists of service providers, service clients, and the provider—client relationships. Service provider is an entity (person, organization, or institution that has skill, technology, and experience, etc.) that makes preparations to meet a need. Service client is an entity (person, organization, or institution) that engages the service of another. Provider and client coproduce the value. Therefore, the definition of IBM Research on service is apt, that is, a service is a provider-to-client interaction that creates and captures value [28].

Services are diversified, reflected in all aspects of various fields. Services have been classified from many perspectives, such as service process, nature of the service, provider—client relationships, availability of service, service demand variation, mode of service delivery, service automation degree, etc. From the manufacturing perspective, this chapter will focus on product service, manufacturing service, and digital twin service.

1.3 Product service and manufacturing service

Pure product manufacturing or assembly is at the bottom of the "smile curve" whose added value is getting lower due to the advancement of science and technology [4]. In the highly competitive global market, many companies are using their resources and capacities to provide services to increase the value of their products, which in turn can bring competitive advantages [29]. Services usually have the characteristics of technology, knowledge, information, and other factors-intensive. It could help improve enterprise productivity by integrating service factors into the whole process of enterprise production and operation activities. It has become a prominent trend in the manufacturing industry from simple manufacturing to service-oriented manufacturing models [3].

The manufacturing servitization has two main directions. Firstly, the enterprise shifted from a pure product provider to a full lifecycle manager or system solution provider. From selling products, to selling the combination of "product + service" and then to providing integrated services for customers, manufacturing enterprises are constantly changing in the process of servitization to create higher value for both themselves and customers [30].

This direction of servitization contributed to product service. Secondly, enterprises focus on the R&D of core technologies or links. Manufacturing activities integrate manufacturing resources through the Internet and service, and closely cooperate with each other, to form a production and manufacturing network [31]. This direction is the servitization of manufacturing resources, that is, manufacturing service. Manufacturing services realize the integration of decentralized manufacturing resources and realize the value-added of stakeholders in the value chain through the reasonable allocation of manufacturing resources, capacity, and service [32]. Manufacturing enterprises provide the customer with comprehensive services in the form of "product + service" and "manufacturing + service."

1.3.1 Product service connotation

As technology advances, products become more complex. When a consumer buys a product, he not only wants to buy the product itself, but also hopes to get reliable and thoughtful service.

From the perspective of product transactions, product services can be divided into three stages according to their sequence with the purchase decision, including presales service, on-sales service, postsales service [33]. Presales service refers to the service provided to customers before the product is sold, which is to stimulate customers' desire [34]. Presales services provide customers with various technical consultations, such as product introduction service, function demonstrating service, shopping guide service, etc. On-sales services refers to the services provided to customers in the sales process, which help the customer utilize the product. On-sales services include services that answer the various queries raised by consumers about the product, installation, and training, etc. After-sales service refers to the service provided to consumers after the product is sold, to ensure that the value of the product is fully displayed and improve the degree of satisfaction [34]. Typical after-sales services include repair and maintenance, etc.

In addition, in order to meet the individual and diversified needs of customers, companies continue to include additional services while providing products, to improve the product's attractions to gain competitive advantages. Thereby, manufacturing companies provide consumers with products and supporting services, which become more highly integrated, resulting in product service system (PSS) [35]. Compared with traditional simple services, the services provided in the PSS serve the entire product lifecycle. According to the different combinations of services and products in the PSS, the services in PSS can be divided into three types, including product-oriented services, use-oriented services, and result-oriented services [30].

Product-oriented services are based on physical products. By selling the product, the producer transfers all the ownership and use rights of the product to the consumer, and provides after-sales services such as installation, monitoring, maintenance, consulting, etc. [30] For example, while providing aircraft engines, Honeywell developed an embedded aircraft information management system that can automatically detect aircraft failures, realizing the value-added [36]. Due to the extended interaction between consumers and enterprises, consumers not only pay attention to the product itself, but also consider the corresponding service when they choose products. For use-oriented services, enterprises retain product ownership and transfer product use rights to consumers [30]. Enterprises develop value-added services based on consumer usage. Consumers purchase services and the right to use products rather than purchase physical products to enjoy its functions, such as rental services [30]. For example, the car rental service allows consumers to pay only a certain usage fee and enjoy the functions brought by the car without purchasing the car [37], so as the sharing bikes. Besides, in order to facilitate maintenance and save costs, airlines no longer buy aircraft engines. Instead, they purchase the effective use time of aircraft engines, namely, airlines purchase the function or the service of these engines in a certain period of time [10]. As for result-oriented services, consumers do not own the product, nor use the product directly, but directly gain the result of using the product [30]. Result-oriented services can exist independently of products, and enterprises can help consumers gain better results by providing some information or services to consumers instead of the physical product itself. For example, some specialized consulting companies meet customer needs by providing customers with some information services. And some 3D printing service enterprises directly deliver the printed objects to the customers, who do not own nor use the 3D printing equipment directly [38]. Result-oriented services enable producers to improve product or equipment efficiency, reduce costs, and obtain higher profits, while consumers can enjoy the results without taking risks.

With the increasing popularity of customization and intelligent interconnection, customers' expectations for product experience are constantly being raised. And the increasing variety adds the complexity of products increases the uncertainty of demands [39]. Therefore, companies need to continuously adjust and optimize products according to changes in customer needs. However, companies face many challenges to do so. Digital twin provides a powerful way for companies to overcome these challenges. By establishing the digital mapping of physical products, and forming closed-loop feedback and optimization based on the data generated during the use of the product, the digital twin can comprehensively improve the full lifecycle management of the product, to improve product experience, reduce costs, and increase efficiency [40].

1.3.2 Manufacturing service and its management

Manufacturing servitization is reflected by the fact that manufacturing is offered as service. Various advanced manufacturing modes and systems adopt service-oriented architecture, such as application service provider, manufacturing grid, PSSs/industrial PSS, cloud manufacturing, etc. [3] Because services have the characteristics of interoperability and platform independence, they pave the way for large-scale enterprise collaboration, in which many services from different companies collaboratively accomplish a complex manufacturing task [3,41]. Through services, the upstream and downstream of the industry chains are connected together. And services and demands are gathered and connected, improving the ability to integrate resources in the industry chain.

1.3.2.1 Scope of manufacturing services

All activities that contribute to the creation of product value and bring convenience to users can become services. Manufacturing services in product lifecycle can be roughly divided into three categories [11] according to the production and use stages of the product:

Production-related service is the service related to the manufacturing process at the front end of the product lifecycle from product design to manufacturing, which is used to process or assemble raw materials or parts into finished products [11]. These services include services provided by different manufacturing companies to each other, such as design services, manufacturing resource/capability services, machining processes, services, testing services, etc. [11,42].

Product-related service is the additional service provided to the users at the back end of the product lifecycle from product sale to recycle/disposal after the product was sold to the users, such as logistics, installation, commissioning, training, repair, maintenance, scrap recycling, and other services [11,42]. These services increase product added value and enrich user experience.

Technical support service mainly refers to the services that provide technical support in all aspects of the product lifecycle [11]. On the one hand, it includes technical or consulting services provided by the specialized service enterprises for the production process of manufacturing enterprises, such as financial services, manufacturing equipment maintenance, information support services (e.g., knowledge, model, algorithm, computing, analysis, decision-making, etc.). On the other hand, it includes technical or consulting services provided by manufacturing enterprises in conjunction with specialized service enterprises for product operation, such as customer consultation, product up-gradation, remote maintenance, online guidance services, etc.

From the above, it indicates that there is some overlap between manufacturing services and product services from a perspective of product lifecycle. But in a narrow sense, manufacturing services in most of the research about smart manufacturing systems refer to the related production-related services and services provided by the specialized service enterprises for the production process of manufacturing enterprises. With the help of services, the smart manufacturing system can conveniently call various algorithms, models, computing resources, etc. required for perception, analysis, and decision-making, as well as the manufacturing resources and capabilities required for production execution.

1.3.2.2 Manufacturing services running workflow

As the analyses in the authors' previous work [3], manufacturing services are sensed and encapsulated from a variety of decentralized manufacturing resources and soft resources. Manufacturing resources mainly refer to material, manufacturing equipment, robots, and transportation equipment and computing equipment and other physical resources, while soft resources are a variety of information systems (e.g., ERP, MES, PDM, CRM, etc.), professional software (e.g., CAD, CAE, CAPP, etc.), experience, models, knowledge, data, and so forth [3]. Manufacturing resources and soft resources are invoked to perform design, manufacturing, maintenance, as well as computing, analysis, simulation, and other activities [3].

As shown in Fig. 1.4, a large amount of data (e.g., basic attributes, real-time status, process parameters, processing progress, and maintenance record, operation data, etc.) generated by these resources are sensed and collected. Based on these data, manufacturing resources and soft resources

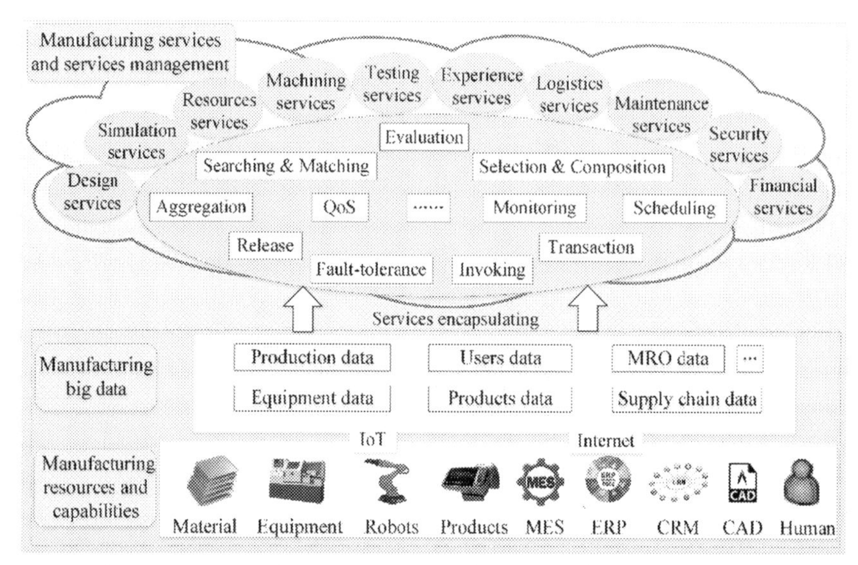

Figure 1.4 Manufacturing services collection and management [3].

are virtualized and encapsulated into different manufacturing services [3]. With the characteristics of interoperability and platform independence of services, various elements in manufacturing become plug-and-play components and can be invoked and used by other participants. Manufacturing participants can conveniently provide their services to their collaborative stakeholders, such as customers, suppliers, partners, etc. [3].

After service encapsulation, manufacturing services are released to the platform to be managed and used by others. Through services, different and decentralized production elements can cooperate with each other to implement synergistic production. Through combination, arrangement, connection of services, it can help manufacturing resources achieve cooperation, interconnection, and communication. In this condition, users and manufacturers not only can manage and use resources remotely but also monitor their performance and operation status [3].

In order to achieve efficient allocation of manufacturing resources and rapid response to user and market, services need to be managed scientifically and effectively to improve the value. Services management mainly includes services release, aggregation, smart searching and matching, comprehensive evaluation, optimal combination, scheduling, fault-tolerance, etc. [3]. Services that are able to complete a subtask decomposed from a complex manufacturing task are searched for invoking. These services are further optimized and combined into a solution to complete the task. In the services execution process, the implementation process and QoS are monitored. When failures occur, fault-tolerance is executed to continue to complete the task. Through the network and services, a variety of resources within manufacturing enterprise can be shared in the form of services to more enterprises around the world to be effectively integrated and used. Standard and shareable manufacturing services can help enterprises achieve internal and external interoperability and cooperation between the upstream and downstream enterprises in supply chain, improving manufacturing efficiency [3].

In conclusion, in order to achieve the effective sharing, real-time scheduling, and optimal allocation of manufacturing services, the running workflow of manufacturing services includes following three stages [43,44].

Manufacturing service generation stage: This stage is mainly to encapsulate manufacturing resource as a service. This stage is the basis for realizing subsequent manufacturing service management and on-demand use. Through perception access technologies such as the sensors, IoTs, and access adapters, multisource information of various scattered and heterogeneous physical manufacturing resources is perceived in real time, which is processed through data filtering, cleansing, integration, and feature extraction to realize the virtualization of physical manufacturing resources. Then the manufacturing resource model is established based on the perception data and knowledge-based manufacturing capabilities data, which is described and encapsulated in service description language (e.g., Defense advanced research projects agency (DARPA) proxy markup language and Web ontology language) to complete the service encapsulation of manufacturing resources [45]. After service encapsulation, manufacturing services are released and aggregated to the platform to be managed.

Manufacturing service management stage: The efficient management of manufacturing service includes the following four aspects: (1) Searching and matching of manufacturing service. Searching and matching is to find suitable services to meet user requests based on the description of manufacturing tasks and manufacturing services [46]. (2) Comprehensive evaluation of manufacturing services. Service quality, trust, and utility evaluation play an important role in the effective management of manufacturing services. Comprehensive evaluation of manufacturing services is to provide users with highly reliable service capabilities and results. Considering the trust problem in the process of manufacturing service management, the concept of service trust-QoS is introduced [47]. (3) Selection and combination of manufacturing services. In the process of using services, multiple services are often required to complete a task together. Selection and combination are to generate service combination plans and select the best service combination plan. (4) Manufacturing service scheduling. Service scheduling is to arrange service execution under time constraints. From the perspective of services participants, there are three service scheduling patterns, that is, provider-centered, consumer-centered, and operator-centered [38].

Manufacturing service on-demand use stage: In the process of using manufacturing services, the matching and mapping between service (supply) and tasks (demand) need to be realized under certain time constraints, that is, the supply and demand matching of manufacturing service. Considering the dynamic characteristics of manufacturing service and manufacturing environment, complex networks are introduced into the supply and demand matching of manufacturing services [48]. After the service is selected and scheduled, the service transaction is executed to complete the set goal. Manufacturing service has the characteristics of high complexity, dynamics, and diversification. As a result, the execution of manufacturing services and the combination of manufacturing services may inevitably experience some failure. Thus, it needs to manage failures related to manufacturing services and their composition [49]. Through the whole process from perception, encapsulation, aggregation, searching and matching, evaluation, selection and combination, scheduling, supply and demand matching, transaction, fault-tolerant, etc., manufacturing services are accessed, invoked, and on-demand used.

According to the traditional manufacturing service models, manufacturing services are generally described for specific functions to finish tasks and output results under certain constraints and fulfill user demands. As information opacity and lack of feedback in traditional manufacturing service frequently lead to information asymmetry between manufacturing service suppliers, low service quality, and unreliable collaboration, it seriously affects the efficiency of manufacturing. With the development of information technology, digital twin is gaining increasing attention as a new method to enhance manufacturing services. The digital twin applications cover the whole lifecycles of product development, process planning, manufacturing, testing, operation, and maintenance, etc., and can help companies promote digital marketing and services.

1.4 Digital twin service

With the further development and in-depth application of New IT, people have put forward more requirements for industrial production and living needs. People want to know what is in the physical world of manufacturing and products, what is happening in different time-space scales, and what will happen in the future? In this context, digital twin has emerged and caused profound industrial changes [50]. Physical entities, their corresponding virtual models, data, connections, and services are the core components of the digital twin, that is, five-dimensional digital twin model [15]. Driven by multidimensional models and fusion data, as well as the interaction between physical objects and models, digital twin can describe the multidimensional attributes of physical objects, characterize the actual behavior and real-time status of physical objects, analyze the future trend of physical objects, and realize the realization of physical objects [50]. The digital twin makes products and manufacturing have a lot of "intelligence" and "information" characteristics, which become complex systems with cyber-physical integration. A large amount of data is generated during the use and manufacturing of the product. With multidimensional models, the use and manufacturing process of products can be monitored, controlled and analyzed, driven by data and models. Digital twin not only can improve production efficiency and reduce downtime of equipment but also continue to improve the product after it is released. As shown in Fig. 1.5, digital twin enables companies to provide services to customers throughout the entire lifecycle of product design, manufacturing, use, maintenance, recycling, etc., and form more closed loops and innovations in application services.

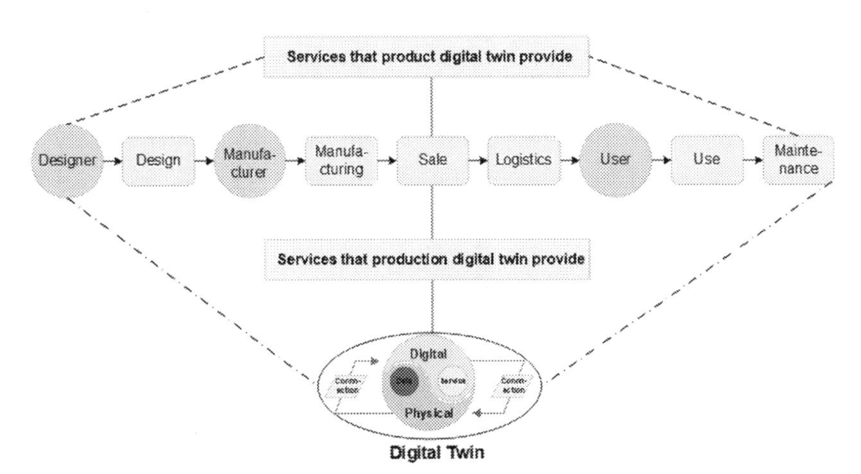

Figure 1.5 Digital twin services applications throughout the product lifecycle.

1.4.1 Digital twin-driven services

Driven by multidimensional models and fusion data, digital twin is to monitor, evaluate, simulate, predicate, optimize, and control the physical world through real-time connection, mapping, analysis, and interaction, so as to maximize efficiencies of the full-elements, all-process, whole-business, and value chains of the physical system. Therefore, as shown in Fig. 1.6, digital twin-driven services can be summarized as digital twin-driven monitoring, digital twin-driven evaluation, digital twin-driven simulation, digital twin-driven prediction, digital twin-driven optimization, and digital twin-driven control [51].

Digital twin-driven monitoring services: The digital models in digital twin are the true equivalent mapping to the physical system. By collecting system status in real time and dynamically updating the digital

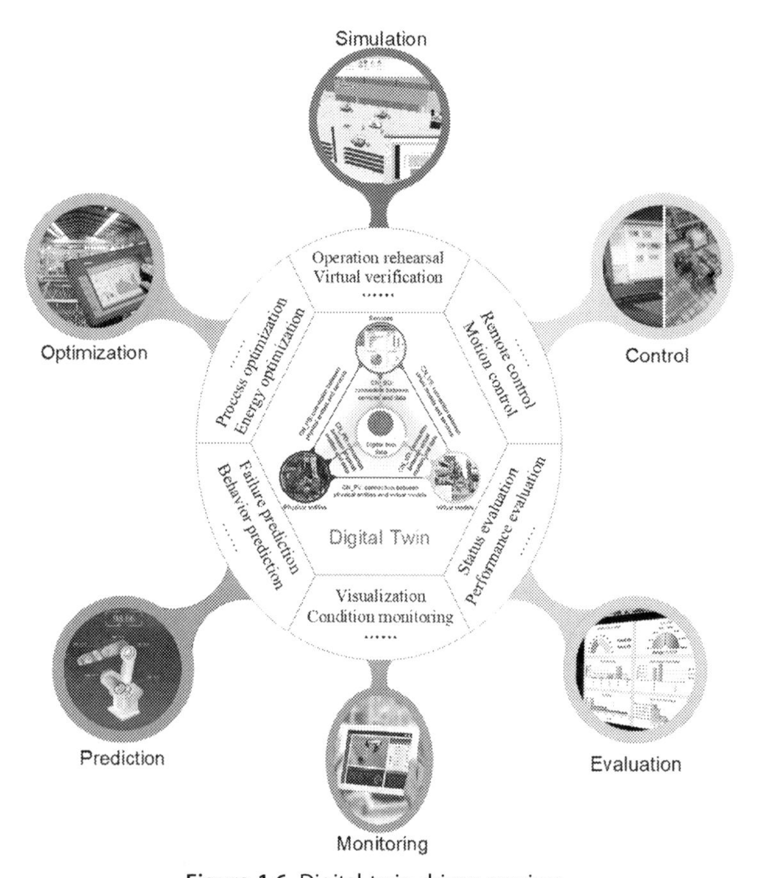

Figure 1.6 Digital twin-driven services.

model, the digital twin can monitor the physical system in real time. On the one hand, digital twins can use virtual reality and augmented reality to visualize geometric models and finite element models. With 3D visualization, digital twins can intuitively monitor the layout of physical systems in real time, changes in the behavior of physical elements, and view the overall operating status of the product or manufacturing system intuitively [52]. On the other hand, digital twins can use charts, graphs, design elements, and other means to visualize data [15]. Complex and abstract data are presented in a more comprehensible form, enabling digital twins to display load changes, energy consumption statistics, analysis results, etc., of physical systems in real time. For example, in the process of product assembly, the assembly process, change in equipment behavior and state, statistical analysis, etc., can be completely and truly reproduced in the digital space of digital twin, so as to realize the monitoring of product assembly quality. In addition, for products or devices operating in extreme environments (e.g., radiation environment), digital twin is an excellent way to monitor the status.

Digital twin-driven evaluation services: Digital twin-driven evaluation is to evaluate the performance and state of a physical system or objects, to determine the availability, feasibility, and need for maintenance of a product or manufacturing system. With data and models, digital twins provide comprehensive evaluation of the dynamic performance of products, machines, and systems. Through running relevant algorithms with the virtual product model (composed of geometric models, material properties, parts linkage coupling model, parts mechanics-temperature-flow coupling model, etc.), combined with real-time state data and history data, product performance and state, as well as energy consumption are provided to users and manufacturer. For example, a digital twin of a vertical transportation system is used to evaluate the system condition and potential corrective solutions [53]. In addition, faults can be diagnosed when digital twin-driven evaluation is carried out. For example, digital twin is used in power transformer for state evaluation and fault diagnosis to ensure safe and stable operation of power system [54].

Digital twin-driven simulation services: Simulation is an integral part of digital twin. Through simulation, digital twins can provide customers with virtual tests (e.g., autonomous driving test), virtual design verification (e.g., structure verification, feasibility verification), operation rehearsal (e.g., virtual commissioning, maintenance plan rehearsal, process planning), and other functions. Virtual testing can reduce the number of physical experiments and reduce the cost of trial production and testing. Virtual design verification can shorten the product design cycle. Operations rehearsal can improve the feasibility and success rate of the scheme, and reduce the risk and error. For example, in automatic driving test [55], while it is more realistic to adopt real vehicle field tests or road tests, there are problems such as high time cost, manpower cost and material cost, low safety, and low repeatability, etc. By building the models of the environment, road, traffic participants, test vehicles, and sensors, digital twin-driven simulation can quickly generate test conditions close to the real traffic environment combined with the scene database data, effectively improving the efficiency and authenticity of the test. For design verification, manufacturers can adjust parameters in a virtual environment to test and validate the structure, functionality, feasibility, safety, and quality of product before actual manufacturing [56]. In addition, while product failure occurs, and the maintenance strategy is provided, users or manufacturers can carry out virtual maintenance based on digital twin-driven simulation to rehearse operation steps and detect and eliminate potential defects, before conducting practical maintenance [57].

Digital twin-driven prediction services: What will happen in the future is a major concern for customers who use digital twin. Based on the above-mentioned monitoring, evaluation, and simulation, digital twin brings predictions for future into reality. Driven by virtual models and fused data, as well as cyber-physical interaction, digital twin can predict operational characteristics and performance of the entire product lifecycle. Digital twin-driven prediction includes fault prediction (such as fan fault prediction), life prediction (e.g., aircraft life prediction), quality prediction (e.g., product quality control), behavior prediction (e.g., robot motion path prediction), and performance prediction (e.g., how an entity will behave in different environments), etc. [51] Fault prediction changes maintenance from passive to predictive. Maintenance is only performed when needed, reducing downtime. Life prediction quantifies the remaining use time of products and systems, reducing the risk of accidents. Quality prediction helps to improve the quality of the products, avoiding high loss from defective products. Behavior and performance predictions can test the operations of products or equipment under different environments and different parameters, to verify the flexibility and plan or guide the use of products and equipment.

Digital twin-driven optimization services: On the basis of monitoring, evaluation, simulation, and prediction, digital twins make the optimization of product lifecycle possible. For example, a digital twin cooperative architecture integrating virtual modeling, process monitoring, diagnosis, and optimized control is used in automatic process applications [58]. Digital twin-driven optimization could be design optimization (e.g., product redesign, product performance optimization, and product structure optimization, etc.), configuration optimization (e.g., layout optimization, equipment scheduling), performance optimization (e.g., equipment parameter adjustment), energy consumption optimization, process optimization, etc. [51] The purpose of optimization is to improve product development, increase system efficiency, save resources, and reduce production costs. Digital twin-driven optimization could be considered as the decision-making process. Digital twin transfers the trial-and-error and decision-making optimization processes originally carried out in the real world to the digital world, so that the physical world always executes the most optimized plans and decisions.

Digital twin-driven control services: The important feature of digital twins is the symbiosis of physical entities and virtual models. The interaction in the digital twin is bidirectional. From the digital world to physical world, the function of digital twin is to control physical entities or systems to execute decision-making plans. After the monitoring, evaluation, simulation, prediction, and optimization of the physical world, digital twin mastered the state and changes of the physical world, and made optimization decisions. Decision results are fed back to the physical world to respond to dynamic changes. For example, in remote surgery, digital twin is used to remote control medical facilities over mobile networks [59]. Besides, digital twins can remotely control the start and stop of equipment, especially in extreme environments (e.g., nuclear radiation, generator system, etc.). Digital twins also can empower the cooperative control of multiple machines.

Digital twins have broken many limitations of the physical conditions. Digital twin-driven monitoring, evaluation, simulation, prediction, optimization, and control could enable continuous innovation of services, immediate response to demand, and industrial upgrading. Based on the advantages of models, data, and services, digital twin is becoming the key to improving quality, increasing efficiency, reducing costs and losses, ensuring safety, as well as energy saving and emission reduction.

1.4.2 Digital twin enhanced manufacturing service

As in Section 1.3.2.1, manufacturing services cover the whole product lifecycle stages such as product design, manufacturing, and after-sales service. According to the difference in physical components, manufacturing services are divided into services that can and cannot be enhanced by digital twins. Especially, in the manufacturing stage, most manufacturing services are highly dependent on physical entities, such as machines and tools, which can be enhanced by digital twins [60]. Therefore, the manufacturing resource services can be enhanced by digital twins. For the manufacturing resources with the high-fidelity virtual entities, they can be encapsulated as a digital twin enhanced manufacturing service.

As shown in Fig. 1.7, manufacturing resources (e.g., machines, robots, AGV cars, etc.) with physical entities and digital twin models, form manufacturing resources digital twins. Service encapsulation is the transformation of manufacturing resources digital twin into services with a unified description. As the analyses in the authors' previous work [60], digital twin enhanced manufacturing services can be denoted as DT-MS = (Func, PI, VI, Input, Ouput, QoS) [60]. Func represents the function of the manufacturing service, and PI and VI are the physical and the corresponding virtual information respectively. Input, Ouput, and QoS are the input, output, and quality of the manufacturing service [60]. The digital twin enhanced manufacturing service can simulate and monitor the service and manufacturing resources execution process and results in advance. As shown in Fig. 1.7, the physical part of DT-MS can be controlled in real time through virtual part of DT-MS to guarantee service accuracy. Demand is input to both the physical and virtual part of the digital twin enhanced manufacturing service. Driven by the integration of data and model, the virtual service quickly obtains possible execution plans through simulation in advance and selects the best plan to be handed over to the physical service [60]. The physical service also transmits real-time data and results to the virtual service to continuously modify the key parameters of the original simulation model. A simulation scheme that satisfies the constraints of physical services can simultaneously guide the actual execution of physical resources [60]. Digital twin enhanced manufacturing services can update manufacturing service information and adjust the corresponding manufacturing resources to a better working state independently through the interaction between physical and virtual parts. Synchronous interaction and real-time coordination between physical and virtual services can ensure

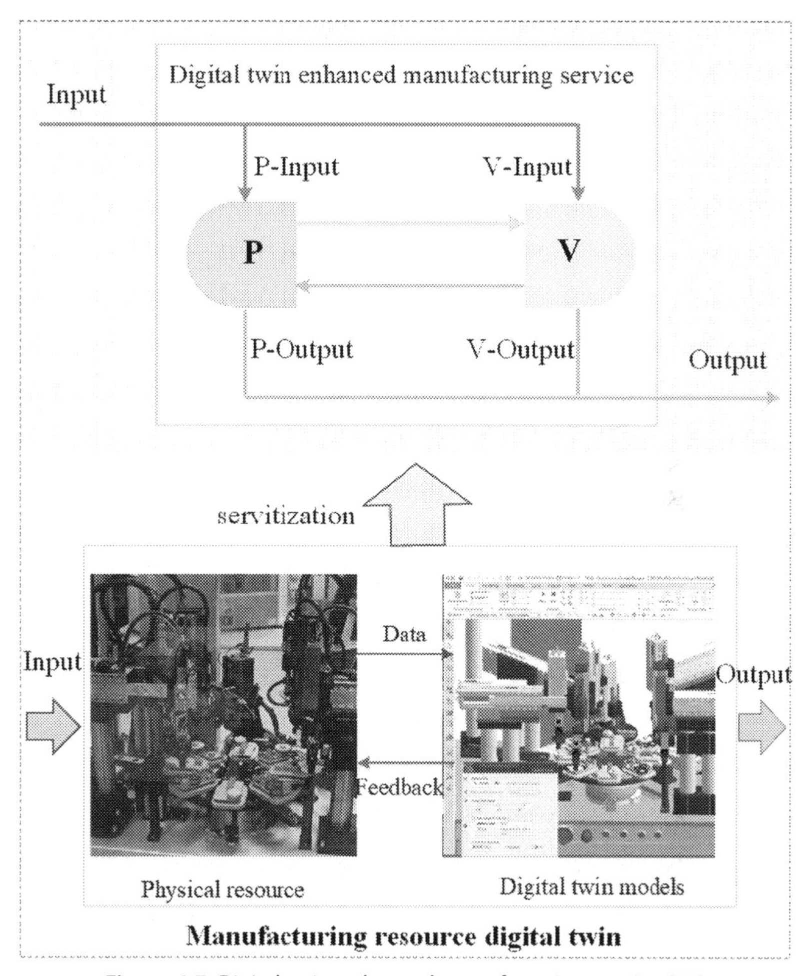

Figure 1.7 Digital twin enhanced manufacturing service [60].

that the digital twin enhanced manufacturing services effectively complete tasks and output better service execution results. Digital twins add new features to manufacturing resources and enrich their functions. Correspondingly, combined with digital twins, digital twin enhanced manufacturing services enrich service functions and improve the efficiency of manufacturing services.

1.4.3 Digital twin as services

As the analyses in the authors' previous work [17], the model creation of digital twins is a complex and specialized project, so are data fusion and analysis. For users who do not have relevant knowledge, it is difficult to build and use digital twins. Therefore, it is imperative that models are able to be shared by users and data analysis is outsourced to the third-party professional organizations [17]. Moreover, in the context of manufacturing socialization, the physical resources involved in manufacturing are geographically distributed [17]. Considering digital twin as a service, others can use digital twin or some elements of digital twin in the form of service. For example, through the Internet, users are able to access and use the various elements to establish their digital twins. Or users can get services provided by someone else's digital twin without having to create their own.

As shown in Fig. 1.8, the first and the most important step of turning digital twins to service is to establish the information template [17], which consists of a variety of information. For the physical objects, the information includes basic attributes (e.g., name, ID, address, etc.), QoS (e.g., time, cost, reliabilities, satisfaction, etc.), capacities (e.g., precision, size, process, etc.), real-time status (e.g., overload, idle, in maintenance, etc.), as well as input and output, which can be described as $PO = \{Basic, QoS, Cap, Status, Input, Output\}$ [17]. *Basic* denotes the basic attributes to identify a physical object. *QoS* (quality of service) denotes the evaluation for the performance of physical object to conduct the user selection. *Cap* denotes the functions that the physical object can do, and *Status* indicates whether the physical object is available.

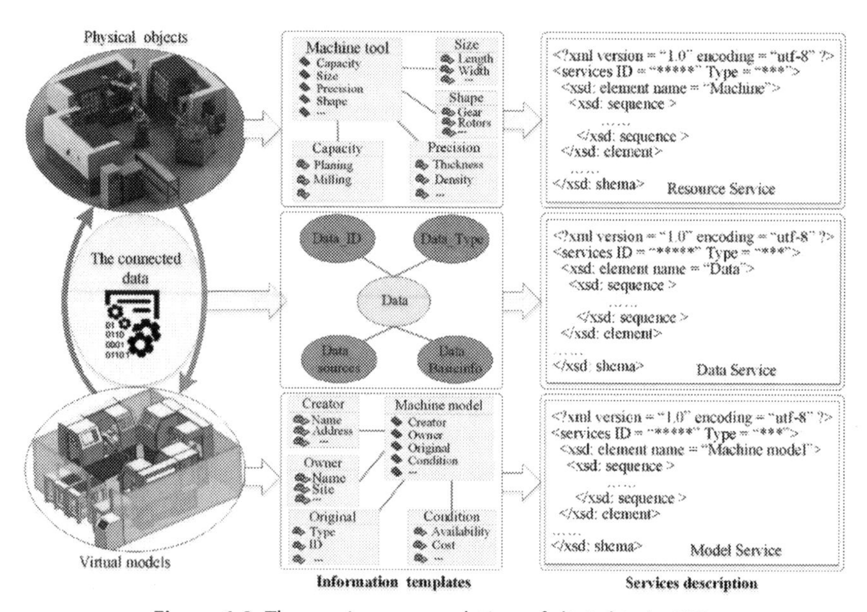

Figure 1.8 The service encapsulation of digital twin [17].

Different from the physical entities, the virtual models can be copied without repetitive creation for a same or equal physical object and used by multiple users at the same time. Therefore, not only creators of models can get benefits from models services shared by users but also users can reduce costs and time. The virtual model can be described as *VM* = {*Ori_phy, Creator, Ori_ID, Cur_ID, Owner, QoS, Online_site, Input, Output,* ...} [17]. *Ori_phy* denotes the original physical object from which the virtual model is created. *Creator* is the one who builds the model based on his specialized knowledge. *Ori_ID* is the original identifying number when the model is first created. *Cur_ID* is the identifying number of current copy. *Owner* is the one who possesses the model. Owner who has copyright or may be a creator, can earn profits through renting out models or selling copies. Similar to the physical object, *QoS* denotes the evaluation for the performance of models, including cost, reliability, functions, etc. *Online_site* denotes the online address where users can access or download the models. *Input* and *Output* may be different according to the specific models.

In addition, data are very important for smart manufacturing [18]. However, because of various standards, and communication protocols/interfaces, data are difficult to acquire and understand. Through services, users can conveniently use data. In general, the attributes of data include the data provider who owns the data, data sources where data are collected, data ID which is used to identify data, data type which denotes the kind of data, and data abstract which is the brief introduction of data value. Therefore, the relationship can be described as *Data* = {*D_prov, D_source, D_ID, D_type, D_abstract,* ...} [17].

After the information templates are established, various kinds of physical objects, virtual models, and data are encapsulated to services based on the service description language such as XML language [61]. After services encapsulation, the digital twin services are published to the services pool and management platform, where they are managed to be shared by various users.

1.4.4 Digital twin services applications

As shown in Fig. 1.9, digital twin-driven services and digital twin as services improved traditional product services. In the era of digital twins, digital twin services not only give users a better experience, but also provide enterprises with valuable insights and systematic thinking.

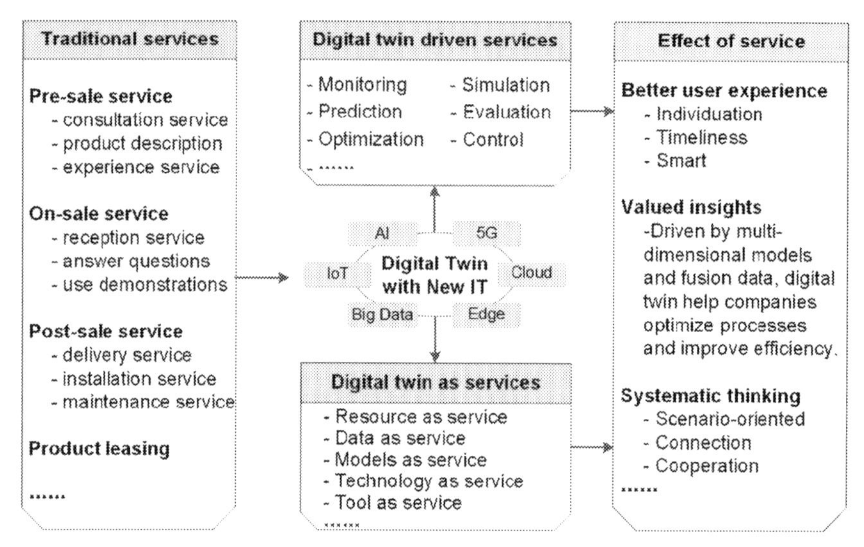

Figure 1.9 From traditional services to digital twin services applications.

For the customer, before the product is sold, users can experience the true use feeling of the product by the product digital twin. Digital twin-driven simulation can simulate various real functions of the product and switch different application scenarios for different users. Product introduction, consultation, and experience services can all be done by product digital twins. The product digital twin can be delivered to the customer when it is traded. After the product is sold, with the high-speed computing in virtual product space, on the one hand, digital twin-driven user operation guide can guide users to operate the product, and on the other hand, it can correct the users' poor habits in real time [57]. In addition, each user has its own operation habits. With the real-time monitoring of digital twins, the influence of the operation habits on product performance and life can be computed [57]. Digital twins endow product services with individuation, timeliness, and intelligence. For companies, driven by multidimensional models and fusion data, digital twins can help companies optimize processes and improve efficiency. Digital twins empower companies to turn their products into a dynamic, connected and cooperated platform, which can continuously feel customer needs. As a result, companies can provide customers with better products and services according to different scenarios.

As shown in Fig. 1.10, digital twin services pool consists of digital twin as services (e.g., equipment services, data services, models services), digital twin-driven services (e.g., monitoring services, evaluation services,

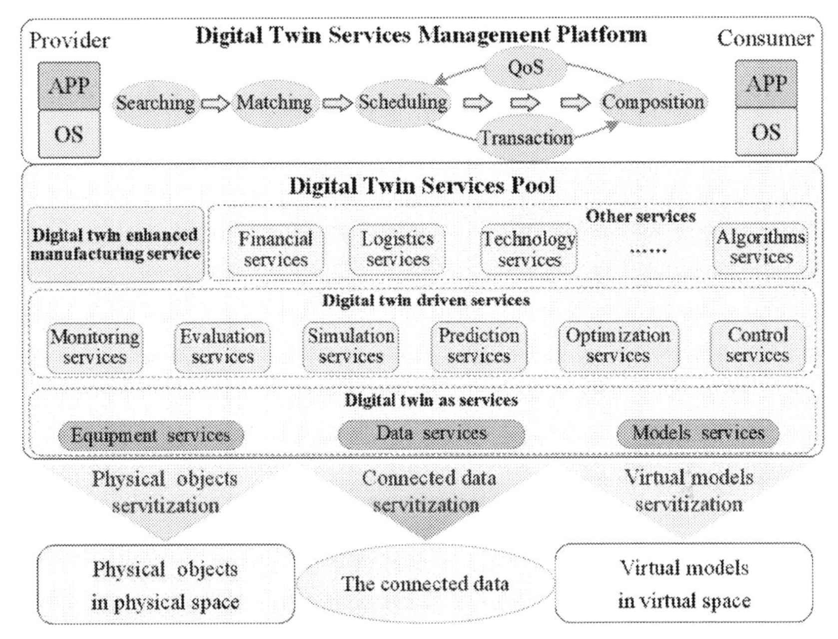

Figure 1.10 The digital twin service management and applications [17].

simulation services, prediction services, optimization services, and control services), digital twin enhanced manufacturing services, and other services (financial services, logistics services, technology services, algorithms services, etc.). The services management includes searching, matching, scheduling, combination, transaction, fault-tolerance, etc. [44]. A task is submitted to the management platform. Then, it is decomposed into subtasks that can be accomplished by a single service. According to the QoS, the manufacturing service supply-demand matching and scheduling is carried out to select the optimal services [41]. After the service transaction, the selected services are invoked and combined to complete the task collaboratively. Finally, the results are fed back to the users.

As the analyses in the authors' previous work [17], from the perspective of product creation, the digital twin services can be used in product design, manufacturing, and equipment or product PHM (Prognostic and Health Management), and other applications [17].

In product design, digital twin can help improve the accuracy of the design and verify the performance of the product. The digital twin-driven design is to turn the expected product in the designer's mind into the digital representation in digital world [56]. Based on digital twin-driven services and digital twin as services, digital twin can break through the limitations of physical conditions, and help designers understand the performance of the product, iterating the product with less cost and faster speed. Designers just simply submit their needs to the services management platform. Services managers will match the various services which designers need. Through invoking, combining, and operating these services, the results are returned to the designers. As a result, designers acquire what they want in the "pay-as-you-go" manner [17]. Moreover, after the function structure and components of the product are designed, the design quality and feasibility need to be tested. In virtue of digital twin-driven simulation, analysis, prediction, and optimization services, designers can quickly and easily test products through virtual verification to guide, simplify, and reduce physical experiments [56]. Through a series of repeatable and variable parameter simulation experiments, the performance and adaptability of the product under different environments are verified. The models of the manufacturing site (e.g., production line or shop-floor, etc.) needed in virtual verification can be invoked and used through services. Through services, digital twins can be easily applied in product design, which can make product design more effectively to reduce the inconsistencies of expected behavior and design behavior, and greatly shorten design cycles and reduce costs [17].

In general, product manufacturing is the whole process from the input of raw materials to the output of finished products. In the manufacturing stage, digital twins can be applied to different levels of the manufacturing processes from the equipment, production line, shop-floor to the factory, through all aspects of process optimization, resource allocation, parameter adjustment, energy efficiency management, and production planning, etc. Through digital twin-driven simulation, evaluation, and optimization services, manufacturers can systematically plan production processes and resources, monitor, discover, and respond to various abnormalities and instabilities in the production process in time. In the phase of production planning and manufacturing execution, a production task is submitted to services management platform and resource services supply-demand matching and scheduling are carried out to find available resources and services. Then, based on the real-time status of physical resources (e.g., machine tools, robot arms, etc.), production plan is drawn up. Digital twin can simulate the plans in virtual space and find out the potential conflicts before even during the actual manufacturing process [62]. However, the digital twin is a complex and specialized work to be built, especially the models including geometry, rule, behavior, dynamics models. With the help of services, these models do not have to be created by manufacturer themselves [17]. For physical equipment and pervasive rules, their models which have been established by other manufacturers, can be bought to use in the form of services. Current manufacturer only needs to create the special models, which is only suitable for themselves [17].

The performance degradation of physical equipment or product is inevitable. When the equipment or product malfunctions, it would result in high maintenance costs and postponement of tasks [17]. PHM is necessary to monitor the equipment or product condition, predict and diagnose equipment faults and component lifetime [17]. In digital twin-driven PHM, digital twin use temperature, vibration, crash, load, environmental, and other data combined with digital twin models to predict equipment or product conditions well in advance so that worn parts can be replaced within scheduled downtime and unplanned downtime can be avoided. A high-fidelity digital mirror for the equipment or product provides access to the equipment or product even out of physical proximity [17]. Besides, the interaction of digital twin can reduce the disturbances from the external environment, improving accuracy. In above process, the models are accessed through services. Moreover, when failures occur, repair services are invoked to repair, or replace the broken-down equipment or product.

Last but not least, digital twin services can be applied to aviation, aerospace, smart city, construction, healthcare and medicine, robotics, ship, vehicles, railway, industrial engineering, engineering management, agriculture, mining, power, energy, environment, and other fields.

1.5 Summary

As digital transformation and intelligent upgrading have become a global hot spot, digital twin has received unprecedented attention and played important values in increasing industry fields and application scenarios. Digital twins are creating huge value for the manufacturing industry in terms of product operation monitoring and maintenance, real-time simulation, and remote monitoring of shop-floor operation status, virtual commissioning of production lines, and the development of complex products. Meanwhile, services could optimize the entire business processes and operation procedure of manufacturing, to achieve a new higher level of productivity. Therefore, digital twin services are beginning to be valued. This chapter first analyzes the development progress of service from the four stages, followed by the connotation of services. Then, from manufacturing perspective, the two main directions of manufacturing servitization are analyzed, that is, product services and manufacturing services. In response to the challenges of product services and manufacturing services, digital twin services are introduced. Digital twin services not only can improve traditional product services but also enrich service functions and improve the accuracy of manufacturing services. Through digital twin-driven services and digital twins as services, digital twin services enable companies to provide services to customers throughout the entire lifecycle of product design, manufacturing, use, maintenance, recycling, etc. Finally, the digital twin services applications are analyzed.

Acknowledgments

This work is financially supported in part by National Natural Science Foundation of China (NSFC) under Grant 52005024, and in part by grants from the Hong Kong Polytechnic University, China (Project No. G-YZ3N i.e., XJ2019057).

References

[1] Tao F, Qi Q. Make more digital twins. Nature 2019;573:490–1.
[2] Legner C, Eymann T, Hess T, et al. Digitalization: opportunity and challenge for the business and information systems engineering community. Bus Inf Syst Eng Vol 2017;59(4):301–8.
[3] Tao F, Qi Q. New IT driven service-oriented smart manufacturing: framework and characteristics. IEEE Trans Syst Man Cyber Syst 2019;49(1):81–91.
[4] Aggarwal S. Smile curve and its linkages with global value chains. J Econ Bibliograp 2017;4(3):278–86.
[5] Hamidi F, Shams Gharneh N, Khajeheian D. A conceptual framework for value co-creation in service enterprises (case of tourism agencies). Sustainability 2020;12(1):213.
[6] Grönroos C, Helle P. Adopting a service logic in manufacturing: conceptual foundation and metrics for mutual value creation. J Serv Manag 2010;21(5):564–90.
[7] Sun L, Li G, Jiang Z, et al. Service-embedded manufacturing: advanced manufacturing paradigm in 21 st century. China Mech Eng 2007;18(19):2307–12 [in Chinese].
[8] Lightfoot H, Baines T, Smart P. The servitization of manufacturing: a systematic literature review of interdependent trends. Int J Oper Prod Manag 2013;33(11/12):1408–34.
[9] https://max.book118.com/html/2021/0204/8075064136003045.shtm; 2021.
[10] Tao F, Cheng Y, Zhang L, et al. Advanced manufacturing systems: socialization characteristics and trends. J Intell Manuf 2017;28(5):1079–94.
[11] Tao F, Qi Q. Service-oriented smart manufacturing. J Mech Eng 2018;54(16):11–23.
[12] McCarthy I, Anagnostou A. The impact of outsourcing on the transaction costs and boundaries of manufacturing. Int J Prod Econ 2004;88(1):61–71.
[13] Wu D, Rosen DW, Wang L, et al. Cloud-based design and manufacturing: a new paradigm in digital manufacturing and design innovation[J]. Comput Aided Des 2015;59:1–14.

[14] Li P, Cheng Y, Song W, et al. Manufacturing services collaboration: connotation, framework, key technologies, and research issues. Int J Adv Manuf Technol 2020;110(9):2573–89.
[15] Qi Q, Tao F, Hu T, et al. Enabling technologies and tools for digital twin. J Manuf Syst 2021;58:3–21.
[16] Lu Y, Liu C, Kevin I, et al. Digital Twin-driven smart manufacturing: connotation, reference model, applications and research issues. Robot Comput Integrated Manuf 2020;61:101837.
[17] Qi Q, Tao F, Zuo Y, et al. Digital twin service towards smart manufacturing. Procedia Cirp 2018;72:237–42.
[18] Tao F, Qi Q, Liu A, Kusiak A. Data-driven smart manufacturing. J Manuf Syst 2018;48:157–69.
[19] Smith A. The wealth of nations. Aegitas 2016:4–10.
[20] Schleich B, Anwer N, Mathieu L, Wartzack S. Shaping the digital twin for design and production engineering. CIRP Ann - Manuf Technol 2017;66(1):141–4.
[21] Mohajan H. The first industrial revolution: creation of a new global human era. J Soc Sci Human 2019;5(4):377–87.
[22] Mokyr J, Robert H. The second industrial revolution, 1870-1914. Storia dell'Economia Mondiale 1998:21945.
[23] Brynjolfsson E, Yang S. Information technology and productivity: a review of the literature. Adv Comput 1996;43:179–214.
[24] Javalgi RRG, Gross AC, Joseph WB, Granot E. Assessing competitive advantage of emerging markets in knowledge intensive business services. J Bus Ind Market 2011;26(3):171–80.
[25] Simeone A, Caggiano A, Boun L, Deng B. Intelligent cloud manufacturing platform for efficient resource sharing in smart manufacturing networks. Proced CIRP 2019;79:233–8.
[26] Janssen M, Joha A. Challenges for adopting cloud-based software as a service (SAAS) in the public sector. In: European conference on information systems (ECIS); 2011. Proceedings. 80.
[27] Rai A, Sambamurthy V. Editorial notes—the growth of interest in services management: opportunities for information systems scholars. Inf Syst Res 2006;17(4):327–31.
[28] IBM Research. Services science: a new academic discipline. 2004. http://www.ssmenetuk.org/docs/IBMReportServicesScience.pdf.
[29] Gremyr I, Löfberg N, Witell L. Service innovations in manufacturing firms. Manag Serv Qual Int J 2010;20(2):161–75.
[30] Beuren FH, Ferreira MGG, Miguel PAC. Product-service systems: a literature review on integrated products and services. J Clean Prod 2013;47:222–31.
[31] Zhe H, Lin-yan S, Zhu-qing H, Gang L. Trend of service-manufacturing and differences between services-manufacturing network and SCM. Soft Sci 2008;4(22):77–81 [in Chinese].
[32] Li BH, Zhang L, Wang SL, et al. Cloud manufacturing: a new service-oriented networked manufacturing model. Comput Integr Manuf Syst 2010;16(1):1–7 [in Chinese].
[33] Frambach RT, Wels-Lips I, Guendlach A. Proactive product service strategies: an application in the European health market. Ind Market Manag 1997;26(4):341–52.
[34] Nie Y, Kosaka M. Customer self-service platform: the next practice for servitization of manufacturing. Int J Serv Oper Manag 2016;25(2):259–73.
[35] Vandermerwe S, Rada J. Servitization of business: adding value by adding services. Eur Manag J 1988;6(4):314–24.

[36] Banning R, Roesch M, Morgan A. Improved embedded flight control system design process using integrated system design/code generation tools. In: Digital avionics systems conference, 1994. 13th DASC. AIAA/IEEE. IEEE; 1994. p. 134−8.
[37] Aurich JC, Mannweiler C, Schweitzer E. How to design and offer services successfully. CIRP J Manuf Sci Technol 2010;2(3):136−43.
[38] Zhang Y, Tao F, Liu Y, et al. Long/short-term utility aware optimal selection of manufacturing service composition toward industrial internet platforms. IEEE Trans Indus Inform 2019;15(6):3712−22.
[39] Ho CF, Tai YM, Tai YM, Chi YP. A structural approach to measuring uncertainty in supply chains. Int J Electron Commer 2005;9(3):91−114.
[40] Macchi M, Roda I, Negri E, Fumagalli L. Exploring the role of digital twin for asset lifecycle management. IFAC-PapersOnLine 2018;51(11):790−5.
[41] Tao F, Cheng J, Cheng Y, et al. SDMSim: a manufacturing service supply−demand matching simulator under cloud environment. Robot Comput Integrated Manuf 2017;45:34−46.
[42] Cheng Y, Tao F, Zhang L, Zuo Y. Supply-demand matching of manufacturing service in service-oriented manufacturing systems. Comput Integr Manuf Syst 2015;21(7):1930−40.
[43] Tao F, Zhang L, Liu Y, et al. Manufacturing service management in cloud manufacturing: overview and future research directions. J Manuf Sci Eng 2015;137(4):040912.
[44] Cheng Y, Qi Q, Tao F. New IT-driven manufacturing service management: research status and prospect. China Mech Eng 2018;29(18):2177−88.
[45] Luo Y, Zhang L, Tao F, et al. A modeling and description method of multidimensional information for manufacturing capability in cloud manufacturing system. Int J Adv Manuf Technol 2013;69(5−8):961−75.
[46] Li HF, Dong X, Song CG. Intelligent searching and matching approach for cloud manufacturing service. Comput Integr Manuf Syst 2012;18(7):1485−93.
[47] Tao F, Hu YF, Zhou ZD. Application and modeling of resource service trust-QoS evaluation in manufacturing grid system. Int J Prod Res 2009;47(6):1521−50.
[48] Cheng Y, Tao F, Xu L, Zhao D. Advanced manufacturing systems: supply−demand matching of manufacturing resource based on complex networks and Internet of Things. Enterprise Inf Syst 2018;12(7):780−97.
[49] Zheng Z, Lyu MRT, Wang H. Service fault tolerance for highly reliable service-oriented systems: an overview. Sci China Inf Sci 2015;58(5):1−12.
[50] Tao F, Zhang H, Qi Q, et al. Theory of digital twin modeling and its application. Comput Integr Manuf Syst 2021;27(1):1−16 [in Chinese].
[51] Tao F, Zhang H, Qi Q, et al. Ten questions towards digital twin: analysis and thinking. Comput Integr Manuf Syst 2020;26(1):1−17 [in Chinese].
[52] Zhao H, Liu J, Xiong H, et al. 3D visualization real-time monitoring method for digital twin workshop. Comput Integr Manuf Syst 2019;25(6):1432−43 [in Chinese].
[53] González M, Salgado O, Croes J, et al. A digital twin for operational evaluation of vertical transportation systems. IEEE Access 2020;8:114389−400.
[54] Yang Y, Chen Z, Yan J, et al. State evaluation of power transformer based on digital twin. In: 2019 IEEE international conference on service operations and logistics, and informatics (SOLI). IEEE; 2019. p. 230−5.
[55] Ge Y, Wang Y, Yu R, et al. Research on test method of autonomous driving based on digital twin. In: 2019 IEEE vehicular networking conference (VNC). IEEE; 2019. p. 1−2.
[56] Tao F, Sui F, Liu A, et al. Digital twin-driven product design framework. Int J Prod Res 2019;57(12):3935−53.

[57] Tao F, Cheng J, Qi Q, et al. Digital twin-driven product design, manufacturing and service with big data. Int J Adv Manuf Technol 2018;94(9):3563–76.
[58] He R, Chen G, Dong C, et al. Data-driven digital twin technology for optimized control in process systems. ISA Trans 2019;95:221–34.
[59] Laaki H, Miche Y, Tammi K. Prototyping a digital twin for real time remote control over mobile networks: application of remote surgery. IEEE Access 2019;7:20325–36.
[60] Tao F, Zhang Y, Cheng Y, et al. Digital twin and blockchain enhanced smart manufacturing service collaboration and management. J Manuf Syst 2020. https://doi.org/10.1016/j.jmsy.2020.11.008.
[61] Tao F, Cheng J, Qi Q. IIHub: an industrial Internet-of-Things hub toward smart manufacturing based on cyber-physical system. IEEE Trans Indus Inform 2017;14(5):2271–80.
[62] Tao F, Zhang M. Digital twin shop-floor: a new shop-floor paradigm towards smart manufacturing. IEEE Access 2017;5:20418–27.

CHAPTER 2

Digital twin-driven service collaboration

Feng Xiang[1], Jie Fan[1], Shiqian Ke[1] and Ying Zuo[2]
[1]School of Mechanical Automation, Wuhan University of Science and Technology, Wuhan, Hubei, China; [2]Research Institute for Frontier Science, Beihang University, Beijing, China

2.1 Introduction

The traditional manufacturing industry is being reshaped profoundly, and many advanced manufacturing systems/models and related policies have emerged successively, such as "Industry 4.0" in Germany, "Industrial Internet" in the United States, "Manufacturing cloud" in the European Union, "Internet + manufacturing" in China, etc., these advanced manufacturing modes all have the common characteristics of "intelligence [1], green [2], service [3]" to meet the global fierce competition situation, complex personalized needs and sustainable development trend, and achieve the coordination and optimization goal of "environment, benefit, and resource". However, manufacturing resources and services of a single enterprise can no longer meet the three characteristics of the above manufacturing mode, and the dynamic acquisition and sharing of manufacturing service collaboration (MSC) to integrate manufacturing resources and manufacturing capacity can pursue more efficient personalized manufacturing and carry out more large-scale collaborative manufacturing [4].

Fig. 2.1 shows a 3D printing cloud platform, which brings together all kinds of manufacturing service providers and their manufacturing services, including 3D model design service, material procurement service, printing equipment use service, etc., to complete 3D printing manufacturing tasks in the way of MSC. In order to complete the MSC for 3D printing, the platform supports (1) provide 3D model sharing and trading, support online processing model library and model management tools; (2) provide 3D printing custom processing capability, support 3D printing equipment sharing integration and cloud online processing methods and tools; (3) the method of intelligent on-demand matching and optimization for personalized customization and 3D printing services; (4) methods, tools, and platforms for managing and optimizing the configuration of social 3D printing resource.

Digital Twin Driven Service
ISBN 978-0-323-91300-3
https://doi.org/10.1016/B978-0-323-91300-3.00002-4

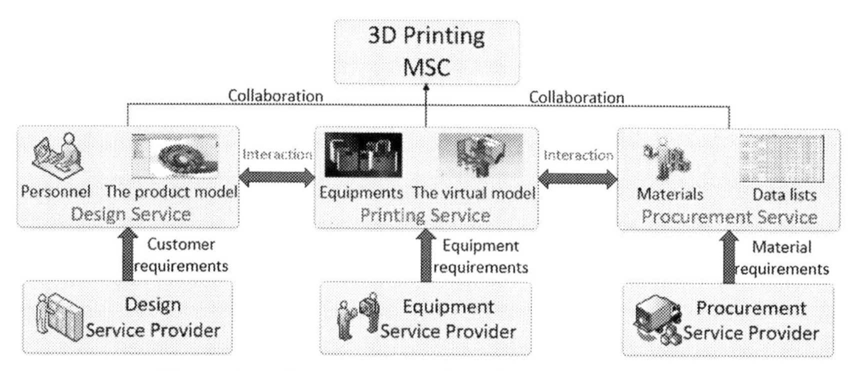

Figure 2.1 3D printing MSC based on cloud platform.

The core idea of MSC is to acquire, share, and integrate manufacturing resources and manufacturing capabilities dynamically, so as to pursue more efficient resource utilization, more personalized on-demand manufacturing and carry out larger scale collaborative manufacturing, and realize value cocreation and benefit sharing among all collaborators. At present, MSC research mainly concentrated in MSC modeling, collaboration optimization, and failure prediction, etc.

In terms of MSC modeling, Xu [5] proposed two kinds of cluster extended matrix factor models to obtain better predictive effects of QoS collaboration. Fan [6] integrated awareness concepts and workflow models and proposed a workflow awareness tracking framework, which promoted collaborative management under multitasks. Shin [7] studied the creation of composite services by linking the inputs and outputs of available services and proposed a composite approach that can explicitly specify and use the functional semantics of Web services. The above researches can only reflect the static collaborative modeling of QoS, workflow, functional semantic constraints, and other aspects between high-level manufacturing services but fail to consider the data—physical entity—virtual entity interaction under the virtual and real space, and fail to reflect that the essence of high-level manufacturing services is the dynamic cooperation of all the production factors under the virtual and real space fusion.

In terms of MSC optimization, Yang [8] used the coordination and optimization mechanism of a multilevel system to optimize the external, internal, and interenterprise resource collaboration. Marvel [9] proposed a new method of human-computer collaborative security by comprehensively considering multiple factors such as tools, nature of contact, and duration. Ren [10] deduced the service collaborative network through weighted aggregation and proposed a service selection model based on the overall synergetic effect on the maximization of collaboration requirements. The above MSC optimization is virtual space model simulation optimization under incomplete information. However, there are insufficient researches on whether the production factors in the physical space can be coordinated, and whether the optimized physical objects can be controlled according to the feedback of the virtual model to reach the optimization goals such as manufacturing task requirements, process optimization, and improvement of the overall operation efficiency of the system.

In terms of MSC failure prediction, Jia [11] proposed a diagnosis method based on the hidden Markov model (HMM), which is used to automatically diagnose faults in the process of web service composition. Bhandari [12] adopted the concept of partially observed random Petri nets, proposed a model-based fault diagnosis method, and converted it into Petri nets and random Petri nets by using existing constructs with Web services. Pang [13] used a spiked nerve P system with colored spikes to simulate faults in available services, which can be used for fault location and correct handling in case of service failure. The above collaborative fault diagnosis of manufacturing services is carried out on the premise of service failure in virtual space, without considering the information of factors of production in physical space and the interaction between virtual and real, it is impossible to determine whether the source of the fault is related to factors of production in physical space.

With the rapid development of industry internet platform [14], the amount and type of manufacturing services have greatly increased, along with the physical world and information world connectivity and integration ability. How to develop from static service collaboration in virtual space (the existing idea) to MSC under the fusion of virtual and real space (the proposed idea in this chapter) is a common research topic in modern manufacturing service industry. DT (digital twin) technology [15] has wide application potential in the field of manufacturing, it establishes a virtual twin model with high fidelity with physical entities in a digital way, and monitors, diagnoses, predicts, and optimizes the behavior and feedback control process of physical objects in the manufacturing environment by means of perception, interaction, and iterative optimization, so as to achieve precise control in various manufacturing application fields. Therefore, this chapter puts forward the DT-driven MSC, DT-MSC, which provides the technical possibility for the new MSC mode of integrated sharing and balanced configuration of various production factors in the physical and cyber space.

This chapter is divided as follows: the first part mainly introduces the concept of DT-driven MSC and describes the differences between DT-driven MSC (DT-MSC) and traditional MSC, as well as the forming frameworks and service processes of DT-MSC. The second part describes procedures and enabling technology of DT-MSC, including (1) MSC state aware monitoring, (2) data-driven MSC Collaborative relationship assessment, (3) MSC failure prediction under virtual model, (4) DT-MSC optimization and reconfiguration. Finally, a case study is provided to test the actual operation process and practical feasibility of DT-driven MSC correlation evaluation and optimal selection.

2.2 The concept of digital twin-driven service collaboration

2.2.1 Features of digital twin-driven service collaboration

MSC has roughly experienced three stages: from intraenterprise [16] and interenterprise process/business collaboration [17] to cross-domain/cross-industry collaboration [18] of manufacturing service platform. Among them, enterprise internal cooperation emphasizes on the basis of internal information integration and process integration, through the local area network to achieve the collection of resources, so as to quickly respond to the market; the process/business collaboration among enterprises pays more attention to the resource sharing and business collaboration among enterprises in the case of technological innovation and fierce competition to make up for the insufficiency of the manufacturing capacity of a single enterprise; the cross-domain/cross-industry collaboration of manufacturing service platform is to realize the on-demand sharing and configuration of distributed and heterogeneous manufacturing resources or capabilities in a distributed environment in the form of services through the manufacturing service platform [19].

Different from the traditional MSC mode, the DT-MSC is supported by the industrial Internet platform and achieves the MSC under the fusion of virtual and real. Its service cooperation features present new changes, which are more in line with the development trend of the manufacturing industry. Their differences are mainly reflected in the core ideas of the enterprise, the degree of information interaction and integration, the scope of collaboration, process management, and common goals, as shown in Table 2.1.

Table 2.1 Traditional MSC and DT-MSC.

Collaboration type	Response way	Degree of interaction	Collaborative way	Resource sharing mechanism	Collaboration scope
Traditional MSC	Passive response	Partial interaction, lack of integration	Static collaboration	Decentralize resources, centralize services	Intraenterprise/interenterprise
DT-MSC	Active response	Multidimensional interaction, virtual, and real fusion	Dynamic collaboration	Pooling resources、Scattered service	Full scope/full process/full factor

(1) Response way. The response way of traditional MSC is mainly driven by the requirements of manufacturing tasks, suppliers respond passively with a fixed number of resources or established solutions to provide services for users. The development of its MSC has a lag, and it is difficult to quickly respond to the manufacturing task demand. DT-MSC is to integrate all kinds of resources needed for product manufacturing into corresponding services under the support of the platform, actively responds to manufacturing service tasks, analyzes product service demand by big data, and carries out MSC through intelligent matching and service recommendation.
(2) Degree of interaction. The traditional MSC mainly relies on the interaction of part of information in the physical space and lacks data fusion in the virtual space. DT-MSC realizes multidimensional information exchange and data fusion between physical space and virtual space through high-fidelity mapping and feedback control between models and physical entities, ensures the certainty and consistency of collaboration development, and improves the stability and global optimization ability of collaboration.
(3) Collaborative way. DT-MSC transforms the static service capability collaboration encapsulated in virtual space into the dynamic collaboration of all production factors in the fusion of virtual and real space. The completion of manufacturing task is no longer a single MSC way, but a new service collaboration way, which is supported by the data, virtual and real object coupling interactive connection, and carries out integration sharing and balanced configuration of all production factors from virtual and real space.
(4) Resource sharing mechanism. The traditional MSC is mainly connect distributed manufacturing resource by network [20], emphasizes how to pool resources to accomplish a manufacturing task, and reflects the thought of the "scattered resources concentrated use" but because of the lack of centralized management and operation of services, manufacturing services work hard to ensure the implementation of the quality and reliability. DT-MSC breaks the traditional way, centralizes the manufacturing resources scattered in different regions, centralizes the management and decision-making of the assembled manufacturing resource services through the cloud platform, and provides services for users in different regions. It not only embodies the thought of "decentralized resources and centralized use" but also effectively realizes the thought of "centralized resources and decentralized use."

(5) Collaboration scope. The traditional MSC is limited by physical resources and timeliness of production factor information, and its scope only involves intraenterprise and interenterprise collaboration in local regions, which is obviously regional and unable to complete large-scale and multiprocess manufacturing tasks. DT-MSC is a full range/full process/full factor MSC model, this model greatly increases the resources involved in MSC. It not only can bring together the manufacturing resources across the organization and across the range, but also improve the ability to adjust and evenly distribute the manufacturing resources, and effectively avoid the problem of low resource utilization rate and low manufacturing efficiency caused by information island.

2.2.2 Framework of digital twin-driven service collaboration

The components of DT-MSC are shown in Fig. 2.2, including manufacturing resources in physical space, virtual model and twin data in virtual space, manufacturing service platform and the interaction and connection between them.

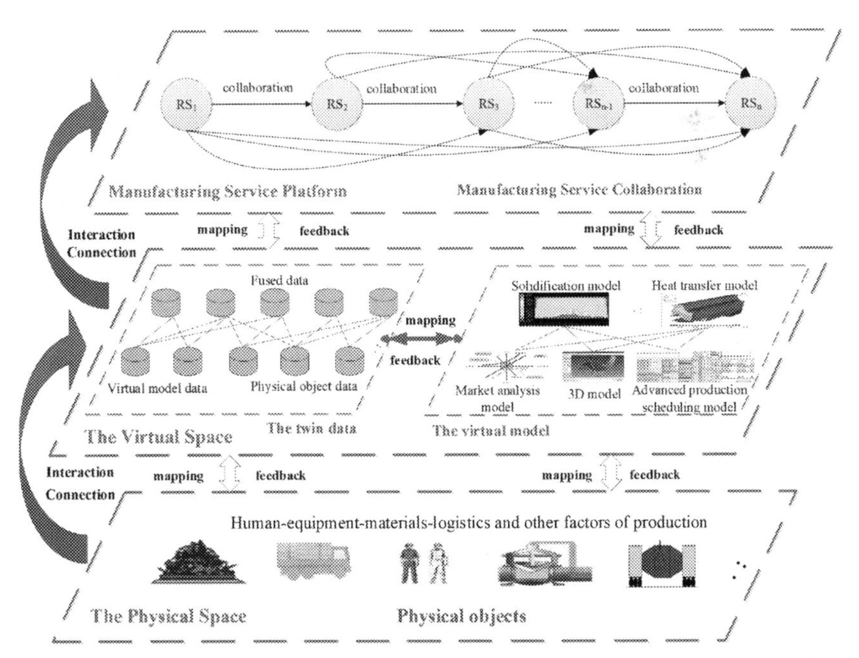

Figure 2.2 The service collaboration mode under the fusion of virtual and reality.

2.2.2.1 Manufacturing resources in physical space

Manufacturing resources in physical space refer to the software and hardware production factors that complete the production activities of the whole life cycle of product service system. It includes all physical objects involved in the process of design, manufacturing, maintenance, and other related activities, which are the material basis for the collaborative execution of manufacturing services, and also the service objects for feedback after virtual space simulation, such as hardware resources, software resources, human resources, material resources, logistics resources, environmental resources, etc.

2.2.2.2 Virtual model and twin data in virtual space

Virtual model

In virtual space, virtual model is the core component of product design, manufacturing and health maintenance, including service application model and service management model. Service application model is to use various modeling and simulation technologies to describe the elements, behaviors, and rules in the physical object, which can completely describe the physical object. Its virtual model involves geometric model, physical model, behavior model, rule model, and other multidimensional, multitemporal, and multiscale models; service management model plays a very important role in MSC decision-making process. It provides a place for MSC process optimization and transfers the optimization instructions to physical objects. Its virtual models include collaborative dynamic recombination model, collaborative multi-objective optimization model, collaborative QoS evaluation model [21], etc.

DT data

In the virtual space, twin data are the basic data support for the MSC, including physical object data, virtual model data, and fused service data. It is a comprehensive, multisource heterogeneous, and complex database set. Among them, physical object data include real-time perception, collection, and storage of physical entity state data and related historical data; the virtual model data are the model data generated by depicting the behavior, geometry, and rules of the physical object, the control parameters of the model and the simulation results of the model; the fused service data are the data generated in all kinds of manufacturing service-oriented high-level applications under the interaction of virtual model and physical object, including service encapsulation, service management, service maintenance, service after-sale, service matching, service search, service monitoring, service collaboration, etc.

2.2.2.3 Manufacturing service platform

The manufacturing service platform gathers the manufacturing resources of physical space and the model services of virtual space and collects the data of the decision-making process of physical object perception, monitoring and analysis and the simulation data of virtual model. It provides platform support for enterprises to participate in MSC and realize value creation, which is also an inevitable trend and one of the common goals of the current advanced manufacturing service model. The manufacturing service platform presents the characteristics of "cross-field, cross-industry, and full-cycle" when providing services [22], and the most typical one is the industrial Internet platform, which can realize the on-demand dynamic sharing and collaboration of distributed and heterogeneous enterprises' socialized manufacturing resources and capabilities in the form of services. And the industrial Internet platform has powerful computing and analysis capabilities, which can be used for equipment failure prediction, manufacturing service optimization, predictive maintenance, product quality analysis, collaborative process monitoring, etc.

2.2.2.4 Interaction and connection

Interaction and connection is the connection of physical objects, virtual models, and manufacturing service management systems into an organic whole, so that information and data can be exchanged and transmitted among various parts. It mainly includes low-level resources, twin data, virtual model, and information exchange and sharing among manufacturing services. It is the guarantee of the consistency and reliability of production factors in virtual and real space and the rapid response to dynamic manufacturing demand, and it is the key to ensure MSC. In the process of collecting and transmitting multisource heterogeneous service process data, the DT-MSC realizes the interconnection between the physical world and the virtual world through an extensible and interoperable interface.

2.2.3 Procedures of digital twin-driven service collaboration

As shown in Fig. 2.3, the DT-MSC process is unrolled in the physical space, and synchronously maps the MSC in the physical space to the virtual space. The virtual and real mapping process of MSC includes static encapsulation and dynamic awareness. Static encapsulation is a static

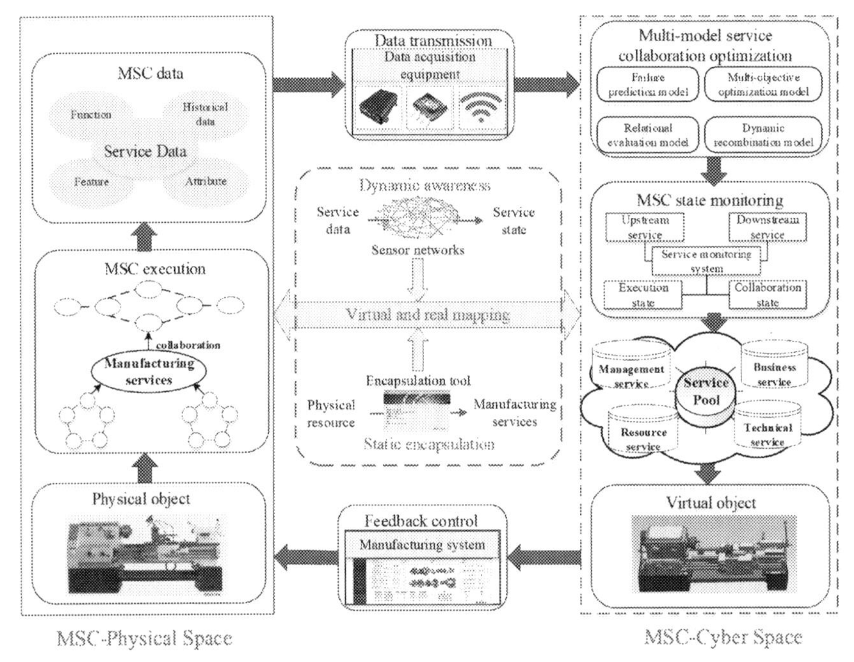

Figure 2.3 The DT-MSC process.

description and service encapsulation of the physical resources, by describing MSC of physical objects and the related information, determine the scenes and objects for the service collaboration, use the tool to encapsulate the physical resources as services, and connect the physical objects and the technology to make service collaborations, and import it into the virtual space service pool to ensure the virtual and real space service collaboration consistency, and finally perform the physical space service collaboration to get the collaboration service data. Dynamic awareness is a process of collecting data from physical space and storing it in a dedicated data awareness network through various data acquisition channels (such as RFID, device data, records, etc.) and then using it for virtual space service collaboration state. According to the service collaboration data obtained from physical space, the service collaboration optimization model is constructed, and the service collaboration management and optimization criteria are established to realize dynamic optimization and reconfiguration. At the same time, in order to verify the effectiveness of optimization, it is necessary to monitor the state of MSC in real time [23], including execution state for subtasks and collaboration state for service chains. It not only can ensure the normal operation of service collaboration, but also provide the basis for subsequent optimization management. On the basis of perception, the collaboration model of sensing data and service is combined to continuously optimize and update the collaboration state and data, transmit them to the manufacturing system, and input them to the physical space, control the allocation of physical resources, and realize the complete DT-MSC process.

2.3 Enabling digital twin service collaboration

2.3.1 State aware service collaboration monitoring

DT-MSC is completed under the support of integration and sharing of low-level multielements and balanced configuration. It needs to ensure the consistency of low-level multielements under the combination of virtual and real conditions, so it needs to obtain the service state in time to confirm (1) are the manufacturing services able to complete the current manufacturing tasks as planned? (2) Are the manufacturing services collaborating together as planned? Therefore, manufacturing service state monitoring includes manufacturing service execution state monitoring and MSC state monitoring.

2.3.1.1 Manufacturing service execution state monitoring

As shown in Fig. 2.4, manufacturing service execution state monitoring is mainly a single service execution state oriented to subtasks, and the execution process is monitored to judge the normal or abnormal completion of subtasks.

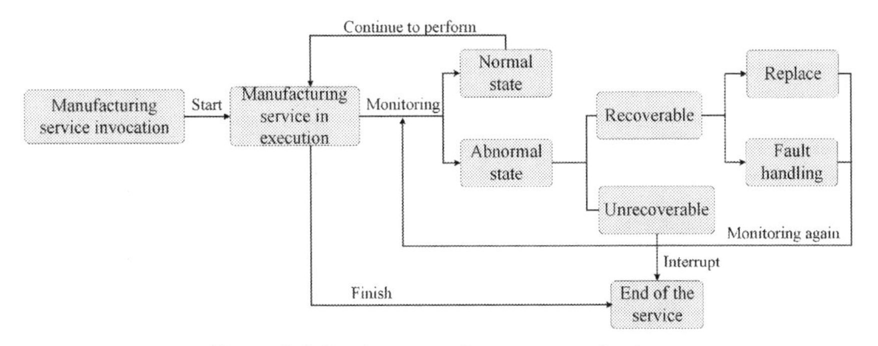

Figure 2.4 Service execution state monitoring.

Table 2.2 Monitored manufacturing service execution state.

Execution state	Description
Normal state	The manufacturing service execution process monitoring state is normal and can continue to complete the task
Abnormal state	The manufacturing service execution process monitoring state is abnormal and needs to be further determined
Recoverable	In case of an exception to the manufacturing service execution state, the task can resume and resume execution
Unrecoverable	In case of an exception to the manufacturing service execution state, the service execution is interrupted
Replace	In case of the manufacturing service execution state is recoverable, replace the current service to continue the task
Fault handling	In case of the manufacturing service execution state is recoverable, repair the failure and continue the task

When the service monitoring judgment for the state to be normal can continue to perform manufacturing tasks; when the service monitoring judgment for the abnormal state, including the service function and service performance of anomaly, etc. The degree of service exceptions is divided into recoverable and irreversible, when it is recoverable, it will be judged again whether it can be replaced or fault handled and repaired, and monitored again to determine whether the service state can continue to complete the task. The monitored service execution state is shown in Table 2.2 below.

2.3.1.2 Manufacturing service collaboration state monitoring

As shown in Fig. 2.5, MSC state monitoring mainly refers to the collaboration state among multiple services oriented according to the overall task, and monitoring the collaboration process is used to judge whether the overall task is completed normally or abnormally.

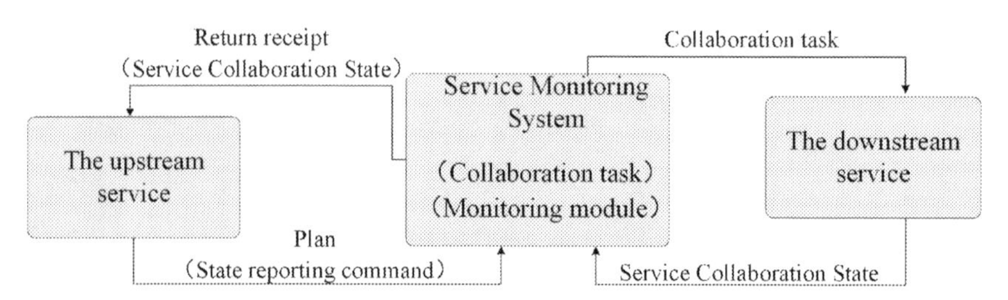

Figure 2.5 MSC state monitoring.

Monitoring the state of MSC is an important step to ensure the stability of the process of service collaboration. By monitoring the process of service collaboration and obtaining the relevant information of collaboration operation, the development of service collaboration can be better supported. MSC process monitoring has three processes, including upstream service, monitoring module, and downstream service. Upstream service: upstream service, as a service, belongs to the service that is executing the task in the collaboration process. The service monitoring process is the same as the execution service state monitoring mentioned above, with normal execution state and abnormal execution state. In addition, the upstream service also needs to send the downstream state reporting command to the monitoring module, and needs the receipt of the downstream service state monitoring, so as to master the state of the downstream service. When the upstream service completes the current manufacturing task, the appropriate downstream service collaboration can be selected to complete the next task to ensure the completion of the entire MSC. After the upstream service completes the manufacturing task, it becomes the downstream service and waits for the task to be reassigned.

Monitoring module: In addition to monitoring the collaboration state of each manufacturing service, the monitoring module also needs to regulate the completion of the entire collaboration process. The monitoring module maintains the progress of the collaboration and updates the corresponding task state according to the task state information sent by the upstream system. If the downstream system is idle, it will be assigned new tasks.

Downstream service: Downstream service collaboration monitoring state is shown in Fig. 2.6 below. When the manufacturing service is idle, it will accept the task to complete the task; when it is in the assigned or

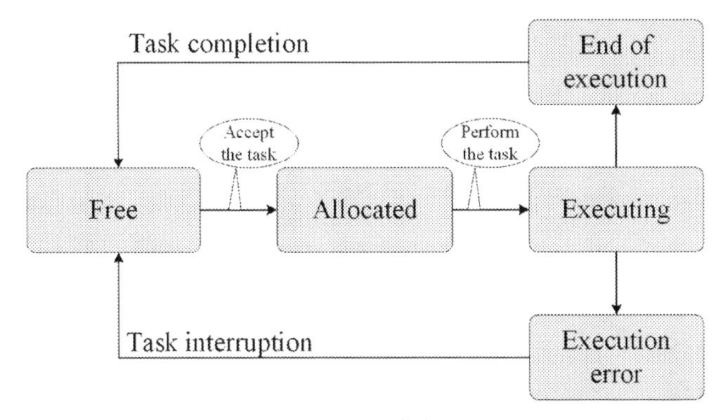

Figure 2.6 Downstream service collaboration monitoring state.

executing state, it waits for the current task to complete, and then it can receive the execution of the next task; when the downstream service receives a task, it becomes the upstream service state.

The MSC monitoring process consists of the following steps:

Step 1: The collaboration task module assigns subtasks to the manufacturing service.

Step 2: Manufacturing services complete subtasks through different workflow structures and collaborate to complete the overall task.

Step 3: The service system regularly sends the state reporting command to the service collaboration monitoring system.

Step 4: The monitoring module monitors the service collaboration state in real time.

Step 5: The service monitoring system collects and collates the information of MSC state.

Step 6: The service system receives receipt of service collaboration state.

Step 7: Complete the collaborative state monitoring process.

The monitored collaboration state is shown in Table 2.3 below:

Table 2.3 Monitoring the MSC state.

Collaboration state	Description
Upstream service normal	The upstream service collaboration state is normal and can complete the current manufacturing task
Upstream service abnormal	The upstream service collaboration state is abnormal and needs to replace other services to complete the current manufacturing task
Downstream service idle	The downstream service is idle and can receive collaboration tasks from the upstream service
The downstream service has been allocated	The downstream service is in the assigned state and needs to wait for the current task to complete before receiving a task
The downstream service is executing	The downstream service is in the executing state and needs to wait for the current task to complete before receiving a task
The overall process of collaboration is normal	The overall collaborative process is in a normal state and can continue to complete the current manufacturing task
The overall process of collaboration is abnormal	The overall collaboration process is in an abnormal state and needs to fix the abnormal service or interrupt the collaboration

2.3.2 Data-driven manufacturing service collaboration collaborative relationship evaluation

The objective is to find the collaborative relationships between manufacturing services. In the process of the actual service collaborations, the collaborative relationships between services are dynamic. Based on the perception of service collaboration state, this chapter uses the service monitoring system to perceive the execution state of a single service and the collaboration state of multiple services and evaluates the collaboration relationship and service optimization among services through the task completion. The judgment process is divided into three stages, as shown in Fig. 2.7.

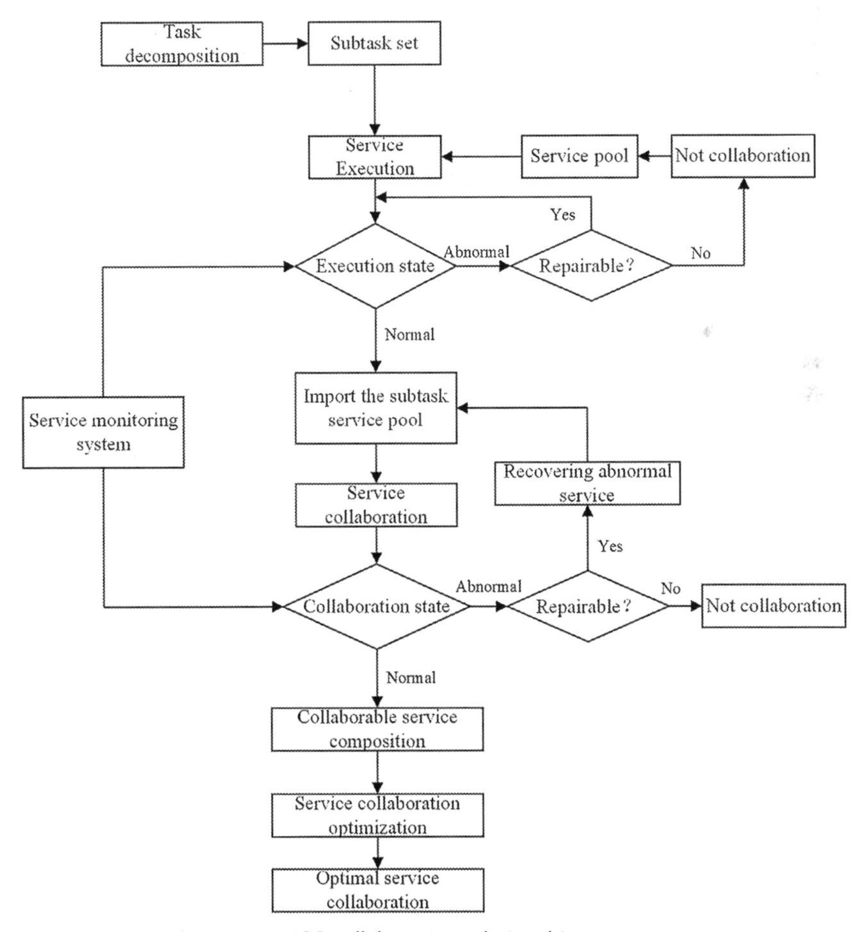

Figure 2.7 MSC collaborative relationship assessment.

The first stage determines whether the selected service can complete the related subtasks. After receiving the manufacturing task, the whole task is decomposed into several tasks to obtain the subtask set, and services are selected from the service pool to complete the subtask. When the selected service performs a subtask, based on the perception of the execution state of the service by the service monitoring system, if the execution state of the service is normal, the service will be imported into the corresponding subtask service pool as one of the solutions for this subtask. If the service execution state is abnormal, it also needs to determine whether the exception is repairable (the repair refers to the fault within the same service, the replacement service is not repairable). If it can be repaired, then continue to test whether the repaired service can complete the subtask, if it can, then import the repaired service into the subtask service pool; if it is not repairable, the service is not available for service collaboration and a new service needs to be reselected from the service pool to perform subtasks. Finally, a subtask service pool for each subtask can be obtained by repeatedly testing the execution state (services are not selected, just used to find out which services can solve the subtasks).

The second stage determines the collaborative relationships between services based on the overall task. According to the composition of the overall task, services are selected from each subtask service pool to form a service collaboration combination, and the collaboration state of the whole process is perceived through the service monitoring system. If the collaboration state is normal, it means that there is a collaborative relationship between the services in the service combination. If the service collaboration state is abnormal, it also needs to determine whether the exception is repairable (different from the first stage, it is a collaboration exception between services), if it can be repaired, the repaired service will be reimported into the service pool of corresponding subtasks and service collaboration will be carried out again; if it is not repairable, it means that the service does not collaborate with other services. Finally, by repeatedly testing the collaboration state, the collaboration service composition of the whole task can be obtained, and the collaboration relationship among the services can be evaluated.

In the third stage, service collaboration optimization is carried out. In general, the collaborative relationships between services obtained through the second stage are not unique most of the time, the efficiency and cost of executing the same task varies among different services. Therefore, in order to complete the manufacturing task more effectively, it is necessary to comprehensively consider the constraints of the manufacturing task (such as time, cost, reliability, etc.), discard the composition of services that do not meet the constraints of task requirements, and select the services that best fit the current plan [24].

2.3.3 Manufacturing service collaboration failure prediction under virtual model

Service failure refers to the state that the service cannot be accessed or obtained. It generally exists in the process of service running and seriously affects the reliability of service collaboration. As shown in Fig. 2.8, this section adopts an MSC failure prediction method of "service collaboration state (simulation data) + prediction model (fault prediction model)." By detecting the state of collaboration, the simulation data are obtained and used for fault prediction model, and the state and configuration of service application are adjusted, so that the service operation can quickly adapt to the change of application environment, which fully reflects the robustness and intelligence of DT-MSC.

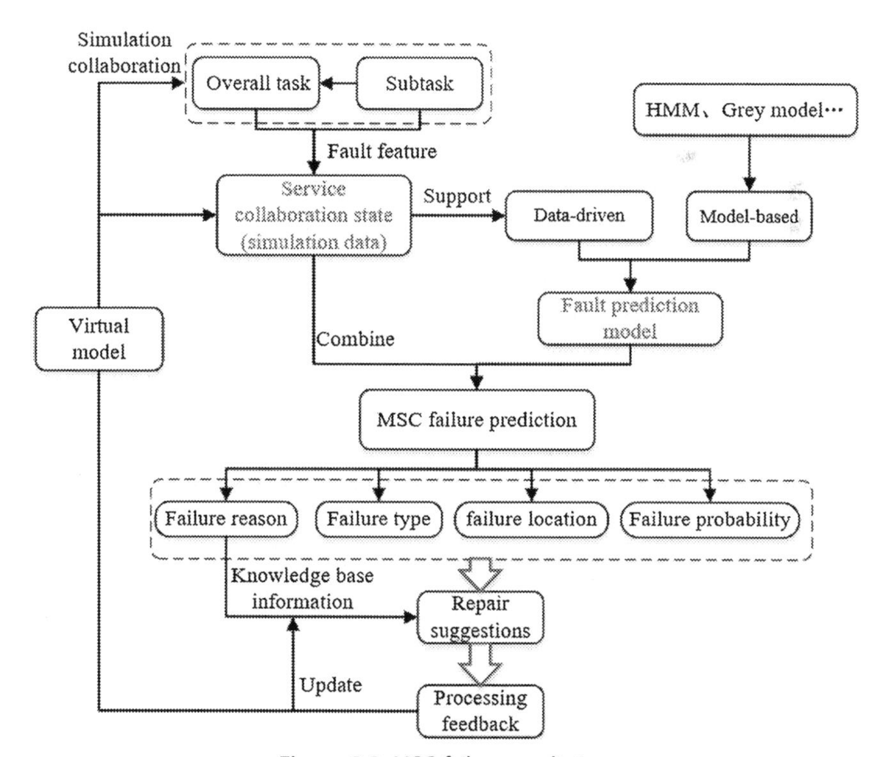

Figure 2.8 MSC failure prediction.

Service collaboration state (simulation data) is execution state data and collaboration state data obtained on the basis of virtual twin. The prediction model includes model-based prediction and data-driven prediction, the model-based prediction method assumes that an accurate mathematical model of the object system can be obtained, such as HMM (which can predict the process state well), gray model (which can predict the short-term fault well), etc. Data-driven prediction method is to generate automated prediction model through training and fitting on the premise of collecting massive data, which can be divided into artificial intelligence technology and statistical technology, such as neural network, decision tree, signal analysis, etc. This excerpt uses the data-driven prediction method to extract the fault features by simulating the operation of the overall task or subtask of the service collaboration, and accurately grasp the service collaboration state, and combine the collaboration state with the data-driven prediction model to predict the failure of the service collaboration [25]. It includes predicting the failure causes of service collaboration (such as poor information availability, dynamic change of service quality, limited service processing capacity, etc.), failure types (internal service failure and external service failure), failure locations and failure probabilities, and forecasting the development trend of service collaboration. In the meantime, synthesize the failure cause and other failure information for preventive maintenance, including repair suggestions and processing feedback. Repair suggestions are based on failure causes and knowledge base information query data, give solutions to eliminate the failure or reduce the probability of failure; processing feedback is to confirm the validity of repair suggestions based on the virtual model and historical service failure instances. If the failure problem is successfully solved, the solution and related information will be stored in the knowledge base to help deal with service collaboration failure [26].

2.3.4 Service collaboration optimization and reconfiguration

For existing MSC, with the unceasing change of user's target demand, the service collaboration usability is also changing, therefore, in order to ensure the effectiveness of the service collaboration, based on the predictions of service collaboration, it needs to decide whether the results of service collaboration can meet expected requirements, optimize, and refactor service collaborations that no longer meet requirements. As shown in Fig. 2.9, by comparing the actual results of physical space collaboration, the simulation results of virtual space collaboration and the user's target

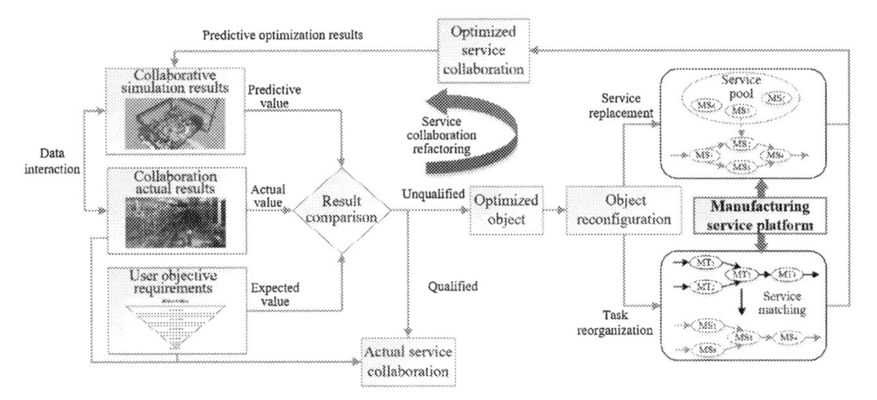

Figure 2.9 Service collaboration optimization and reconfiguration.

demand, when there is a big difference between the results and the results or between the results and the demand, that is, the service collaboration does not conform to the requirements, analyze the related services included in the service collaboration process, determine the object of service collaboration optimization, reconfigure the service collaboration object in the virtual space, and reconstruct the new service collaboration meeting the requirements. Service reconfiguration can be divided into service replacement and task reconfiguration:

(1) Service replacement means that the service selected in the current service collaboration cannot meet the requirements, and the service pool already has services that can meet the requirements. The service invoked by the manufacturing service platform should replace the service that does not meet the requirements. The invoked service should not only meet the new needs of users but also meet the collaboration requirements among services.

(2) Task reorganization means that the selected services of the current service collaboration cannot meet the requirements, and there is no service in the service pool that meets the requirements. According to the characteristics of decomposition and reconstruction of the manufacturing task, the whole manufacturing task is decomposed into different subtasks, and the services are matched and restructured into new service collaboration through the manufacturing service platform.

After the optimized service collaboration is obtained by reconstructing the manufacturing service through the manufacturing service platform, the optimization results are detected by the virtual space simulation collaboration, and the data are synchronized to the physical space and compared again. When the result of service collaboration meets the requirements of users, the actual service collaboration can be carried out and the optimization and reconstruction of service collaboration can be realized [27].

2.4 Case study

This section takes 3D printing as an example to illustrate the application process of MSC. Its application is mainly divided into five parts, including manufacturing task decomposition and demand, construction of service collaboration scheme, selection of service provider, implementation monitoring of service collaboration scheme, and evaluation of service collaboration. Through the analysis of 3D printing process of manufacturing resources, determine the demand for services, which decides the collaborative relationship between the service, obtain the corresponding optimal service collaboration of manufacturing tasks, and test the reliability and availability of service collaboration scheme. Finally, the data of service collaboration are stored on the network platform and shared to the users in real time to realize the application of service collaboration, Fig. 2.10 shows the overall process of 3D printing service collaboration.

(1) Take 3D-printed car lamp parts as an example, the overall printing task can be decomposed into five subtasks: scanning, design modeling, slicing, 3D printing, and postprocessing. At the same time, in order to verify whether the printed headlight parts can meet the product design requirements, it is generally required to add two subtasks: entity test and model correction. Guided by the functional needs of consumers and combined with subtasks to determine the manufacturing resources and capabilities needed in the production process of parts, including materials, design, software, scanning equipment, processing technology, equipment, path, and environment, etc. Then, based on these resource requirements and capability requirements, one can deduce what conditions the selected service should have, that is, to determine what kind of service is needed by the subtask requirements.

(2) As can be seen from Section 2.3.2, the use of services is a selection process, and the steps for obtaining the optimal service collaboration scheme are divided into three stages. The first stage is to select the subtask service on the basis of the subtask requirements of the car lamp parts mentioned above. Under certain subtask requirements and

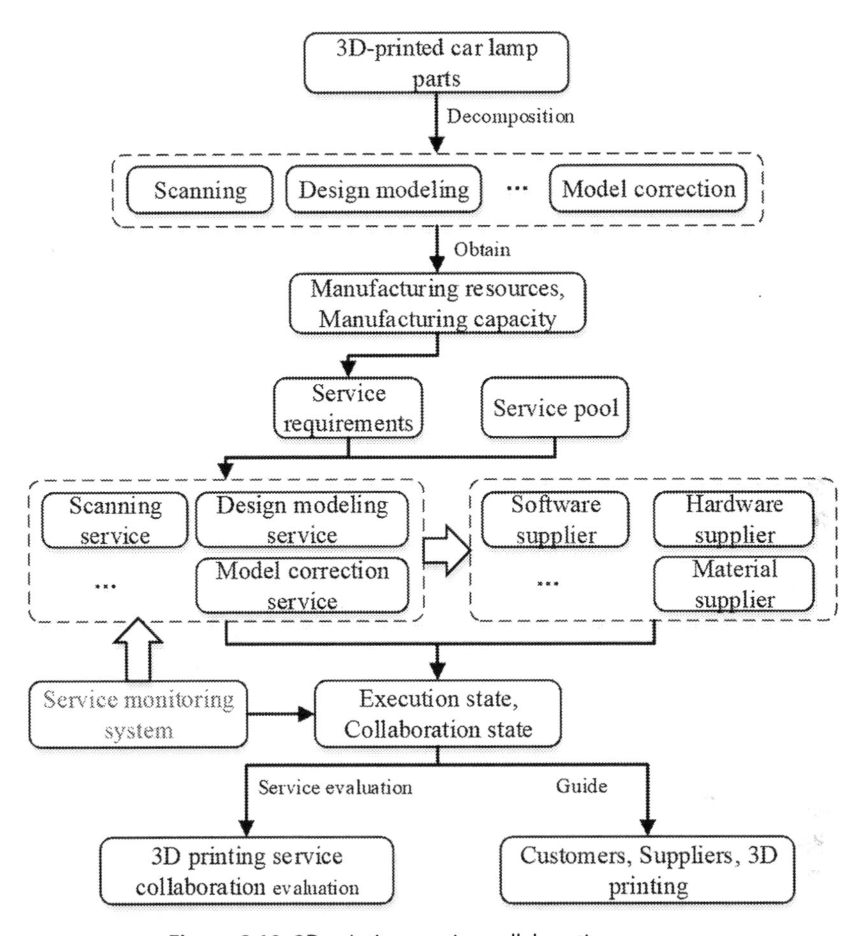

Figure 2.10 3D printing service collaboration process.

constraints, the manufacturing services in the service pool are searched to match the conditions. With the support of the service monitoring system, the execution state of the services is perceived to verify the correctness of the selected services. Finally, the seven subtask service pools of 3D-printed car lamp parts are obtained. In the second stage, it continues to rely on the service monitoring system to perceive the overall process service collaboration state, verify the correctness and reliability of the collaborative relationship between services, and obtain the collaborative service combination for the manufacturing of auto lamp parts. In the third stage, the optimal service collaboration is selected from the collaborative service combination. By comprehensively considering the constraints of the manufacturing task, establish service quality evaluation indicators and assign weights to reflect their importance. Finally, a service quality evaluation model is established to select the optimal service collaboration scheme.

(3) Supplier selection is used to determine whether a service provider is qualified to perform the currently required services. Due to the characteristics of diversified service contents, its choice of service objects is also greatly increased, and the upstream and downstream service providers also play dynamic games at all times, which makes stable service relationships more flexible and changeable. Enterprises need to make decisions in a more complex information environment. The construction of evaluation index system of supplier selection is an important guarantee of the supplier selection, under the premise of meeting the demand of service, to comprehensively consider the product competitiveness of service providers, service ability and quality, environmental performance, cost flexibility and enterprise credit ability, and the service requirements of subtasks are used to evaluate what kind of suppliers to choose.

(4) On the basis of Section 2.3.1, the implementation monitoring content of service collaboration scheme includes state awareness of service collaboration scheme and follow-up processing. After dealing with the collaboration solution and suppliers, verify the availability of the service by actually executing the solution, simulate the process of 3D printing of car lamp parts, use the service monitoring system to perceive the real-time state data of the service collaboration execution process (such as equipment data, personal data, environmental data, etc.), and obtain the service execution state, service collaboration state, and service completion progress. The Table 2.4 shows the service state and service progress at a given time. For ease of processing and use, the services in the optimal service collaboration scheme are directly named after the subtask name (① scanning service, ② design modeling service, ③ slicing service, ④ 3D printing service, ⑤ post-processing service, ⑥ physical testing service, ⑦ model correction service).

The following processing part is divided into service normal state processing and service abnormal state processing. In the normal state, that is, the selected service runs normally and the service collaboration is normal. However, in order to obtain a more secure and reliable service collaboration scheme, the service collaboration failure prediction is carried out with the support of the service collaboration state (data),

Table 2.4 3D-printed car lamp parts service collaboration state.

Service name	Service relationship		Service state		Service progress
	Upstream service	Downstream service	Execution state	Collaboration state	
Scanning service	—	②	Normal	Normal	100%
Design modeling service	①	③	Normal		100%
Slicing service	②	④	Normal		60%
3D printing service	③	⑤	Normal		20%
Postprocessing service	④	⑥	Normal		0%
Physical testing service	⑤	⑦	Normal		0%
Model correction service	⑥	—	Normal		0%

and the failure situation of the MSC is simulated to ensure that the service collaboration can be repaired quickly and effectively. In the abnormal state, according to the source of abnormal problems (including the service collaboration process and service providers, the correctness of the service and the cooperative relationship between services has been verified), to detect, locate, optimize, and restructure service collaboration and to adjust service collaboration or change service providers.

(5) After the completion of the 3D printing manufacturing task, the overall evaluation of the service collaboration, such as product quality, service quality, requirement completion, etc., can be realized through the feedback information of product use. The service monitoring system plays an important role in the process of service collaboration. Combined with the service state (data), it realizes the selection of services, the perception of the collaborative relationship between services, and obtains the optimal service collaboration scheme. At the same time, by observing the state of the follow up actual service collaboration process, the MSC control of the whole life cycle of the product is realized, the resource service management system is established and the data are stored, which provides a valuable reference for the future 3D printing customers and service providers.

2.5 Summary

This chapter starts with the MSC under the virtual-real fusion and explores the DT-MSC, this chapter introduces the development process and characteristics of MSC, and based on the shortage of the existing service collaboration builds the DT-MSC method, illustrates the mode of collaboration in the form of digital twins. The main research contents are as follows:

(1) The introduction of the concept of digital twin-driven service collaboration. This chapter first describes the differences between DT-MSC and traditional MSC, and explains its characteristics of service collaboration. Secondly, this chapter describes the DT-MSC framework and introduces the components of the DT-MSC. Then, this chapter analyzes the processes of DT-MSC and briefly explains the operation mode of its service collaboration through the flow chart.

(2) The introduction of the enabling technology that supports DT-MSC. Firstly, the state-aware service collaboration monitoring technology is analyzed to ensure the element consistency under the virtual-real fusion. Secondly, the Data-driven MSC Collaborative relationship assessment is described to guarantee the reliability of service collaboration. Then, the failure prediction of MSC under the virtual model and the optimization and reconstruction of DT-MSC is explained to provide methods and suggestions for the failure and maintenance of service collaboration. Through the description of each enabling technology, the perfect DT-MSC technology system is constructed.

(3) Taking 3D automobile lamp parts as an example, the proposed method of DT-MSC is demonstrated through a complete description of service collaboration process.

References

[1] Tao F, Qi QL. Service-oriented smart manufacturing. J Mech Eng 2018;054(016):11–23.

[2] Xiang F, Huang YY, Zhang Z, et al. New paradigm of green manufacturing for product life cycle based on digital twin. Comput Integr Manuf Syst 2019;025(006):1505–14.

[3] Sholihah M, Maezono T, Mitake Y, et al. Formulating service-oriented strategies for servitization of manufacturing companies. Sustainability 2020;12(22):1–30.

[4] Wang H, Liu L, Fei Y, et al. A collaborative manufacturing execution system oriented to discrete manufacturing enterprises. Concurr Eng 2016;24(4):330–43.

[5] Xu Y, Yin J, Li Y. A collaborative framework of web service recommendation with clustering-extended matrix factorization. Int J Web Grid Serv 2016;12(1):1–25.

[6] Fan S, Kang L, Zhao JL. Workflow-aware attention tracking to enhance collaboration management. Inf Syst Front 2015;17(6):1253–64.

[7] Shin DH, Lee KH, Suda T. Automated generation of composite web services based on functional semantics. Web Semant Sci Serv Agents World Wide Web 2009;7(4):332–43.

[8] Yang X, Shi G, Zhang Z. Collaboration of large equipment complete service under cloud manufacturing mode. Int J Prod Res 2014;52(2):326–36.

[9] Marvel JA, Falco J, Marstio I. Characterizing task-based human–robot collaboration safety in manufacturing. IEEE Trans Syst Man Cyber Syst 2015;45(2):260–75.

[10] Minglun R, Lei R, Hemant J. Manufacturing service composition model based on synergy effect: a social network analysis approach. Appl Soft Comput 2018;70:288–300.

[11] Jia ZC, Lu Y, Li X, et al. HMM-based fault diagnosis for web service composition. J Comput 2020;31(1):18–33.

[12] Bhandari GP, Gupta R. Fault diagnosis in service-oriented computing through partially observed stochastic Petri nets. Serv Orien Comput Appl 2020;14(1):35–47.

[13] Pang SC, Wang M, Qiao SB, et al. Fault diagnosis for service composition by spiking neural P systems with colored spikes. Chin J Electron 2019;28(5):1033–40.

[14] Li J, Qiu JJ, Dou KQ. Research on the reference architecture, core function and application value of industrial internet platform. Manuf Autom 2018;040(006):103–6.

[15] Liu Q, Leng J, Yan D, et al. Digital twin-based designing of the configuration, motion, control, and optimization model of a flow-type smart manufacturing system. J Manuf Syst 2020;58:52–64.

[16] Lewis MA. Lean production and sustainable competitive advantage. Int J Oper Prod Manag 2000;20(8):959–78.
[17] Jiang P, Zhang Y. Visualized part manufacturing via an online e-service platform on web. Concurr Eng 2002;10(4):267–77.
[18] Tao F, Zhang L, Venkatesh VC, et al. Cloud manufacturing: a computing and service-oriented manufacturing model. Proc IME B J Eng Manufact 2011;225(10):1969–76.
[19] Li P, Cheng Y, Song W, et al. Manufacturing services collaboration: connotation, framework, key technologies, and research issues. Int J Adv Manuf Technol 2020;110(9–10):1–17.
[20] Liu W-Q, Ji Y. Construction of industrial design resource sharing mechanism based on cloud service. Int J New Dev Eng Soc 2020;4(2).
[21] Li XJ, Yuan YP, Sun WL, et al. QoS-driven evolution model of manufacturing service collaborative networking cloud manufacturing mode. Modu Mach Tool Autom Manuf Tech 2017;(01):46–49+53.
[22] Sheng BY, Zhang CL, Lu QB, et al. Research and implementation of cloud manufacturing service platform's intelligent matching for supply and demand. Comput Integr Manuf Syst 2015;21(03):822–30.
[23] Chaabane M, Krichen F, Rodriguez IB, et al. Monitoring of service-oriented applications for the reconstruction of interactions models. Spring Int Publ 2015;7(13):172–86.
[24] Hou Y, Cao ZJ, Yang SL. Cloud intelligent logistics service selection based on combinatorial optimization algorithm. J Eur Systèmes Automatisés 2019;52(1):73–8.
[25] Sharma M, Glatard T, Gélinas É. Service failure prediction in supply-chain networks. In: Proceedings-2018 IEEE international conference on big data; 2018. p. 1827–36.
[26] Li XB, Rao F, Guo L, et al. Research on prediction of logistics equipment based on cloud service. Manuf Autom 2020;42(07):131–4.
[27] Zhang Y, Zhang P, Tao F, et al. Consensus aware manufacturing service collaboration optimization under blockchain based Industrial Internet platform. Comput Ind Eng 2019;135:1025–35.

CHAPTER 3

Digital twin-driven production line custom design service

Hao Li and Haoqi Wang
College of Mechanical and Electrical Engineering, Zhengzhou University of Light Industry, Zhengzhou, Henan, China

3.1 Introduction

With global market competition, some companies have developed customized systems for their customers, covering various products such as clothing, computers, integrated kitchens, and mobile phones. Many leading high-tech companies are implementing personalized custom design services to maintain a competitive advantage. Among them, the "Dell model" is a relatively successful custom case. The Dell Company in the United States has realized the customer's independent product configuration design on the Internet, thereby reducing sales links and reducing costs. Japan's Matsushita Electric Co., Ltd. customizes the refrigerator on the Internet, and the customer can choose the style, color, door handle, and other personalized customization. European bicycle companies have launched typical personalized customization services, including parts configuration, frame painting, etc. Buick Motors North America website provides users with personalized customization services, including the choice of a series of parameters such as engine model, color, car tires, etc., and even the customer's name can be indicated on the back of the car. One month later, a unique car in the world will be delivered to the customer. In China, Haier Group, a large-scale home appliance company, provides an order-based personalized customization system. Customers can customize a personalized refrigerator that meets their requirements according to their individual needs for capacity, style, color, etc., and then send the customized information to the manufacturer as a production order, the user thus can get the customized product within a week [1,2].

Digital twins provide a new approach for the implementation of custom designs for production lines. Digital twin technology can realize the integration and iterative optimization of the physical model and information model of the intelligent production line, thereby shortening the production line customization cycle and reducing design costs [3,4]. When the state of

Digital Twin Driven Service
ISBN 978-0-323-91300-3
https://doi.org/10.1016/B978-0-323-91300-3.00001-2

the physical system changes, the state of the corresponding virtual model in the digital world also changes simultaneously. The optimization and adjustment of the virtual model will also be fed back to the physical system through automatic and manual adjustment of industrial Internet so that the physical system can also make corresponding adjustments.

Many scholars study how to use digital twin technology in production line design. For example, Zhang et al. proposed a personalized rapid design method for insulating glass production lines based on digital twins, which combines physics-based system modeling with distributed real-time process data to generate the digital design of the system during the production phase. Different from the traditional simulation architecture, the virtual twin based on distributed simulation establishes the relationship between the physical production line and its virtual model, which can verify whether the production line meets the design intent and customer needs [5]. Yi et al. proposed a digital twin reference model for intelligent assembly process design [6]. Referring to the five-layer architecture of CPS, Sun designed the overall architecture of digital twin system of aircraft assembly line by using the improved CPS grid model and Moore finite state machine [7]. Zhang focused on the modeling method of the digital twin stamping production line based on physical data [8]. On the basis of analyzing the current situation of the motor assembly line and aiming at some problems existing in the process of production management, Ding proposed the construction and application scheme of the digital twin model of the motor assembly line based on the concept of real-time mapping of digital twin physical elements [9]. After analyzing the concept and connotation of digital twinning technology and discussing the application scenarios of digital twinning technology in aero-engine design, manufacturing, service, and other fields, Cui studied and verified the application of digital twinning technology in the intelligent production line [10]. Kyu et al. proposed a cyber-physical production system architecture framework based on digital twins. The framework includes five services and operates based on the information of the proposed product, process, plant, factory, and resource information model. The application of CPPS based on digital twins in microsmart factories provides advanced solutions for the personalized production of various products [11].

From the above analysis, the custom design of the production line driven by the digital twin involves the modular design, parametric design, virtual modeling and simulation technology, and virtual debugging. The following chapters, respectively, introduce in detail the connotation, framework, key theories, and technologies of the digital twin-driven production line custom design.

3.2 Framework for digital twin-driven production line custom design service system

3.2.1 Connotation of digital twin-driven product design

In the traditional design approach, a design process is completed after the design is verified by the virtual prototype, the design model and documents are then delivered to the production department. Manufacturing-oriented design, assembly-oriented design, and maintenance-oriented design are all completed in the design phase. Complex product design based on digital twins is different from traditional design methods, which requires analysis from the concept of the digital twin.

The earliest definition of the digital twin is proposed by Prof. Michael Grieves of the University of Michigan, who believes that a digital twin forms a close loop between design and implementation [12]. Airbus proposed the concept of industrial digital mock-up (iDMU), pointing out that iDMU is a complete definition and verification of virtual product manufacturing, which contains interactive product design, process design, and manufacturing resource data [13,14]. Based on the analysis of the above two concepts, the product design based on the digital twin studied in this chapter is an integrated method of design and manufacturing, which forms a tight closed loop in the design and manufacturing stage and can deliver products and actual design and manufacturing data to users [15]. Fig. 3.1 shows the complex product design process based on the digital twin. This process goes from analyzing user needs to function/performance-oriented product design. At the same time, design models and documents need to

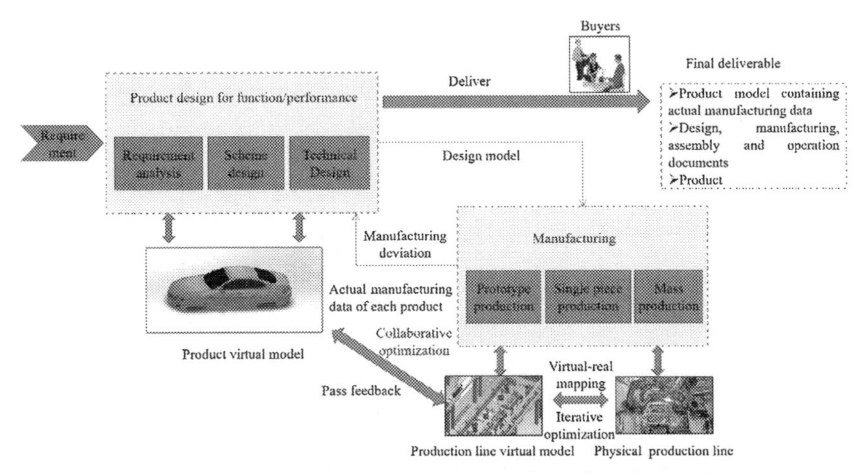

Figure 3.1 Product design process based on digital twins [15].

be transferred to the manufacturing stage. After processing and assembly, the actual size, assembly parameters, and defective product information of the sample product are formed and fed back to the design stage. It can be delivered to customers' custom products and their unique product design parameters and models to form a large-scale personalized provision of user needs.

The digital twin-driven product design method differs from the traditional design method in the design focus, design process, design and manufacturing cross-phase submission, design cycle, and delivery of user information, as shown in Table 3.1.

Based on the above analysis and comparison, the connotation of complex product design based on digital twin can be summarized as follows:

(1) Complex product design based on digital twins is a design method oriented to large-scale individual needs of customers, which can realize the submission of unique single-piece instance products to users. An instance product includes the product itself, virtual prototypes, process information, actual manufacturing parameters of the product, operation manuals, etc.

(2) After product processing and assembly are completed, the actual manufacturing data are returned to the digital twin design, and then the complex product design based on the digital twin is completed. It realizes the integration of design and manufacturing and forms a tight closed loop between the design and manufacturing stages.

(3) The design of complex products based on digital twins is a design process of virtual-real iteration and loop optimization.

(4) After the design stage, the designer provides a complete design digital twin to the manufacturing stage, including a high-fidelity virtual prototype, and integrates it into the digital twin of the processing, process, and production line in the manufacturing stage to realize the design virtual model and manufacturing coordinated operation of virtual models.

3.2.2 Basic structure of digital twin-driven production line custom design service

Modular design and parametric design provide theoretical and technical support for production line custom design services. Modular design can quickly match key equipment from the module library, and parametric design can drive customized modeling of virtual production lines according to users' individual needs [16,17]. Digital twin technology can realize the

Table 3.1 The relationship between complex product design methods based on digital twins and traditional design methods [15].

Design method	Design concerns	Design process	Design and manufacture cross-phase submissions	Design cycle	Deliver user information
Traditional design methodology	Function/performance	Serial iterative optimization during product design	Providing drawings and lightweight 3D models to manufacturing	Serial or parallel design in stages such as user requirements, program design, and technical design.	Product main design parameters, installation, and operation manual
Complex product design method based on digital twin	Customer	Virtual and real iteration, loop iteration	Providing a complete design of digital twins and integrate them into manufacturing digital twins	On the basis of the traditional design stage, it is extended to prototype trial production and product manufacturing stage. A tight closed loop is formed between design and manufacturing, forming an integrated collaboration of design and manufacturing.	The deliverables of a single instantiation include product design, process design, manufacturing parameter data, operation manual, etc.

integration and iterative optimization of the physical model and virtual model of the production line, thereby shortening the production line custom design cycle and reducing design costs. When the state of the physical production line changes, the state of the virtual model also changes simultaneously. The optimized adjustment of the virtual model is fed back to the physical production line through the industrial Internet automatic and manual adjustment so that the corresponding design changes can be made. The custom design service framework of the production line driven by the digital twin is displayed in Fig. 3.2, including four main components:

(1) Modular design system: This system is oriented to the individual needs of users, these needs are divided into modules which are assigned to the key equipment in the production; then, through module configuration and optimization, a personalized, customized production line that meets user needs is generated.

(2) Parameterized customization system: The system can automatically complete production line layout design, key equipment design, and environmental condition design according to the design parameters of the production line design plan, which lays to a foundation for the establishment of a digital model of the production line.

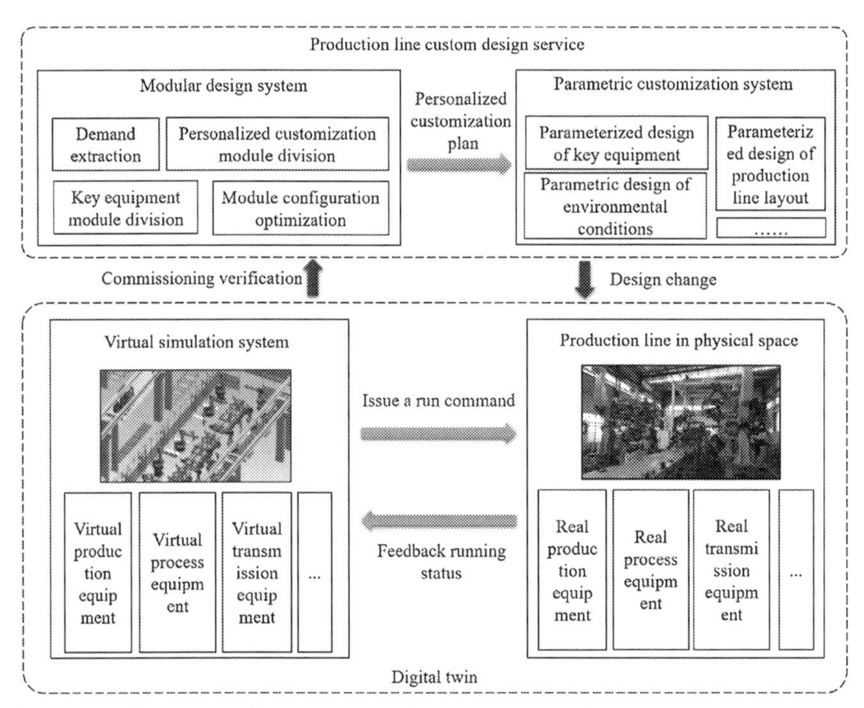

Figure 3.2 The composition structure of the production line custom design service driven by the digital twin.

(3) Virtual simulation system: This system is a digital mapping of the real production line in the physical space, which includes the virtual production equipment model, virtual process equipment model, and virtual conveying equipment model corresponding to the actual production line. It can be compared in the virtual space—production line capacity analysis, process planning, design optimization, etc.
(4) The actual production line in the physical space: It mainly refers to the physical entity of the production line, including actual production equipment, process equipment, conveying equipment, and other equipment, as well as a collection of production activities formed by related equipment of the production line.

The modular design system and the parametric customization system constitute the production line customization design service, and the virtual simulation system and the real production line in the physical space constitute a digital twin. When the design needs to be changed, the production line custom design service sends a change service request to the digital twin system in time. After receiving the request, the digital twin system performs simulation analysis in the virtual simulation system. After the simulation result meets the requirements, it can be used to guide the adjustment of the real production line in the physical space and complete the design change.

The virtual simulation system and the real production line in the physical space constitute a digital twin. On the one hand, the virtual simulation system can perform virtual debugging and optimization of the physical production line before it is established, which can reduce the construction and save time of the production line and saving design costs; on the other hand, the running status of the real production line can be fed back to the virtual simulation system. Through the linkage between the virtual production line and the real production line, the problems in the operation of the production line can be found in time and the maintenance cost of the production line can be reduced.

3.2.3 Key theory and technology of digital twin-driven production line custom design service

Under the digital twin-driven production line custom design service framework, the realization of custom design services requires different theoretical and technical support. Fig. 3.3 describes the four-layer structure required to realize the customized design service of the production line. From bottom to top, they are the theoretical layer, the technical layer, the business layer, and the demand layer.

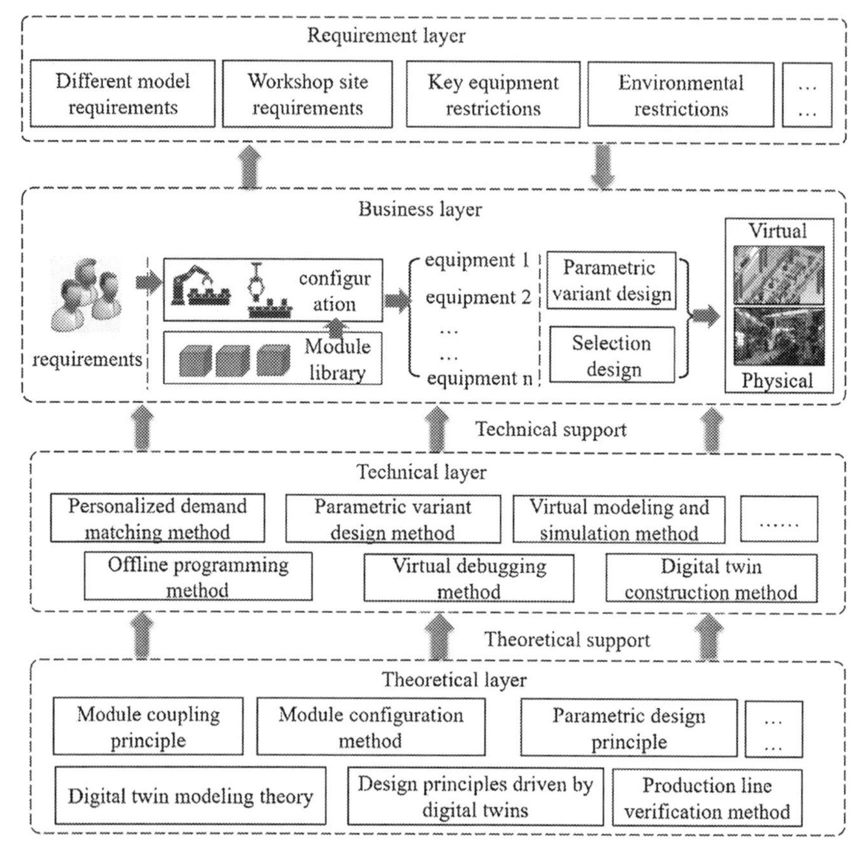

Figure 3.3 The key theory and technical framework of production line custom design service driven by digital twin.

(1) The theoretical layer: It provides theoretical support for the digital twin-driven production line custom design services, which mainly includes three theories, namely, modular design theory, parametric design theory, and digital twin-related theories. Modular design theory includes module coupling principles and module configuration methods; digital twin-related theories include digital twin modeling theory, digital twin-driven design principles, and production line verification methods.

(2) The technical layer: It provides technical support for the business layer, mainly including personalized demand matching methods, parametric variant design methods, virtual modeling simulation methods, offline programming methods, virtual debugging methods, and digital twin modeling methods.

(3) The business layer: Under the warrant of related theories and technologies, designers complete the construction of the digital twin production line at the business layer. First, configure the production line according to the user's personalized needs defined by the demand layer. During the configuration process, the equipment that meets the needs and functions can be matched from the existing module library; second, the detailed design plan of the production line is obtained through parameterized variant design and selection design; third, the virtual production line is established and interacts with the physical production line to form a digital twin system.

(4) The requirement level: As the top level of the design, it reflects the personalized customization requirements which can be divided into functional requirements and nonfunctional requirements according to different types. According to the objects, these personalized customization requirements can be divided into requirements for different product models, workshop site requirements, key equipment restrictions, environmental conditions restrictions, and policy and regulatory restrictions. The construction of the digital twin production line is to meet these individual needs.

3.3 Development method of digital twin-driven production line custom design service system

3.3.1 Design and development of modular system

The modular design of the production line is to divide the production line into multiple modules. Each module has independent function, consistent geometric connection interface, and consistent input and output interface units. The same types of modules can be reused and interchanged in the product family. The permutation and combination of related modules can form the final production line. Through the combination and configuration of the modules, it is possible to create product production lines with different requirements to meet the individual customization needs of customers.

The modular system construction of the production line is the basis for realizing customized design. The modular system consists of a set of modules (physical modules and service modules) and corresponding rules. They are adding or replacing a group of special modules in the modular system (the special modules can be physical modules or service modules) and derive a series of personalized production line design methods, called modular design based on product platform. A series of products derived from the same product platform constitutes a product family.

The production line modular system design and development process is mainly divided into three steps, that is, production line modular scheme design and planning, production line modular configuration design, and modular realization, as shown in Fig. 3.4.

In the stage of modular design and planning of production line, personalized production line data can be obtained according to customers' customized requirements, scheme design, module division, and module design can be carried out to form a series of general equipment and personalized and customized design, and the production line architecture can be planned and the production line design platform can be established. In the design stage of the modular configuration of the production line, personalized recommendation and experience are carried out according to

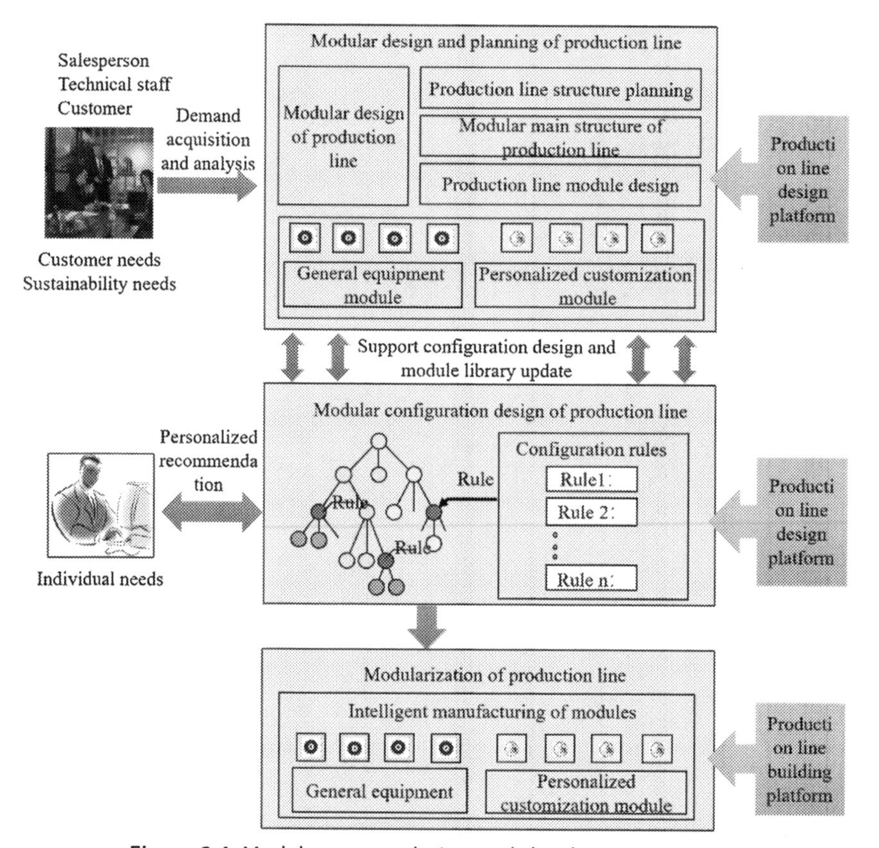

Figure 3.4 Modular system design and development process.

customer needs, and the modular instance structure of the production line is established to make the configured production line meet the unique needs of some customers as far as possible. In the modular stage of the production line, through the intelligent manufacturing of the general equipment module and the personalized customization module, users can track and experience the construction process of the customized production line, so that users can get a unique experience of the construction and manufacturing process of the production line.

Compared with the traditional product design process, the customized design of the production line driven by digital twins emphasizes the personalized recommendation of service solutions and the personalized customer experience. Acquisition of customer needs and functional design is the key to modular design. The needs of different customers vary greatly, and there are a large number of uncertain requirements in the actual design. Therefore, it is necessary to analyze the unknown uncertainties related to customer needs and find out the main design parameters related to them. The uncertainty is classified as a part of the demand, and the personalized demand hierarchy model of the production line is built by using the Kano model graph and the Kano model demand classification matrix to determine the importance ranking of the personalized demand items [18]. According to the different categories of needs analysis and the importance of demand items, the function mapping relationship model is established to complete the function planning, and the reasonable function design is realized through effective methods. Kano's quality model divides customer needs into basic needs, expected needs, and exciting needs. Basic needs are the functions or needs that customers think the product should have. Expected needs are what customers discuss in market research. It is usually expected demand. The more the desired demand is realized in actual products and services, the more satisfied the customers will be. Exciting needs are out of customers' expectation in market research. If the production line meets such needs, the customer is very satisfied with the product [19].

The division of general modules and personalized customization modules is an interactive design process. That is, based on the general needs of customers, the general modules are first divided, then the personalized customization modules are divided based on the customer's individual needs. Finally, the module is improved based on the divided customized module. This process can be regarded as a master—slave two-level planning and decision-making problem based on customer needs and service provider design needs. The idea of the master—slave decision-making problem is proposed by Stackelberg [20], which is used to study the output decision-making problem in the market competition mechanism. Two-level programming is a mathematical model of a two-level decision-making problem. It is a system optimization problem with a two-level hierarchical structure. Both the upper and lower problems have their objective functions and constraints. The objective function and constraint conditions of the upper-level problem are not only related to the upper-level decision variables but also depend on the optimal solution of the lower-level problem. At the same time, the optimal solution of the lower-level problem is affected by the upper-level decision variables. After preliminary research, it is found that the main layer is the personalized customization module division layer, which is based on the user's maximum personalization and service experience. The general module division layer takes performance, quality, cost, and other factors into consideration and aims at the most cost-effective and flexible manufacturing from the point of view of service providers [19].

3.3.2 Design and development of parametric customization system

In order to quickly design a customized production line that meets the needs of users, the parametric design has become one of the effective methods to ensure its high-quality completion. Parametric design constrains certain key geometric elements through the three-dimensional (3D) model provided by the modular system. The geometric model can be automatically derived without precise graphics. The entire 3D model can be modified by adjusting the values of some key parameters. For parts with exactly the same shape or relatively similar shapes, new and satisfactory designs can be generated by adjusting the relevant parameters [21]. Parametric design greatly simplifies the design process of the production line, avoids repeated design by engineers, shortens the design cycle, and reduces manufacturing costs.

The standard equipment in the production line corresponds to the general modules in the modular design system and can be used directly. However, for the customized modules corresponding to nonstandard equipment, parameterized custom design is required due to different customer needs. Based on computer-aided design (CAD) software, a parametric customization system is developed. For example, use the Visual Basic programming language to develop an interface program in the form of EXE or build a parameterized customization system based on the COM interface.

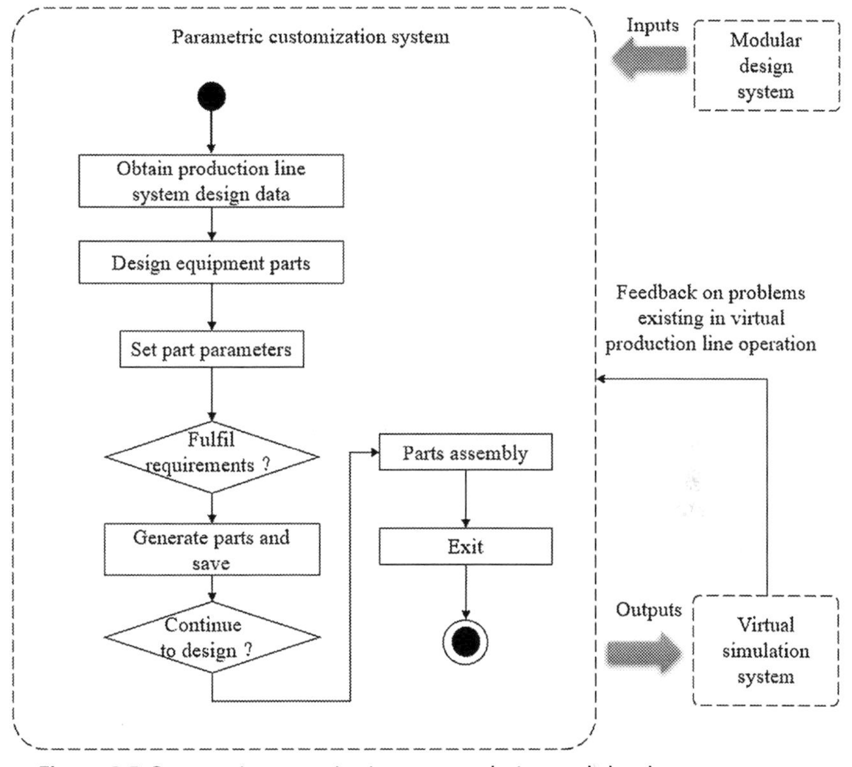

Figure 3.5 Parametric customization system design and development process.

As shown in Fig. 3.5, in the parametric customization system, individual customized design and automatic assembly of parts can be completed. First, by inputting dimensions, a part model of production line equipment or equipment is generated, and the parts are classified and saved. Then, the automatic assembly program is developed using the secondary development technology of the CAD software, and the assembly datum of each part is established and saved in the database of the CAD software CATIA to realize the automatic assembly. The automatic assembly program can confirm the integrity of the parts by comparing the shapes with the existing parts in the parts library. The server and the host communicate according to the IP protocol, and the server sends assembly instructions to the automatic assembly program on the host to complete the assembly design.

3.3.3 Virtual simulation and modeling method of production line

Virtual simulation modeling is one of the key technologies to realize the digital twin system of the production line. The virtual production line can reflect the operation, work changes, and production processes of each production unit of the physical production line. Virtual simulation modeling is applicable to the design and operation of production line. Before the production line is set up, the virtual production line policy modeling technology can quickly understand the production process, beats, bottlenecks, and other factors of the preset production line, realize the previous verification of the production line design, and finally provide ideas for improvement of the design. After the production line is running, the virtual production line simulation modeling technology can be used to make rapid changes to the production line and run to verify the rationality of the changes, identify problems, and reduce production risks [22].

One can use the Siemens Process Simulate to establish a production line simulation environment. Process Simulate software is a digital manufacturing solution that uses a 3D environment for manufacturing process verification. The Robotics function provides a virtual environment that integrates the planning and verification of robots and automatic equipment, which can simulate the working conditions of the robot in the real environment. It has logic driving equipment technology and integrated real robot simulation technology, which greatly improves the efficiency and quality of robot offline programming. Set the model to a JT format file dedicated to Process Simulate. After importing into Process Simulate, the limit motion range of the model joints and the constraint parameters between the joints were set. After the definition, the device can move according to the established idea and build a reference coordinate system and provide a reference for equipment placement and tool installation. For example, Fig. 3.6 depicts the use of the continuous process generator function to define the robot's weld path in a bicycle welding production line.

To improve the sense of immersion, virtual reality technology (Virtual Reality) can be used to model and simulate the production line. For example, the method of scene model premodeling is used to replace the technically more complex real-time 3D reconstruction [23]. In addition, a hybrid modeling method can be used to complete the virtual production line construction by combining the advantages of different CAD software. For example, the combined use of Pro/E, 3DMax, and Unity3D software to complete virtual production line modeling (Fig. 3.7) can give full play to the advantages of Pro/E software in modeling mechanical equipment, 3DMax's advantages in model material and texture editing, and Unity3D Advantages of model-driven and virtual reality development. First, Pro/E software is used to set the location of production and process equipment,

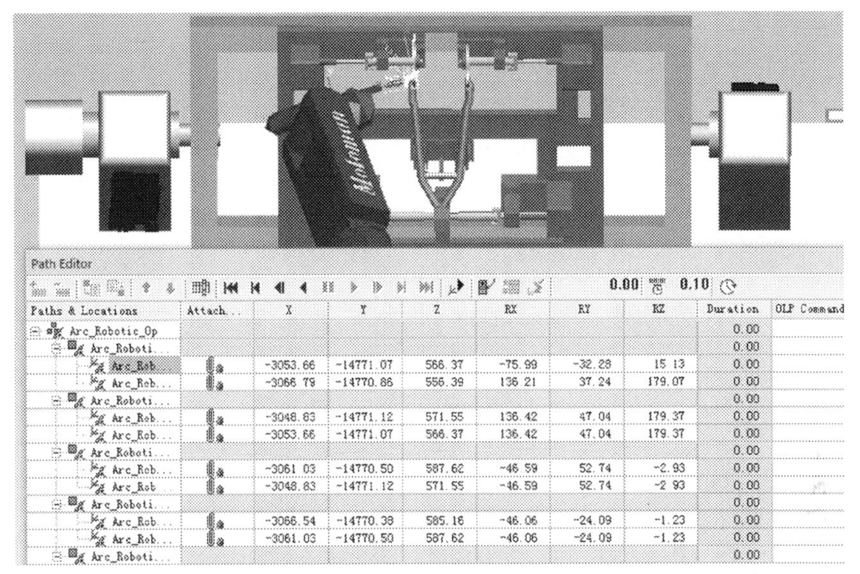

Figure 3.6 Definition of the welding path.

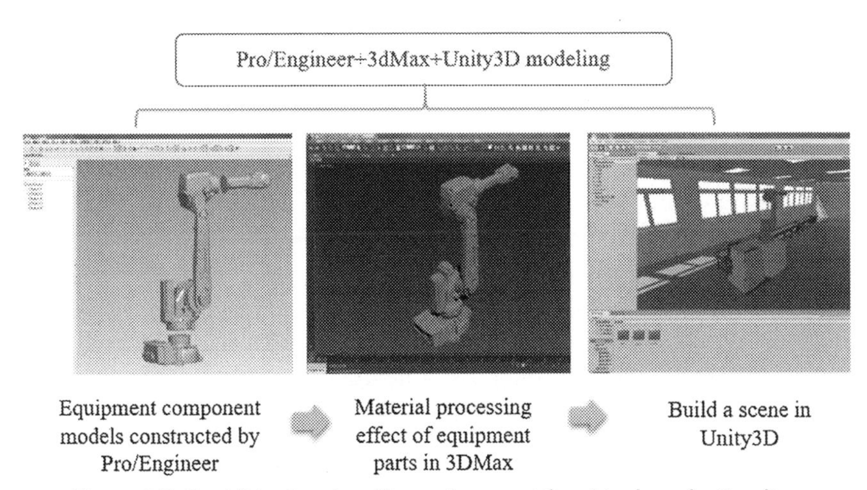

Figure 3.7 Establish virtual reality environment for virtual production line.

obtain the space layout of the production line, install the established production and process equipment models on the welding production line, and export these models to obj format files that can be processed by 3DMax software; second, 3DMax software is used to edit the material of the model to make the model have better fidelity; third, the model processed by 3DMax is exported to a file in FBX format, which can be interpreted by Unity3D software [23].

However, the virtual production line model established in Unity3D is static. In order to enable the static model to be updated in real-time according to the state of the physical production line, a physical driver script program needs to be added. First, programming language C# is used to write the production operation program of the production line. Unity3D provides multiple sets of instructions to control the movement of the model, which can be used to control the position of the model in each frame by calling these instructions in the script program. For example, the Transform component in Unity3D can control the spatial position, angle, zoom, and other states of the model. The rigid body component can control the physical properties of the object's gravity, elastic coefficient, and movement after impact, and the Character Controller component can control the position of the worker. Second, a control system is established outside the virtual simulation system. The Unity3D-based virtual simulation system communicates with the PLC controller through the C# language and controls the PLC controller to send instructions by inputting instruction codes to realize the control of the virtual simulation system by the control system and realize the virtual simulation of the automotive welding production line.

For instance, the robotic arm is a link structure, and the motion drive of the robotic arm is realized through the joint angle between the links. In Unity3D, joint connections between models can be established. The joint connections include Hinge Joint, Fixed Joint, Spring Joint, Character Joint, and Configurable Joint. The base of the mechanical arm is a fixed connection, and each revolving joint is set as a chain connection. Fig. 3.8 shows how to add Hinge Joint components to the joints of the robotic arm and how to set the rotation anchor point, rotation axis, and connection object of the joint in Unity3D. To enhance the reliability of the robotic arm movement, the rigid body component provided by Unity3D is used to give each part of the robotic arm, and the relevant parameters can be adjusted to make each part of the robotic arm have rigid body properties. When a collision occurs between models, the collision algorithm can achieve the rebound effect of objects with different materials.

3.3.4 Construction and debugging method of digital twin system

Industrial field equipment such as robotic arms, motors, machine tools, and other equipment are all electrically controlled by PLC (programmable logic controller). However, production line design and simulation are usually

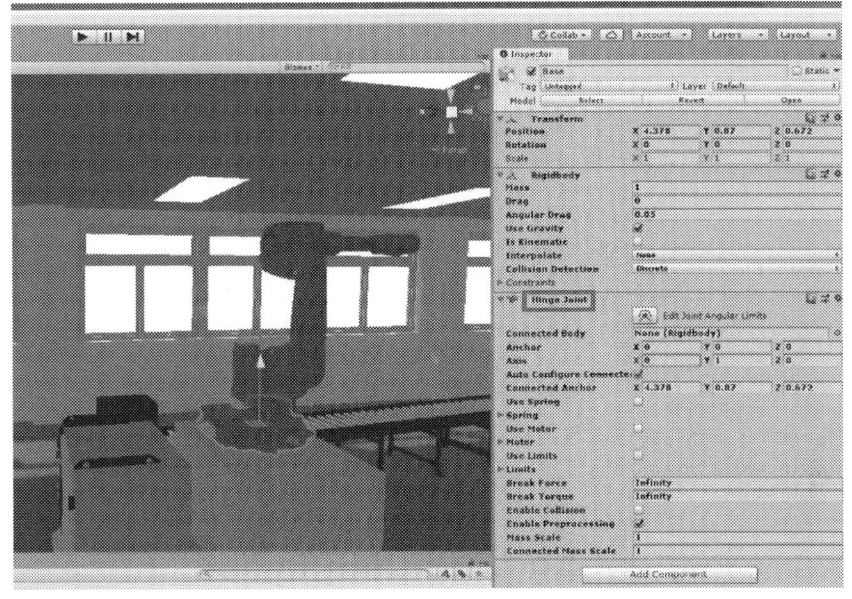

Figure 3.8 The physical drive of the virtual production line [23].

carried out in standard mode, and the simulation process is separated from PLC control. When the production site and the simulation results are different, engineers need to perform on-site debugging. It can be seen that the simulation performed in this standard mode is a timing simulation, and there is no control intervention. Virtual debugging is different, and it can effectively reduce the time spent in the on-site debugging process in actual production and greatly save costs. As shown in Fig. 3.9, the connection between PLC control and simulation environment can be established. In the simulation environment, the use of the PLC program to achieve logic control, which is close to the actual production environment, to achieve virtual and real synchronization.

To realize virtual debugging, it is necessary to realize the interconnection of various modules and the mutual transmission of signals. Signals can be transferred to each other through HMISim (virtual HMI software), PLCSim (virtual PLC software), and Process Simulate software without hardware, as shown in Fig. 3.8. When the hardware conditions are satisfied, the real PLC and HMI control panel and the Process Simulate software communicate with each other to build a virtual debugging simulation platform. In addition, the specific communication interconnection methods are different according to the different types of communication between

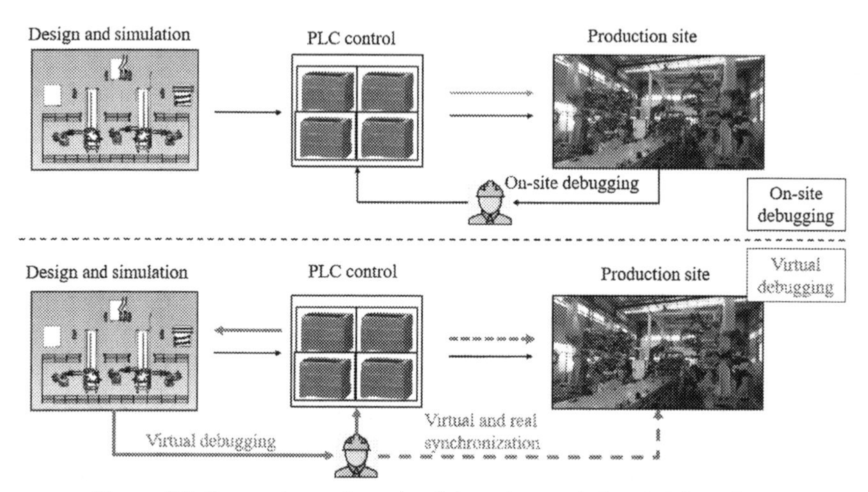

Figure 3.9 Comparison of on-site debugging and virtual debugging.

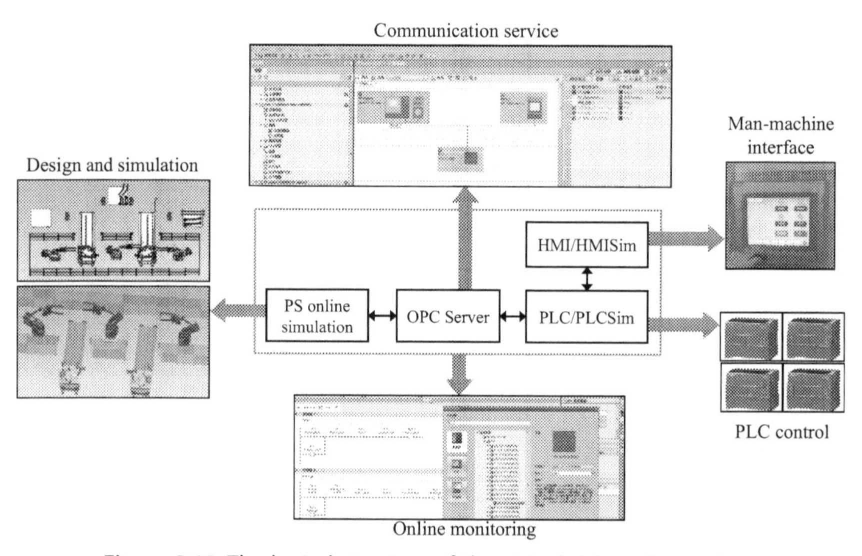

Figure 3.10 The logical structure of the virtual debugging system.

PLC and Process Simulate. Fig. 3.10 displays the connection method based on the OPC communication protocol. Virtual commissioning includes the following three main steps.

(1) Establishing information exchange through OPC (OLE for Process Control). Although the application of PLC driving the physical production line through OPC communication has become very popular, the OPC-based communication between PLC and the virtual production line has specific software requirements on the PC side. Process Simulate integrates the OPC interface, and SIMATIC NET software can be used as the OPC server to realize the virtual production line's PLC and Communication.

(2) Writing production line control program. The control program is based on event logic program simulation, and the execution of all actions is driven by the received signal [24]. The process of writing a control program is as follows.

- **(1)** Assigning input and output I/O variable points according to each event to complete the event mapping;
- **(2)** Establishing sensor models and corresponding sensor signals. The triggering of certain events requires the intervention of sensor signals;
- **(3)** Completing the material flow setting, which is different from the material flow direction planned by the simulation model. In the process of debugging the logic control program, the materials are continuous and related to events. According to the assigned I/O variables, one can write the PLC program in the TIA Portal software.

(3) Virtual debugging based on Process Simulate. The programmed PLC program is first imported into the PLC hardware, which is then combined with Process Simulate software to debug the virtual production line's control program. Second, check the authenticity of the program logic and adjust the program repeatedly until the logic program meets the process requirements of the production line. Finally, the debugged logic program is imported into the actual production line to make the control logic of the virtual production line consistent with that of the actual production line.

In addition, the process of establishing communication between the virtual simulation system based on the Unity3D production line and the PLC controller is shown in Fig. 3.11. The communication is primarily based on character string data. The PLC controller receives the data and converts the data into corresponding control instructions. First, a new script TCP Client is created in Unity3D for sending data instructions, and a Socket server is created for listening to incoming connection requests to realize the data sending function. The data are then transmitted to the server through the TCP protocol, and the PLC controller executes the corresponding control commands. Finally, the Unity3D-based virtual simulation system communicates with the PLC controller through the C# language

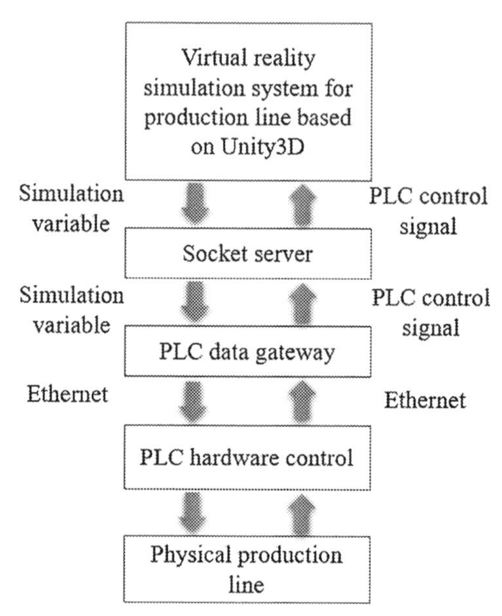

Figure 3.11 Communication between virtual reality system and programmable controller.

and inputs the instruction code to control the PLC controller to send instructions. If the debugged PLC program is used to drive the operation of the physical production line simultaneously, the production actions of the virtual production line and the physical production line can be consistent, and the virtual and real linkage can be achieved.

3.4 Case study

3.4.1 Design and simulation of intelligent welding production line for bicycle rear frame

The bicycle frame includes a front triangle frame and a rear triangle frame. The automatic welding of the triangle part is taken as an example to illustrate the production line's virtual simulation modeling method. The bicycle rear triangle frame includes an upper fork, lower fork, hook claw, and reinforcing tube. The frame components are all made of aluminum alloy with a wall thickness of about 2 mm. The welding surface is required to be free of slag inclusion, nonfusion, porosity, welding bead, etc., and the shape is uniform. The weld edge is well integrated with the base material. After a comparative analysis, the most suitable welding process method is fine-wire mixed gas (Ar + CO_2) shielded welding, which can achieve the required high welding speed and produce high-quality welds. The bicycle frame structure is shown in Fig. 3.12.

The first is the plan design of the welding production line. The welding scheme design is related to the transportation operating parameters, layout, and cost of the entire production line. The whole production line consists of two necessary inspection links, batch processing of the two unqualified workpieces, and then welding. The designer conducts statistical analysis on the welding quality problems that have occurred, finds out the welding quality problems that are prone to appear, adjusts the welding process, and forms a benign feedback system. The technological revolution of the scheme is provided in Fig. 3.13.

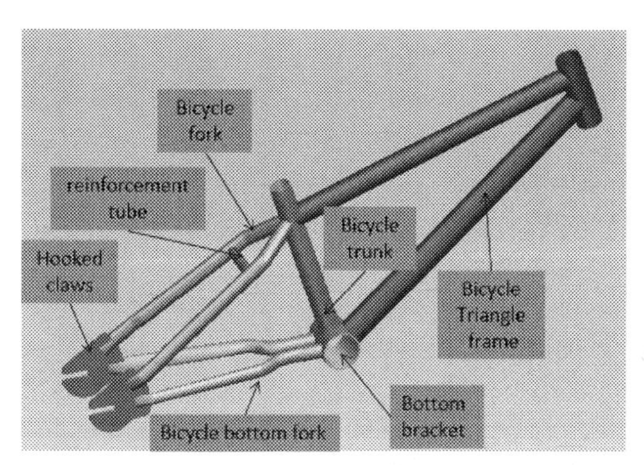

Figure 3.12 Bicycle frame structure.

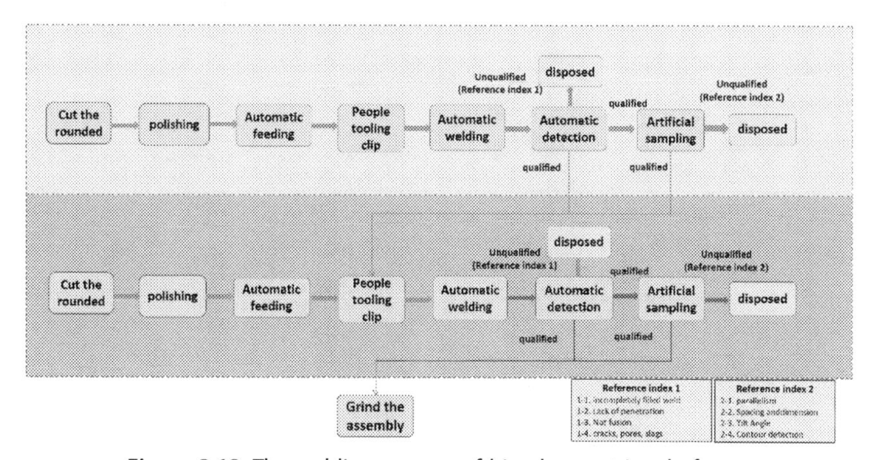

Figure 3.13 The welding process of bicycle rear triangle frame.

According to the balance calculation of the production line, a reasonable space layout scheme can be made for the equipment used in the production process, as shown in Fig. 3.14. The welding system's working area includes the welding area, the inspection area, the sampling area, the loading workpiece storage area, the unloading product placement area, and the waste recycling area. The advantages of adopting this layout scheme are as follows: the welding system adopts the layout method of feeding and discharging on both sides, which improves the transportation efficiency and agility of logistics. The overall compact structure of the system can make full use of the workshop area of the factory. The use of parallel welding robot greatly improves the production efficiency.

A CAD software NX is used to create 3D models of welding robots, flip positioners, testers, welding fixtures, conveyor belts, and test benches required by the welding production line. The output JT models from NX can then be imported into Process Simulate, in which we set the limit motion range of the model joints and the constraint parameters between the joints. Moreover, the defined equipment can move according to the established scheme. The reference coordinate system is established, which provides a reference for the determination of robot model, welding path, workers and position planning of 3D model, etc.

Fig. 3.15 describes the welding robot model's definition after importing the robot model into the simulation space. First, two 3D reference coordinate systems are created, one is the robot chassis coordinate system, and the other is the coordinate tool system. The origin of the chassis coordinate system is the center of the robot base. The purpose of the creation is to

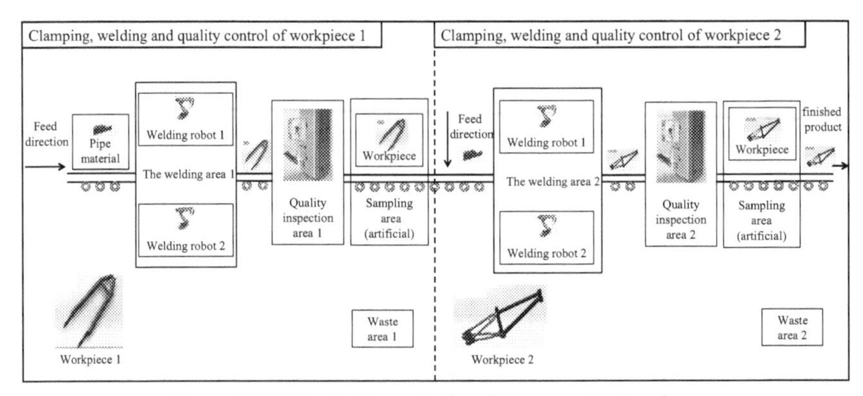

Figure 3.14 Spatial layout of welding production line.

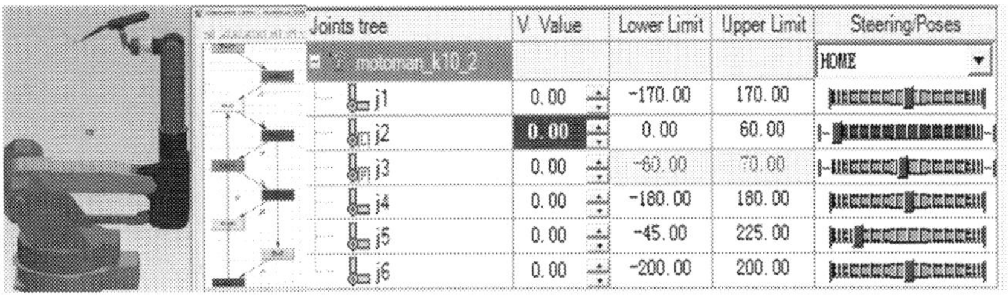

Joints tree	V Value	Lower Limit	Upper Limit	Steering/Poses
motoman_k10_2				HOME
j1	0.00	-170.00	170.00	
j2	0.00	0.00	60.00	
j3	0.00	-60.00	70.00	
j4	0.00	-180.00	180.00	
j5	0.00	-45.00	225.00	
j6	0.00	-200.00	200.00	

Figure 3.15 Define the robot model in Unity3D.

install the ground press in the set space layout accurately. The base coordinate system of other equipment such as turning positioner, X-ray detector, inspection table, conveyor belt, etc. is similar. The origin of the coordinate tool system is the center of the flange. Creating the coordinate tool system accurately installs tools such as welding torches and grippers on the flange. Then, the robot joints' limited motion range is defined, and the connection between the joints and the joints is established. The definition of other equipment, such as flexible pneumatic clamps, flip positioners, etc., is similar to the definition of robot joints.

After completing the 3D model, the 3D digital models are accurately positioned one by one according to the designed spatial layout. The complete virtual simulation model of the bicycle rear tripod welding production line is shown in Fig. 3.16.

The accessibility and interference of robot are important factors affecting welding efficiency, which have been verified in advance in the virtual simulation model of welding production line. The robot's reachability refers to whether the robot can reach the target welding spot according to the required pose, that is, whether the robot TCP frame and the welding spot coordinate system can overlap in the simulation model [25]. There are two reasons why the robot cannot reach the target welding spot in the

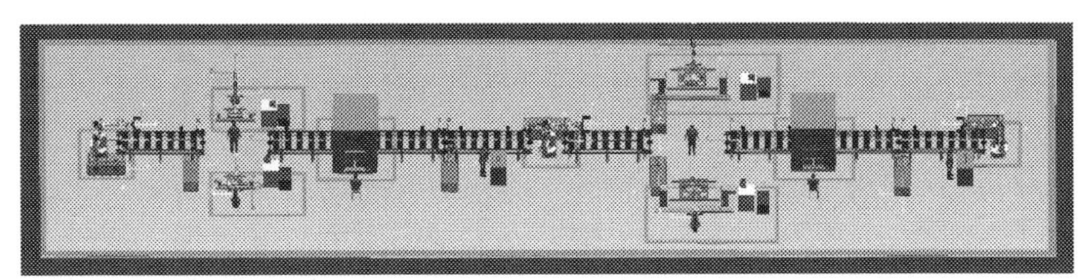

Figure 3.16 Virtual simulation model of welding production line for bicycle rear tripod.

welding process, one is that the motion range of the robot exceeds the designed limit, another is that the welding posture of the robot is not reasonable. The range of motion of the robot is beyond the limit, which makes it impossible for the robot to reach. In this case, the layout of the design space can be avoided. When the robot is designed, it has its own range of motions. The material to be welded is placed in an appropriate position to ensure that each welding point is within the range of motion defined by the robot. Another kind of unreachability is caused by improper welding position. When the robot welds two adjacent welds, the welding posture sometimes reverses, as shown in Fig. 3.17. Therefore, it is necessary to define the welding pose of the robot, respectively, on the basis of the welding pose of the robot in the previous section and take it as the welding pose of the next region.

The spatial structure of the bicycle frame is complex, and most of the welding paths are arc-shaped. The robot can interfere with the fixture during automatic welding. Due to interference, the solder coordinate system is not oriented properly on the workpiece, and the welding process lacks transition points. The welding interference caused by the welding point coordinate system's improper direction on the workpiece can be solved by optimizing the welding point's position and adjusting the direction of the welding point coordinate system. If the direction of the solder joint coordinate system is changed, interference will still occur during the welding process. One should observe the interference between the robot and the fixture and manually program the transition avoidance point. After debugging and running, the robot can smoothly complete the welding, which verifies the robot's accessibility and interference. At the same time, the rationality of fixture design is proved.

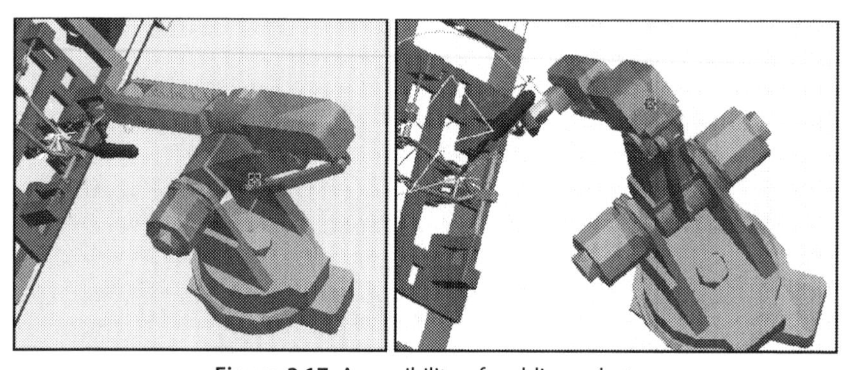

Figure 3.17 Accessibility of welding robot.

3.4.2 Design of personalized fixture for automobile welding production line based on digital twin

The welding of automobile body-in-white has problems such as long design cycles and low welding quality. To quickly design high-quality and personalized fixtures, the personalized furniture of the automotive welding production line is designed based on the digital twin, as shown in Fig. 3.18. In the parametric customization system, the entire 3D model can be modified by adjusting some key parameter values. Also, by changing the relevant parameters, a new and satisfactory design can be generated. The virtual production line provides customers with a real-time and straightforward interactive method. Through the virtual operation of the automotive body-in-white welding production line, it is possible to complete the analysis of the welding production line fixture opening and pressing status, interference inspection, virtual operation, etc., find out problems existing in the design, complete the collection of customized data and feedback to the parameterized customization system for modification.

First, designers can use the modular system to modularize the general unit of welding fixtures. Personalized automotive sheet metal data are obtained based on customer customization requirements. According to the automotive sheet metal data, the fixture clamping position, clamping method, and fixture type are determined. The modular principle is adopted to classify the parts in the fixture, including standard parts (Fig. 3.19):

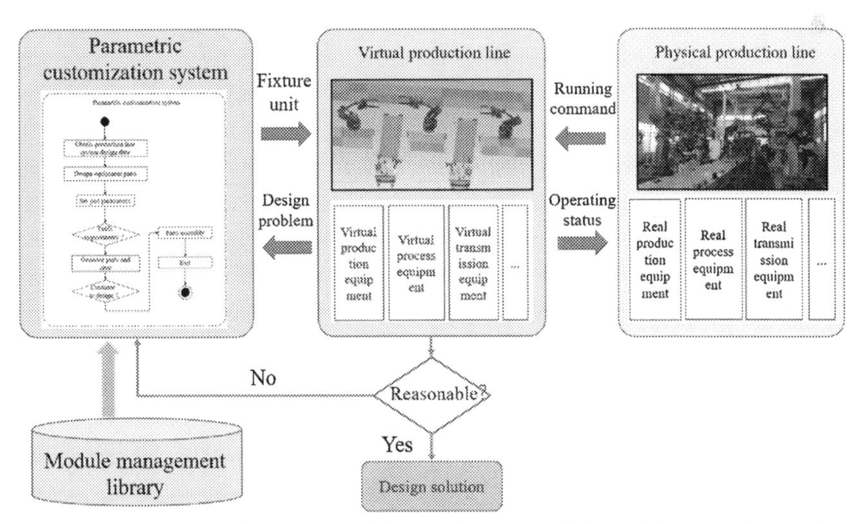

Figure 3.18 Design of personalized fixture for automobile welding production line based on digital twin.

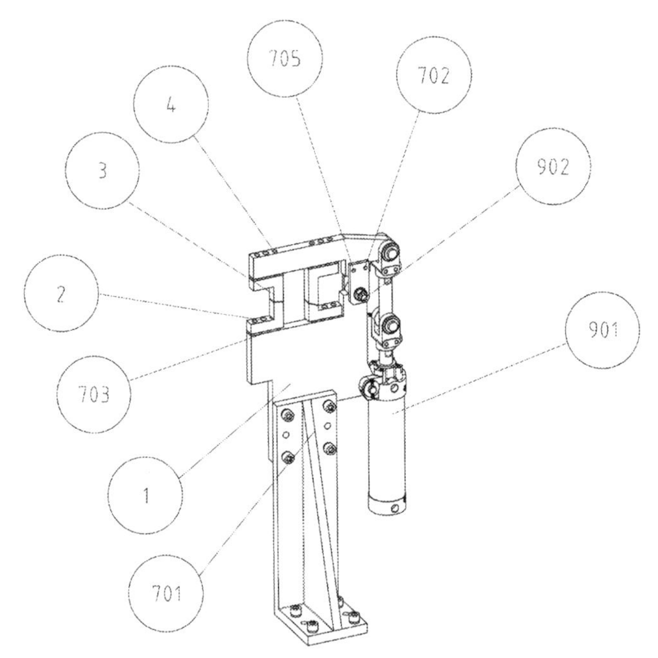

Figure 3.19 Automotive welding fixture standard unit.

casting angle seat 701, hinge plate 702, adjusting washer 703, limit block 704, hinge pin 705, nonstandard parts: connecting plate 1, support block 2, pressing block 3, pressing arm 4, and purchased parts: cylinder 901, bushing 902. Among them, one needs to develop special function modules for nonstandard parts that are generated parametrically. As for standard parts and purchased parts, a series of 3D models should be established and saved in the database for recall at any time. Nonstandard parts are generated by inputting dimensions in the parametric customization system, and the fixture parts are assembled through automatic assembly procedures.

Second, the size of the step fixture parts is input into the parametric customization system to obtain the fixture parts. The automatic assembly unit in the 3D modeling software assembles the fixture parts to form a set of automotive sheet metal welding fixtures. The core is to establish the assembly datum of each part. The cast corner seat is assembled through the bolt hole and the connecting plate. The connecting plate is assembled through the plane and the supporting block. The cylinder is assembled through the hinge hole, and the pressure arm, and the connecting plate, and the pressing block is through the plane. It is assembled with the pressing arm, the limit block is matched with the connecting plate and the pressing arm through the plane, and the hinge plate is assembled with the pressing arm and the connecting plate through the hole and shaft coordination.

Third, in the virtual simulation system, the existing automotive sheet metal welding fixture, fixture clamping position, and automotive welding virtual production line are programmed through C# to develop a virtual assembly program, and the virtual production line components are realized by starting the assembly program. The final virtual welding production line of the car body in white is obtained, as shown in Fig. 3.20.

Fourth, a physical car welding production line is built based on the car body-in-white virtual welding production line. The car body-in-white virtual welding production line is connected to the car welding physical production line through a PLC controller, and PLC drives the movement of the production line through OPC communication. The virtual simulation system communicates with the PLC controller using string type data, and the PLC controller communicates with the TCP protocol of the automobile welding physical production line. The programming software is used to write the PLC control program and import the control program into the PLC control hardware. After receiving the data, the PLC controller converts the data into corresponding control instructions, to drive the virtual welding production line and the physical automobile welding production line.

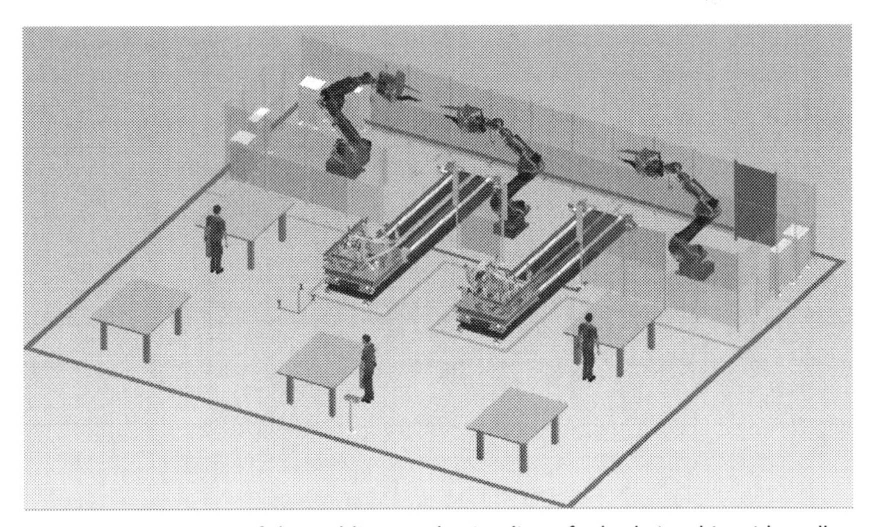

Figure 3.20 Layout of the welding production line of a body-in-white side wall.

Fifth, the PLC controller's control logic is set according to the simulation production action of the virtual welding production line of the automobile body-in-white. The control logic of the PLC controller is used to drive the automobile's physical welding production line and perform the production action. One can check whether the designed fixture model meets the welding requirements through the linkage of the physical model and the virtual model of the automobile welding production line. If the simulated production action is consistent with the production activities, the fixture model designed for automobile sheet metal parts meets the welding requirements. Otherwise, one should adjust the parameterized customization system.

Sixth, the virtual simulation system's communication is established based on Unity3D software and the PLC controller. When the PLC program is running in the PLC hardware, and the I/O signal is used in the PLC control hardware, the car welding virtual production line and the corresponding physical production line are consistent with the production actions. It is convenient for engineers to check the linkage operation status of the physical model and the virtual model of the automobile welding production line, check whether there is an interference phenomenon, and check whether the designed fixture model meets the welding requirements.

3.5 Summary

This chapter first analyzes the research progress of digital twin-driven production line custom design. Then, after comparing the relationship between digital twin-based complex product design methods and traditional design methods, the framework of a digital twin-driven custom design service system is presented. The framework consists of four parts: the modular design system, parameterized customization system, a virtual simulation system, and a real production line in physical space. On this basis, Section 3.2 introduces the fundamental theories and technologies of the digital twin-driven production line custom design service. Section 3.3 focuses on the development methods of digital twin-driven production line custom design service system, including modular system design and development, parameterized custom system design and development, production line virtual simulation modeling methods, and digital twin system construction and debugging techniques. Finally, two examples are given to illustrate the effectiveness of the aforementioned theories and practices.

References

[1] Chen W. Web-based bicycle customization system design. Guangdong University of Technology; 2014.
[2] Ye Z. Research on Customization application design: a case study of Jiangling motor App. Huazhong University of Science and Technology; 2014.
[3] Wang HQ, Li H, Wen XY, et al. Digital twin-based product design process and the design effort prediction method. Comput Integr Manuf Syst 2020:1−15. http://kns.cnki.net/kcms/detail/11.5946.TP.20200506.1719.008.html.
[4] Wang HQ, Li H, Wen XY, et al. Unified modeling for digital twin of a knowledge-based system design. Robot Comput Integrated Manuf 2021;68:01−13.
[5] Zhang H, Liu Q, Chen X, et al. A Digital twin-based approach for designing and multi-objective optimization of hollow glass production line. IEEE Access 2017;5:26901−11.
[6] Yi Y, Yan YH, Liu XJ, et al. Digital twin-based smart assembly process design and application framework for complex products and its case study. J Manuf Syst 2020:01−14.
[7] Sun MM. Research on several key technologies of digital twin system for aircraft final assembly line. Zhejiang University; 2019.
[8] Zhang Q. Research on modeling method of digital twin press lien based on physical data. Wuhan University of Technology; 2018.
[9] Yu XF. Construction and application of digital twin model for motor assembly line. Hebei University of Science and Technology; 2019.
[10] Cui YH, Yang BT, Fang Y, et al. Application of digital twin in aeroengine smart production line. Aeroengine 2019;45(5):93−6.
[11] Park KT, Lee J, Kim H, et al. Digital twin-based cyber physical production system architectural framework for personalized production. Int J Adv Manuf Technol 2020;106(5):1787−810.
[12] Grieves M. Virtually perfect: driving innovative and lean products through product lifecycle management. Cocoa Beach: Space Coast Press; 2011. ISBN: 0982138008.
[13] Mas F, Menéndez JL, Oliva M, et al. Collaborative engineering: an airbus case study. Procedia Eng 2013;63:336−45.
[14] Dai S, Zhao G, Yu Y, et al. Trend of digital product definition: from mock-up to twin. J Computer-Aided Des Comput Graph 2018;30(08):1554−62.
[15] Li H, Tao F, Wang HQ, et al. Integration framework and key technologies of complex product design-manufacturing based on digital twin. Comput Integr Manuf Syst 2019;25(6):1320−36.
[16] Li H, Wang HQ, Cheng Y, et al. Technology and application of data-driven intelligent services for complex products. China Mech Eng 2020;31(7):757−72.
[17] Zhu SS, Lou XF, Li WJ, et al. Product customization method based on extension design. Comput Integr Manuf Syst 2020;26(10):2661−9.
[18] Kano N, Seraku K, Takahashi F, et al. Attractive quality and must be quality. J Jpn Soc Qual 1984;14(2):39−48.
[19] Li H, Tao F, Wen XY, et al. Modular design of product-service systems oriented to mass personalization. Zhongguo Jixie Gongcheng/China Mech Eng 2018;29(18):2204−14.
[20] Stackelberg H. The theory of the market economy. Oxford University Press; 1952.
[21] Oxman R. Thinking difference: theories and models of parametric design thinking. Des Stud 2017;52:4−39.
[22] Lin WB. Research and application for key technology of simulation systems of production line based on virtual reality. Zhejiang University; 2019.

[23] Liu G. Research and application of key technologies of human-computer interaction and equipment remote control in virtual reality environment. Zhengzhou University of Light Industry; 2020.
[24] Cheng ZY, Li L, Li XC, et al. Joint virtual debugging research based on TIA and TECNOMATIX. Auto Technol Mater 2020;02:66–71.
[25] Mantel RJ, Schuur PC. Order oriented slotting: a new assignment strategy for warehouse. Eur J Ind Eng 2007;1(3):301–16.

CHAPTER 4

Digital twin-enhanced product family design and optimization service

Kendrik Yan Hong Lim[1,3], Pai Zheng[2] and Dar Win Liew[1]
[1]School of Mechanical and Aerospace Engineering, Nanyang Technological University, Singapore, Singapore; [2]Department of Industrial and Systems Engineering, The Hong Kong Polytechnic University, Hung Hom, Hong Kong, China; [3]Advanced Remanufacturing and Technology Centre (ARTC), Agency for Science, Technology and Research (A*STAR), Singapore, Singapore

4.1 Introduction

Following the rise of mass customization approaches to meet changing customer sentiments and demand, manufacturing industries have strived to offer greater product and service varieties by enhancing production capabilities [1]. Faced with the paradox of higher variability and lower revenue, industry players such as Volkswagen [2] and Intel [3] have continuously applied the product platform concept to accelerate design processes and reduce development costs in their bid to raise competitive advantage. With product platforms being a key fundamental toward enabling product family paradigms, this chapter adopts the platform definition as a standardization of products with common modules to create user value and facilitate economies of scale [4]. As a prevailing concept, the product family approach expands on this platform concept to efficiently develop and evaluate new product configurations with customized attributes [5]. Existing research in this field has yielded a range of innovative solutions, which includes knowledge-based systems [6], variation-based models [7], and even indices frameworks such as the Generational Variety Index and Process Customizability Index [8,9] for product family design and optimization. Other product family paradigms such as modular design [10], platform-based scalable design [11], design structure matrix [12], and adaptable design with open architecture product [13] offer effective methodologies to expedite design and optimization approaches, extending to production facilities and shop-floor operations. Nevertheless, these proposed solutions are typically conducted either from the onset of the conceptual design stage where functional requirements and design parameters are mapped (e.g.,

Digital Twin Driven Service
ISBN 978-0-323-91300-3
https://doi.org/10.1016/B978-0-323-91300-3.00003-6

quality function deployment) or during the usage stage where relevant user feedback and system information are collected for reverse design processes. With complex product family design and optimization constituting a major challenge in present-day mass customization and personalization paradigms, these techniques serve to manage product component variation with emphasis on extending product life cycles. However, as the proposed solutions primarily hinge on qualitative marketing strategies leveraged before product development and quantitative data collection and analysis obtained during operations, there is a lack of context-aware testbeds to support product family design and optimization with consideration to both design and reverse design throughout the product life cycles.

With new entrants leveraging on smart manufacturing technologies to lower economic barriers to entry, along with heightened social-environmental awareness toward sustainable production, industry incumbents are faltering in this new market environment brought forth by increased competition and stricter manufacturing policies. To remain competitive and recapture market share, more manufacturers are embarking on digitalization roadmaps based on Industry 4.0 framework to enhance production capabilities. By utilizing advanced information and communication technologies (ICT) to digitize physical resources within smart, connected environments, value-adding solution bundles can be offered [14] while smart product-service systems can facilitate co-creation design paradigms for next-generation product development in a robust manner [15]. Among these technologies, digital twin (DT) serves as a fundamental to enhancing manufacturing operations and can bolster specific engineering product life cycle management (PLM) aspects. Since its inception in 2003 [16], DT research has gained momentum in recent years due to advancements in communication technologies and low-priced sensors. Empowered by cutting-edge techniques and approaches to realize high-fidelity simulation and cyber-physical interconnectivity, DT can be utilized to realize complex product family design and optimization holistically via context-aware systems across many sectors. Furthermore, its ability to simulate real-time working conditions and provide intelligent decision support allows stakeholders to visualize and evaluate potential outcomes [17]. This chapter regards DT as a high-fidelity virtual replica of the physical asset with real-time two-way communication for simulation purposes and decision-aiding features for product service enhancement [11] and details a generic

DT-enhanced approach to support in-context virtual prototyping and reconfiguration/redesign of complex product families with extensive life cycle considerations. This DT-driven system is capable of bolstering both modular and scalable product design strategies in the early design stage while aiding stakeholders in existing product family optimization during the usage stage. The subsequent subsections in Section 4.1 offer a review of key engineering product family approaches and DT-driven solutions and the rest of the chapter is as follows: Section 4.2 highlights a DT-enhanced ambient-based benchmark and interaction mechanism to aid product family design. Section 4.3 showcases a DT-enabled in-context product family optimization for asset reconfiguration and reverse design. Section 4.4 provides a case study featuring the detailed establishment of a DT-enhanced service environment. Lastly, Section 4.5 summarizes the core findings in this chapter and highlights potential trends.

4.1.1 Engineering product family design and optimization

Product family design is an established methodology featuring solution platform architectures based on module, component, or subsystem commonalities to satisfy growing user expectations from a wide range of market segments [18]. Throughout past decades, product family design approaches were frequently examined to overcome obstacles relating to mass customization, especially in business-driven frameworks. These platform-based products possess common traits, and development can be supported via formal logics [13], and cost-driven additive manufacturing product family design [19]. One may refer to Simpson et al. [18] and Jiao et al. [20] for more details. Meanwhile, recent studies have ever-increasingly transformed into the data-driven clustering/module partition approaches with Ma and Kim [21] predicting user preferences or functions, owing to large amounts of available online data and rapid development of AI techniques. For instance, Takenaka et al. [22] leveraged 600 users' smart appliance logs and their survey responses on their lifestyles to design smart grid service. Zheng et al. [23] introduced a novel cyber-physical co-design approach for smart product family design in today's smart environment. Fig. 4.1 shows a conventional product family design approach that is initiated from the very beginning of a design process. A typical product family design process considers customer requirements in both functional and physical domains. Modular design defines and selects optimal design solutions while scalable

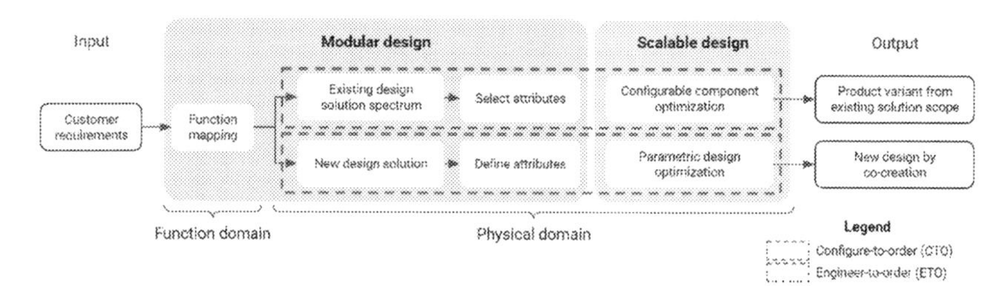

Figure 4.1 A generic framework for product family design.

design improves asset configurations to enhance existing solutions and aid design co-creation. This approach facilitates a step-by-step mapping process to derive final design solutions, but without any existing benchmark tool or graphical information, the mapping process lack visualization and does not actively consider ambient information during the design process.

Product family optimization, on the other hand, represents the change, upgrade, and reconfiguration of existing product families during the usage stage [24]. One typical type of work emphasized on redesign management. Design structure matrix (DSM), as a classic engineering management approach, has been widely adopted [25] with other heuristic or bio-inspired algorithms (e.g., adaptive ant colony [26]) to enable minimized design propagations with redesign efficiency. Meanwhile, by embracing the cutting-edge smart enabling technologies, DSM-based learning [14], and smart reconfiguration approach [27] have been brought up recently to close the design iteration cycles. Another type of work considered the joint optimization of product configuration and reuse/remanufacture. For instance, Du et al. [28] proposed the Stackelberg game theory-based approach for modular and scalable design optimization. Wu et al. [29] introduced a joint decision-making approach to support product family optimization by considering both design and remanufacturing processes. Product family optimization in the usage stage generally involves various engineering management processes such as the design structure matrix to identify links between entities. Like the design stage, the integration of technologies such as DTs can provide process visualization capabilities and, the use of ambient information can aid designers in enhancing solution redesign and reconfiguration.

4.1.2 Digital twin-enabled services for engineering solution re-/design

DT as a prevailing topic has emerged as a trend in recent years due to rapidly evolving simulation and modeling capabilities, better interoperability, and low-cost Internet of Things (IoT) sensors. It integrates high fidelity CAD models, knowledge representation, and computational tools to provide real-time two-way connectivity, simulation, and decision-aiding functionalities. Since the DT concept was proposed as an inexpensive means to investigate and evaluate complex systems, many research works have been done in the DT-driven engineering life cycle management field [30,31]. This work selects the ones related to the design and inverse design stages, and further categorizes them into three aspects: (1) Geometry assurance. Schleich et al. [32] introduced a holistic DT reference model for design geometry inspection based on the Skin Model Shapes concept. Biancolini and Cella [33] further proposed a DT-enabled mesh morphing workflow for geometric model validation. (2) Value co-creation (codesign) aspect. Zheng et al. [23] presented a data-driven platform-based value co-creation process in the smart, connected environment by leveraging the massive product-sensed and user-generated data. They also introduced a DT-enabled codesign approach for smart wearable development [15]. (3) Model-based design and simulation aspect. Schluse et al. [34] integrated DT into the existing model-based systems for context-aware simulation purposes. Moreover, Damjanovic-Behrendt and Behrendt [35] established an open-sourced DT demonstrator for high-fidelity simulation of physical products.

Table 4.1 highlights a list of DT-driven applications designed to tackle challenges in specific PLM stages, showing that DT has the potential to enhance PLM aspects, especially during the design and usage stages. Based on a comprehensive review conducted as shown in the table, there is no existing study that considers both product family design and optimization holistically to facilitate both context-aware design and redesign processes. As such, this study aims to demonstrate a DT-enhanced product family design and optimization with the implementation of a real-life case study. To achieve a functional DT for complex product family design, the work follows the hierarchical data-information-knowledge wisdom model which leverages product-related data modeling and mapping to achieve design efficiency. This proposed DT model will be further described in Section 4.4 to highlight its effectiveness and capabilities.

Table 4.1 DT applications to enhance product life cycle management stages.

Design stage	Manufacturing stage	Distribution stage	Usage stage	End of life stage
Geometry assurance Guo et al. [36] Schleich et al. [32] **Model-based design simulation** Alam and Saddik [37] Damjanovic and Wernher [35] Schluse et al. [34] Tao et al. [30] **Value co-creation** Zheng et al. [38] Zheng et al. [15]	**Production digitalization** Bao et al. [39] Ding et al. [40] Liu et al. [41] Lu and Xu [42] Tan et al. [43] Zhang et al. [44] **Modeling strategies** Guo et al. [45] Luo et al. [46] Sharif Ullah [47] Zheng et al. [14] **Production optimization** Coronado et al. [48] Liu et al. [49] Liu et al. [50] Moreno et al. [51] Park et al. [52] Söderberg et al. [53] Tabar et al. [54]	**Robot–human collaboration** Bilberg and Malik [62] Nikolakis et al. [63] Petković et al. [64] **Warehouse management** Baruffaldi et al. [65] Bottani et al. [66] **Supply chain optimization** Defraeye et al. [67]	**Knowledge reuse and evaluation** Arafsha et al. [68] Liu et al. [50] **Workflow improvement** Schneider et al. [69] Haag and Anderi [70] Iglesias [71] **Shop-floor enhancement** Tao and Zhang [72] Zhuang et al. [73] **Plant management digitalization** General Electric [74] He et al. [75] **Energy and resource efficiency** Coraddu et al. [76] Ferguson et al. [77] Kannan and Arunachalam [78]	Lu et al. [82] Popa et al. [83] Wang and Wang [84]

	Xu [55] Zhao [56] **Individualized production** Leng et al. [57] Liu et al. [41] Soderberg et al. [58] Zhang et al. [59] **Situational adjustments** Lu and Xu [60] Sierla et al. [61]		MacDonald C. et al. [79] **Prognostics health management** Tao et al. [30] Tao et al. [84a] Wang et al. [80] Xu et al. [81]	

4.2 DT-enhanced services for ambient-based product family design

To overcome the above-mentioned research gaps regarding existing engineering product family design approaches, an ambient-based DT system is proposed to facilitate design processes via the use of both benchmarking and interacting mechanisms. As product family design processes frequently start at the conceptual design stage, functional product modules are mapped to stakeholder requirements corresponding to the respective design parameters. As such, the absence of graphical information relating to physical product modules will result in counterintuitive and inefficient design processes that can be improved with DT utilization. DT acts as a fundamental tool toward supporting the creation, planning, and evaluation of engineering product configurations by reflecting intricate details of the physical product. With simulations, key functionalities can be assessed and matched for compatibility throughout product development processes. In addition, smart recommendations derived from entity relationships and knowledge graph frameworks assist designers and planners in making informed decisions regarding new product ideation and configuration selection. Although DT systems possess the capabilities to enhance product design by providing feasible and transparent recommendations to support design operations, the lack of relevant context information may result in a flawed design that is unsuitable for use in the intended environment. Hence, collecting and transmitting relevant ambient information is crucial for establishing context awareness in smart products. This ambient information is generally classified into static and inconstant parameters, whereby static parameters refer to rigid and immobile machinery or structures that are mirrored in the form of 3D models while inconstant parameters involve acquiring continuous manpower movement and atmospheric conditions. To obtain the appropriate inconstant information, data types and sources are identified at the onset with specific predefined properties relevant to the intended use case, following the collection and transfer by embedded sensors and IoT systems. These parameters are then translated and mapped onto virtual layouts to display a realistic environment model to facilitate product configuration recommendations. With such an environment layout, stakeholders can plan and design suitable product applications in a user-friendly manner with clear visualizations of potential applications.

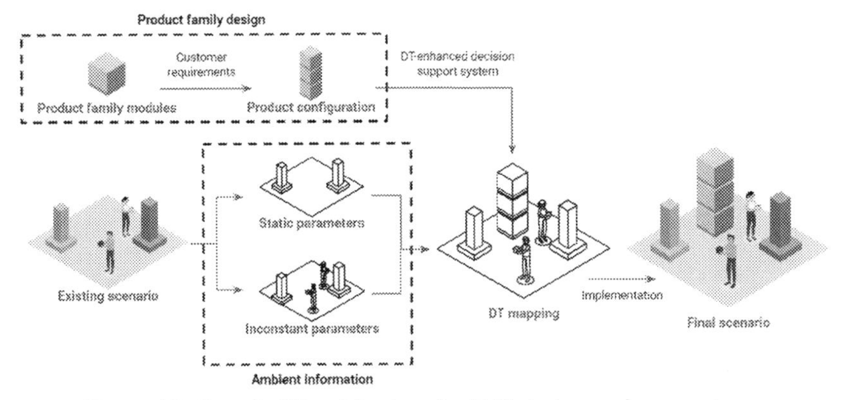

Figure 4.2 Generic DT architecture for PLM design and usage stages.

An overview of the design process is depicted in Fig. 4.2, featuring the use of ambient information to advance product family design approaches. Centered on established customer requirements, new product designs and configurations are first conceptualized with considerations for intended utilization scenarios. Next, an existing scenario reflecting realistic representations of predefined problem statements would be identified, after which a virtual model consisting of both static and inconstant parameters would be created. The virtual model may be modified to include potential product application concerns and serves as an optimal testbed for product usage before integrating the product configuration and virtual environment into the simulation system. Meanwhile, the virtual environment exhibits ambient information with static and inconstant parameters mapped onto the simulation application. As such, various configurations and situations can be deployed to evaluate design effectiveness based on a simulation-verified process that includes positioning assessment and budget estimation while accounting for disruptive aftereffects. Furthermore, the use of simulation-verification techniques offers detailed representation and analytics on expected usage conditions to evaluate potential products features via virtual prototypes before implementation. The use of ambient information in product family design approaches ultimately decreases reliance on expensive third-party experts and expedites design processes.

As large volumes of semistructured and unstructured ambient data are accumulated over time, there is a need to establish a NoSQL database management structure for big data storage and retrieval. A knowledge graph framework would also be required to convert the heterogeneous data types into meaningful insights. These data processing sequences are integrated within the proposed DT system and further described in Section 4.4. Meanwhile, this section highlights the advantages of incorporating ambient information for product family design and showcases the various methods for capturing and mapping ambient information into a generic DT system. A systematic process utilizing both benchmark and interaction mechanisms is also showcased to highlight the importance of context aware capabilities to aid stakeholders throughout design and planning processes. In a manufacturing shop-floor context, the benchmark mechanism provides a static environment layout, which serves as a standard to identify optimal asset configurations and support installation planning processes. Likewise, the interaction mechanism factors in various inconstant parameters to evaluate the effectiveness of the intended asset configuration within realistic working environments via the use of simulation-based verification workflows. With both mechanisms integrated within a generic DT-enhanced service built on top of existing design frameworks, a consistent and reliable decision support system can be realized to evaluate new product configurations in a practical and transparent manner.

4.2.1 Benchmarking service for product family design

To enhance existing product family design approaches, a benchmark mechanism is proposed to expedite design processes in a user-friendly manner. As digitization of ambient information serves to reflect realistic application environments and aid designers to visualize new product configurations, this benchmark mechanism can be integrated into existing DT systems to provide an automated and effective service. This benchmark service will enable planners to achieve multifactor considerations to accommodate stakeholder expectations, including optimal asset configuration and installation. Moreover, the integration of such services into DT systems can facilitate in-depth analysis to identify potential snags in functional compatibility, product integration, and requirement satisfaction while also generating insights not limited to setup cost estimations, product installation, and execution. Product design and planning typically relies on third-party experts to provide suitable recommendations and utilizes existing user feedback to formulate new design features. These inefficient methods are not only resource consuming but also do not guarantee reliable and consistent results. With the aid of this novel benchmarking service, a virtual testbed can be mapped to provide visualization aid to assess product configurations for distinct product application scenarios.

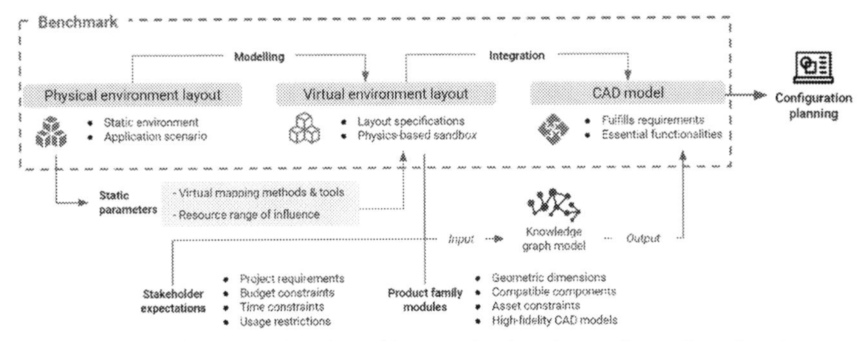

Figure 4.3 Three-step benchmarking mechanism for configuration planning.

The benchmark mechanism forms the foundation toward establishing an ambient-based product service system and stimulates designers during design generation with considerations to embedded functional constraints such as graphical and mechanical, etc. Fig. 4.3 presents an approach toward establishing a three-step benchmarking mechanism starting with digitization of the physical environment to create high-fidelity simulations with environment layouts representing specific scenarios. Firstly, the target scenario for product usage is selected with a predefined range of static and inconstant parameters. For example, a shop-floor upgrading project to provide 3D printing capabilities would require a layout of the environment with considerations for static structures and exit points as well as inconstant attributes such as resource flow paths and ambient temperature to provide designers and planners with context.

Next, the physical environment is reproduced onto a virtual layout where 3D replicates of the immobile installations are imported into the simulation module within the DT system. Aided by recent progress in cloud data processing, advanced techniques in point cloud scanning such as LiDAR and high-resolution radar can be used to map out the 3D environments in a convenient manner. Alternatively, commercial modeling tools such as CATIA and AutoCAD, offered by established engineering technology companies Dassault Systèmes and Autodesk, respectively, allows specific zones of interest to be recreated on a virtual platform. Faced with growing industrial appetite for DT-driven solutions, industry players are starting to provide design solutions that facilitate the creation of DT systems and other emerging forms of technology. PTC, for instance, offers a suite of solutions among which the CAD design tool, CREO, assists designers to construct engineering models which serves as a foundation for DT establishment.

The acquisition of ambient information is conducted through embedded sensors and tracking devices set up at vital locations to capture real-time datasets corresponding to a predefined set of inconstant parameters. The real-time ambient data are then forwarded to the data lake for storage and retrieval through IoT systems and industry-standard communication protocols. These protocols include OPC-UA and MQTT to ensure reliability and consistency in data transfer. As the amount of data increases exponentially over time, cloud and edge computing techniques can be applied to analyze and process these heterogeneous data before storing and retrieving in the NoSQL database systems. Following that, knowledge reasoning frameworks within ontologies can support entity relationship inference to obtain insights and solution generation based on the established data-driven graph models and predefined project constraints.

To create the virtual environment layout, 3D models of the static parameters are uploaded onto a simulation testbed to exhibit potential application scenarios while the sensor data is mapped onto the same simulation testbed to reflect real-time asset status and ambient conditions. While these application scenarios can be recreated using commercial simulation software such as Arena and Plant Simulation by Rockwell Automation and Siemens, respectively, it is important to note that these ready-made testbeds do not necessarily support cyber-physical integration with real-time control and sensor data input. Due to the constraints of these mass-market tools, which are not originally designed to support advanced cyber-physical-related functionalities, existing research typically utilizes custom-made simulation testbeds to establish dynamic DT systems. In addition to this, a thorough assessment needs to be conducted for the selection of an optimal simulation type to visualize both static and inconstant parameters in a comprehensive manner that meets project objectives and requirements. Well-established simulation types include discrete-event, dynamic, and agent-based simulations have varying attributes such as level of details and event-system response, which ultimately affects the outcome of decision support systems.

Lastly, high fidelity CAD models of product family modules based on manufacturers' specifications are necessary for product design and configuration. These compatible components, along with relevant attributes, are imported into an ontology framework along with stakeholder requirements as entities to enable smart solution generation. Using these modular assets, various product configurations can be scrutinized by visualizing the functional CAD model. Meanwhile, modular asset details such as dimensions, functionalities, and pricing provide reference points for configuration selection based on feasibility and preferences. Coupled with real-time sensor input, this knowledge graph model then provides knowledge inference capabilities for smart solution generation which are elaborated in subsequent sections.

The three-step benchmark mechanism established shows that ambient information can be effectively utilized to create, plan, and evaluate engineering product configurations by utilizing in-context information throughout the design stage. The use of benchmarking in DT systems not only facilitates new product creation based on a range of product family modules but also assists stakeholders in identifying optimal installation locations. With multiple restrictions from project objectives, management expectations, and even social and environmental policies by authorities, optimal asset configuration often involves tedious analysis and comparison between various product family combinations. Moreover, miscalculations in the design stage will result in inefficiency and unnecessary costs which will reverberate throughout asset usage. Lim et al. demonstrated the acquiring of essential ambient information to support a product family tower crane design [85]. Other examples include the point cloud of construction site ambience for target tower crane product family design [86].

4.2.2 Interacting service for product family design

Interacting mechanism represents a higher-level design process in today's smart, connected environment, where product family design should also consider the interaction with ambience. Building on top of the benchmark mechanism, complex design evaluations in smart, connected environments can be achieved via a collective virtual platform that considers how various potential product configurations interact within a dynamic environment. Hence, instead of solely designing the physical components of the product family, the embedded hardware should also be considered holistically, as shown in Fig. 4.4. In such context, sensory data fusion, bi-directional communication, data exchange protocol should all be involved dynamically. Hence, the designer should make the product family design compatible with the ambient objects in an ecosystem manner [9]. For instance, the point cloud of city roads with real-time sensed road condition information from intelligent traffic facilities can be leveraged to design better unmanned vehicles in the virtual space.

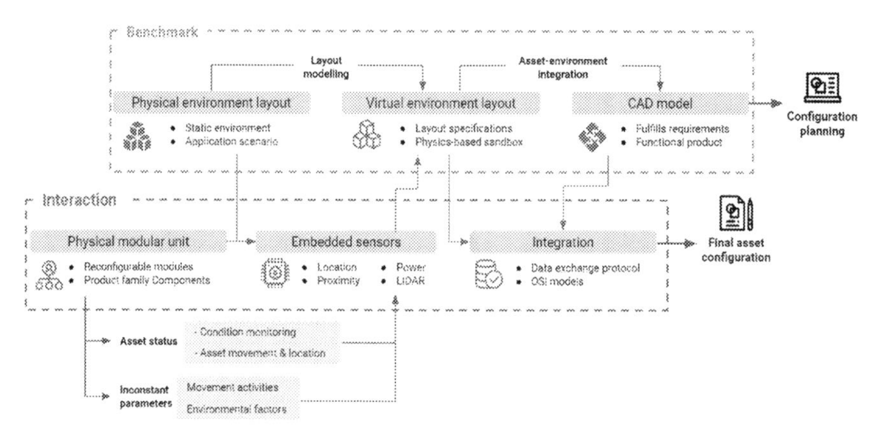

Figure 4.4 Benchmark and interaction mechanisms for asset configuration.

Once the configuration planning is completed, the interaction mechanism utilizes sensor data from the physical modular units and environment layout to obtain information related to the asset status as well as ambient information. With sensors installed at strategic locations throughout the environment layout, the sensor parameters are predetermined before project commencement to identify potential factors affecting asset performance. These real-time sensor data relating to both asset status and environment layouts are mapped onto the simulation platform. The simulation platform should have the capability to monitor real-time asset performance and control the assets as well. By testing out the proposed asset configuration, designers will be able to evaluate the feasibility of such design and perform comparison analysis more efficiently. By taking advantage of ambient information to enable a higher-level design process, the benchmark and interaction mechanisms will reduce reliance on third-party experts and expedite the design process.

4.3 DT-enabled services for in-context product family optimization

The use of DT solutions has been researched extensively to overcome challenges within engineering PLM usage components ranging from shopfloor enhancements to workflow efficiencies. Despite the success of DT-driven solutions to provide decision support capabilities, a key drawback is that the DT systems used are rigid and only serve to optimize a limited number of issues. By leveraging the ambient information to create a context-aware DT system, stakeholders will be able to maximize asset utilization by raising workflow efficiency and reducing downtime risks throughout the usage stage via a product family optimization process. This process aims to improve existing product family asset capabilities to meet new stakeholder requirements as well as manage potential disruptions.

The purpose of using context-aware DT systems for process optimization is to assist the creation of next-generation products as well as improve existing operation processes. To advance new product development, huge amounts of user data and feedback are required for analytics processes to determine the right factors and pain points. While many studies are conducted toward formulating continuous improvement strategies, such as the six sigma concepts to provide statistical identification of key improvement points, these approaches are time and resource consuming. As such, the considerable effort spent toward the collection and analysis of relevant information to generate insights can be reduced with the use of a context-aware DT system. Relying on low-cost sensor input and IoT systems, functionalities of product family modules can be mapped onto different application scenarios to allow potential configurations to be evaluated by the DT-driven simulation. This inexpensive work-saving approach to realize inverse design methodologies enables management to narrow down on well thought out market penetration approaches and incorporate business strategies to spur growth.

Fig. 4.5 illustrates an overview of the in-context usage optimization phase whereby arising situations would prompt the context-aware DT system to evaluate potential disruptions and inefficiencies and derive solutions to avoid or mitigate them. As large amounts of usage data and

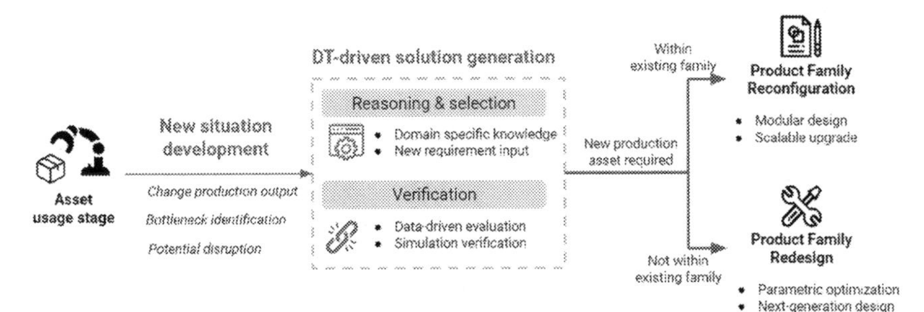

Figure 4.5 Product family reconfiguration and redesign approaches for design optimization.

product status information can be acquired during this stage, a graph-based context-aware approach can be utilized if the solution is within the product family scope for the reconfiguration process to take place. Otherwise, the parametric optimization or next-generation design for a new product family can be considered to meet these evolving challenges.

4.3.1 Reconfiguration services for product family optimization

Reconfiguration refers to the change or upgrade of configurable components of an existing product to meet new requirements [27]. DT-enabled systems that connect both physical and cyber assets, are able to rely on existing reconfigurable components, adaptable interfaces, and predefined constraint sets to execute inference and reasoning procedures. Hence, potential alternatives can be easily obtained from the existing product family based on the product's self-awareness passively (e.g., prediction of remaining life cycle) or via an active user request. For example, the monitoring of tire pressures and retrieval of a recommended car wheel module from the existing knowledge base.

Fig. 4.6 showcases the process flow toward the product family reconfiguration process. Based on the existing ambient-based CPS, new situation developments such as new project requirements would require the use of the DT-driven solution generation to identify optimal modules and come up with new asset configurations. These potential configurations would have to be evaluated for feasibility and compatibility as well as tested using the simulation verification approach. The product family approach facilitates minimal disruptions to existing operations by replacing asset modules to fulfill new asset capabilities. This can be applied in a bike-sharing system whereby replacing a module, for instance, the wheels could cater to a better user experience in an area where road networks are not as good.

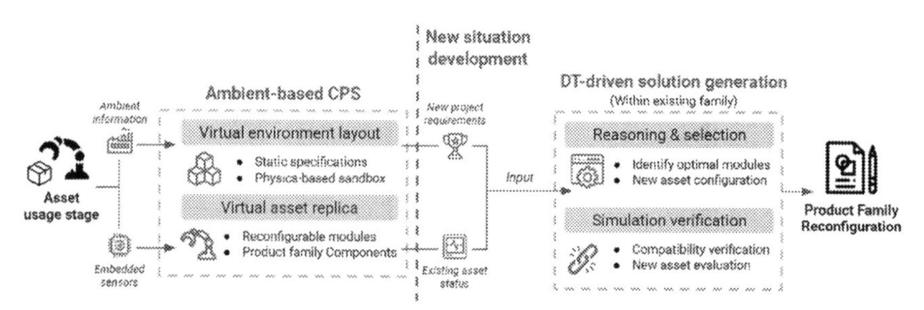

Figure 4.6 DT-driven product family reconfiguration approach.

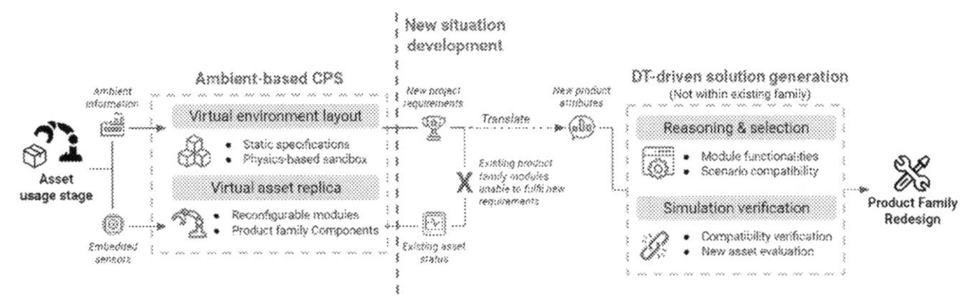

Figure 4.7 DT-driven product family redesign approach.

4.3.2 Reverse design services for product family optimization

On the other hand, reverse design denotes the redesign of existing components/products based on its conditions in the context of design improvement or next-generation design purposes. DT undertakes the real-time monitoring of the critical components of the physical objects, where design optimization results can be further obtained based on the models established. For instance, user's in-context riding information can be used to optimize the next-generation structural design of the bicycle.

Fig. 4.7 showcases the process flow for product family redesign if existing product modules are unable to meet new requirement changes. The requirements are mapped onto specific module features using a graph-based context-aware approach to the customer requirements elicitation. The DT-driven solution generation component would then facilitate the design of new asset modules with 3D visualization while, the simulation-verification approach would evaluate the new design based on compatibility and interactions with the new environment. With both benchmark and interaction mechanisms, both the design and usage workflow processes can be expedited with increased outcome visibility to ensure a positive design and optimization outcome.

4.4 Overall system framework

To enable the in-context product family design and optimization service, an overall framework of the cyber-physical system is depicted in Fig. 4.8 featuring a four-tier structure, as elaborated below.

4.4.1 System architecture

Cyber-physical tier. The goal is to establish a two-way communication flow between the physical and virtual components consisting of both the product

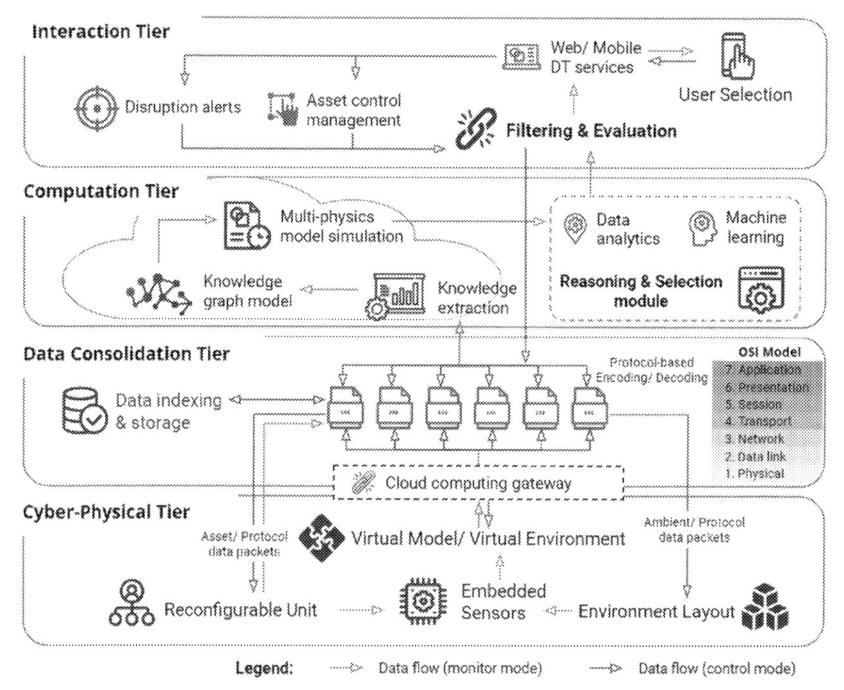

Figure 4.8 Generic DT architecture incorporating ambient information input.

family asset and its current environment layout. Starting with the DT modeling process, high fidelity CAD models of product family assets and accurate dimensions of environment layout are created with 3D modeling software. Next, sensors are embedded at strategic locations on physical assets and layout to obtain relevant data, which is transferred via industrial communication protocols such as OPC UA to achieve consistent and reliable data transfer. These cyber asset and layout components are mapped to reflect the real-time status of their physical counterparts. The arrows marked in green highlight the data flow to establish connection to the data consolidation layer. These data are stored in NoSQL databases which are suited to handle large amounts of heterogeneous sensor data.

Data consolidation tier manages these heterogeneous data passed on by IoT networks, filters out irrelevant data, and performs both storage and retrieval roles to ensure information transfer in a reliable and consistent manner. Data conversion, indexing, filtering, and storage are handled on this layer before being extracted into a predefined ontology model as entities in the computation layer.

Computation tier, information is translated into valuable insights via knowledge graphs and optimization algorithms to enhance existing workflows. The ontology model, which includes stakeholder requirements, safety rules, and operation details, is then populated with real-time or forecasted data to create an effective knowledge graph. The solutions inferred from this graph model are evaluated via a multiphysics model simulation to verify their feasibility. Data analytics and machine learning tools enable key statistics to be derived, providing operators with transparent solutions to make decisions.

Interaction tier forms the final stage whereby the recommended solutions are ranked and presented to end-users in a comprehensive layout and features an interface for asset control. The red arrow highlights the flow from operator input or disruption alert detection back to the physical layer to provide asset control capabilities. This DT system is capable of in-context solution generation for disruption management and process optimization while facilitating asset re-/configuration processes.

4.4.2 Process flow

The process flow details the crux of the DT system, with Fig. 4.9 presenting a process flow model depicting DT-enhanced disruption handling and operator control to facilitate process optimization. The DT system initiates the classification processes to identify the situation type based on predefined parameter thresholds when controller inputs and disruption alerts are triggered. This information is recorded in the cloud database

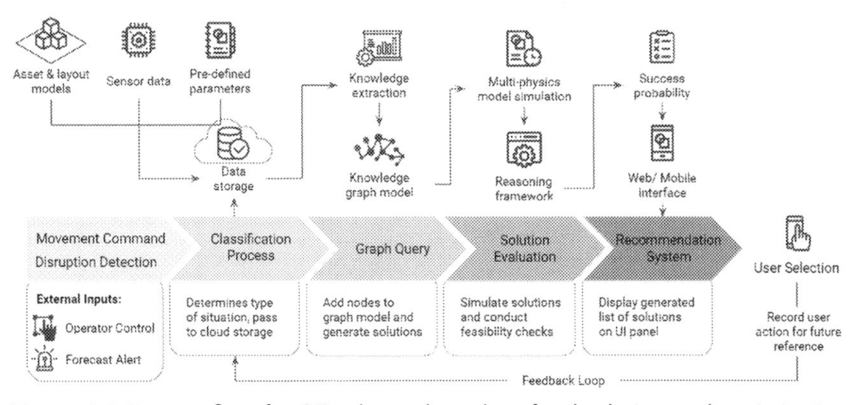

Figure 4.9 Process flow for DT-enhanced product family design and optimization service.

before passing through the knowledge extraction unit, where it is converted into an event entity within the knowledge graph model. Using graph queries to derive potential solutions for asset optimization and minimize disruptive repercussions, the solutions are evaluated via a multiphysics model simulation which eliminates options that exceed predefined safety and financial risk thresholds. Next, the reasoning framework determines the feasibility of these solutions based on parameters such as deadlines, resources, and space constraints that affect shop-floor capabilities. Finally, the choices are categorized according to time and cost factors with other preset priority modes and featured via the system interface. The featured recommendations display the required input and affected parameters, thereby providing transparency which is vital for stakeholders to make informed decisions. This allows stakeholders to inspect the logic and reasoning behind a recommendation and instill user confidence in the DT-enhanced recommendation system over time. A feedback loop records selected recommendations to facilitate knowledge reuse for similar situations in future.

The process model is capable of two different automation modes with varying levels of supervision, thus allowing operators to manage multiple equipment during the usage stage. With a human-in-the-loop approach, user authorization is required before any form of solution implementation. This process is suitable for high-risk equipment and requires extended oversight, although bottlenecks might occur when managing continuous operations such as assembly lines. This poses a challenge during shop-floor disruptions whereby multiple assets would be affected. Alternately, a human-on-the-loop approach enables the DT system to implement solutions without human approval, thus facilitating autonomous processes based on the highest-rated recommendations unless operator interventions occur. These approaches support dynamic shop-floor events and grant a degree of robustness throughout the manufacturing process. In the subsequent section, a case study featuring asset design and optimization is demonstrated by leveraging real-time sensor input, data representation, and evaluation models.

4.5 Case study

To demonstrate the benefits of the benchmark and interaction mechanisms, a case study featuring a 3D printing service family is conducted to aid manufacturing throughout the asset design and usage processes. With 3D printing technology increasingly applied to enhance existing manufacturing capabilities, there is an urgent need to consider asset design and optimization aspects to maximize production efficiency with the capacity to manage disruptions. This section primarily showcases the incorporation of these concepts via the use of an Anet A8 3D printer and highlights key elements necessary for the creation of a context-aware DT system.

4.5.1 Advantages of context awareness in 3D printing services

With evolving manufacturing environments, a hybrid production shopfloor consisting of advanced 3D printers is expected to be integrated into factory assembly lines with typical equipment such as CNC machines and universal robots. Henceforth, in the design stage, an appropriate 3D printer configuration and optimal layout planning are some of the critical factors to be considered as any undue actions will incur efficiency loss which reverberates throughout the production cycle. Without relying on time-consuming and expensive third-party experts, a context-aware DT-enhanced system is proposed to meet these design challenges using a product family approach.

Throughout the design and planning processes, context aware capabilities derived from both ambient information and external knowledge, graph models are essential toward deriving optimal solutions due to the massive amount of heterogeneous data compiled and a sense of flexibility required. Ambient information refers to surrounding details within the shop-floor such as resource flow, which maps out potential bottleneck areas and environment cues such as conditions for 3D printing, including temperature, humidity, etc. These two factors influence the quality of production and manufacturing efficiency extensively and are fundamentals in the design stage with considerations to other existing assets. Next, the external knowledge refers to order requirements and generated forecasts from inventory or demand, allowing the DT system to accurately pinpoint and infer specific constraints to achieve customer satisfaction. The role of this context-aware DT system expedites the selection and verification of suitable 3D printer configurations based on product family components while providing a detailed visualization based on realistic scenarios to evaluate the intended functionalities. In addition, the setup sequence and costs can be computed to allow designers and stakeholders to make informed decisions.

In the usage stage, the context aware DT system would be able to perform 3D printer optimization based on dynamic production scenarios such as customer requirement changes and manufacturing disruptions. Dealing with new events and constraints, the system translates these parameters into new requirements and reevaluates new solutions required for optimizing the existing 3D printer setup. A set of recommendations would be provided to stakeholders, detailing specific actions either for reconfiguration or redesign processes to minimize adverse disruptive effects. For instance, a change in the desired production output would require certain 3D printer modules to be replaced with compatible components, resulting in the aforementioned reconfiguration process. In another typical scenario, disruptions within the shop-floor, such as separate shop-floor equipment downtime, would trigger alternate product printing, reducing the risk of production delays. Besides the conventional DT features such as prognostics and health monitoring in this case study, ambient information is critical toward encompassing multiple PLM aspects.

4.5.2 Establishing a context-aware DT system

The foundation of the 3D printer DT system is based on the technology stack congruent to the generic framework discussed earlier. Starting with asset and context digitization layer, high-fidelity CAD models of all the modular components within the 3D printer family are designed with accurate dimensions in Unity. These modules must be compatible with functionalities such as movement, rotation, and range reflected on the virtual environment. With many original equipment manufacturers (OEM) providing asset CAD models, these models can be disassembled into their modular components for the configuration process. The static layout model of the shop-floor layout can be created using the same modeling tool with same emphasis on the equipment dimensionalities and placement accuracy. Alternatively, the model can be created via point cloud mapping of existing sites with both the final asset configuration and layout models uploaded onto the same simulation system. With existing structural requirements known, identification of suitable areas for 3D printer placement can take place based on manufacturing stipulations. To map the inconstant environment parameters, relevant sensor data can either be obtained from installing new sensors on the 3D printer or from existing sensors embedded within the printer. Utilizing the simulation platform, designers would be able to conduct benchmark and interaction tasks to plan and evaluate potential assets configurations with considerations to various environment constraints. A cyber-physical 3D printer model with preexisting and external sensors installed can be seen in Fig. 4.10, with two-way communication established to provide real-time monitoring and control functionalities. The external ultrasonic sensors are controlled by Arduino

Figure 4.10 A cyber-physical model of the 3D printer with real-time monitoring and control capabilities.

boards which is in-turn connected to the virtual model. The ultrasonic sensors serve to gather output data, while other sensors such as location and temperature sensors embedded within the printer serve to collect asset and ambient data for disruption detection.

Next, the representation model models the relationship between entities which are categorized according to predefined parameters and project requirements. This model holds essential information relating to asset module specifications, layout boundaries and restrictions, stakeholder requirements, sensor input details, and many more. Fig. 4.11 depicts a 3D printer knowledge constructed using Neo4j with nodes containing information related to printing technology, materials, OEM, functionalities, etc. Neo4j is an open-source ACID-compliant labeled property graph that allows the unique identification of same type relationship instances. Meanwhile, additional graphs related to sensor type and other shop-floor equipment

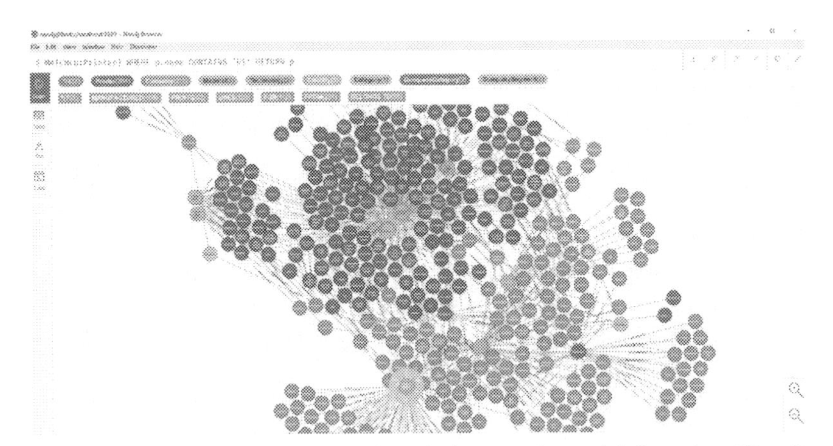

Figure 4.11 Overview of the Neo4j knowledge graph model featuring 3D printer nodes and relationships to enable automatic solution generation.

information, as well as their corresponding relations, can be added separately to form a dynamic reusable system adaptable to different types of manufacturing scenarios. The entities, attributes, and their interrelationships provide the domain-specific knowledge of the tower crane product family. Project-specific ontologies consisting of customer expectations can also be mapped onto an ontology framework via a requirement elicitation approach [87]. With the acquisition of real-time quality sensor data input into the knowledge graph, a range of in-context solutions can be derived to expedite design and optimization workflows. Thus, the 3D printer knowledge graph relies on interactions between asset, environment, and other schemas such as project objectives and constraints to realize knowledge inference. Modeled after a learning pedagogy proposed by David et al. aimed at facilitating a learning environment for asset designing and optimization [88], this enables optimal asset configurations suitable for the predefined layout to be generated and the constant stream of data input further enhances future asset PLM reconfiguration and process optimization. Proposed configurations can be evaluated via simulations as mentioned earlier, thereby achieving context-aware cognitions through information storage and relationship linkage to fulfill product family design and optimization. When integrated with different AI models embedded in the DT computational layer, predictive capabilities can be enabled for maintenance, usage optimization, asset re-/configuration, and layout planning.

Lastly, in the evaluation layer, the 3D printer DT system follows the process flow in Section 4.4.2 to generate a list of recommendations relating to both design and usage stages. In the design stage, family asset configurations provide potential module combinations that are compatible with each other deemed capable of carrying out the customer order requirements with considerations to existing environment constraints. Using the benchmark and interaction mechanism, configurations can be evaluated via the simulation platform to ensure optimal results while alternative suggestions highlight a modular switch to balance setup time and costs. As for the usage stage, besides providing support for asset status monitoring and predictive maintenance, the DT system can track changes in both project and stakeholder requirements as well as disruptions such as material shortage or delayed supply schedules, rerouting production output to maximize printer utilization rate. To meet production-level changes, the new requirements are mapped onto tangible entities in the knowledge graph, after which the system reevaluates realistic remedies via the redesign and reconfiguration methods depending on the product family module availability. The three layers are connected via APIs to provide flexibility for software switch, offering a modular DT framework to encompass wider variety of users and scenarios. Likewise, the use of such DT systems enables stakeholders to scale up shop-floor production by adding additional equipment into the model. This 3D printer case study offers a comprehensive overview on the benefits of the proposed theoretical models and provides an insight into the use of such DT-enabled systems to facilitate product family design and optimization.

4.6 Summary

As a prevailing concept, DT can be leveraged to establish the context-aware testbed to support product family design and optimization as a digitalized service through its life cycle. This chapter introduces a generic DT-enhanced approach to support the in-context virtual prototyping in the early design phase and optimization (e.g., reconfiguration/upgrade) of complex product family during the usage phase. The main contributions can be concluded in three aspects: (1) Introduce a generic tri-model-based approach for product family DT establishment, which includes the digital model, representation model, and computation model. These models interact with each other to provide the fundamental basis (i.e., the cyber-physical environment) for enabling the context-aware product family design and optimization. (2) Propose a novel benchmarking and interacting mechanism for product family design. Unlike the conventional design approach conducted in a mapping process or simulation only process, the DT environment established offers an ambient-based testbed for better virtual prototyping with physical interactions. A case study of a smart 3D printer solution family design and optimization service was further exploited to validate its feasibility. It is hoped to attract more open discussions and promising research works in the DT-enhanced design service field.

References

[1] Tseng MM, Jiao RJ, Wang C. Design for mass personalization. CIRP Ann - Manuf Technol 2010;59(1):175–8. https://doi.org/10.1016/j.cirp.2010.03.097.
[2] Karlsson C, Sköld M. Counteracting forces in multi-branded product platform development. Creativ Innovat Manag 2007;16(2):133–41. https://doi.org/10.1111/j.1467-8691.2007.00432.x.
[3] Cusumano MA, Gawer A. The elements of platform leadership. MIT Sloan Manag Rev 2002;43(3):51–8. https://doi.org/10.1371/journal.pone.0015090.

[4] Harland PE, Uddin Z, Laudien S. Product platforms as a lever of competitive advantage on a company-wide level: a resource management perspective. Rev Manag Sci 2020;14(1):137–58. https://doi.org/10.1007/s11846-018-0289-9.
[5] Li JH, Lin L, Chen DP, Ma LY. An empirical study of servitization paradox in China. J High Technol Manag Res 2015;26(1):66–76. https://doi.org/10.1016/j.hitech.2015.04.007.
[6] Zha XF, Sriram RD. Platform-based product design and development: a knowledge-intensive support approach. Knowl Base Syst 2006;19(7):524–43. https://doi.org/10.1016/j.knosys.2006.04.004.
[7] Nayak RU, Chen W, Simpson TW. A variation-based method for product family design. Eng Optim 2002;34(1):65–81. https://doi.org/10.1080/03052150210910.
[8] Martin MV, Ishii K. Design for variety: developing standardized and modularized product platform architectures. Res Eng Des 2002;13(3):213–35. https://doi.org/10.1007/s00163-002-0020-2.
[9] Jiao J, Tseng MM. Customizability analysis in design for mass customization. CAD Comput Aid Des 2004;36(8):745–57. https://doi.org/10.1016/j.cad.2003.09.012.
[10] JB D, JP G-Z, KN O. Modular product architecture. Des Stud 2001;22:409–24.
[11] Simpson TW. Product platform design and customization: status and promise. Artif Intell Eng Des Anal Manuf 2004;18(1):3–20. https://doi.org/10.1017/S0890060404040028.
[12] Browning TR. Design structure matrix extensions and innovations: a survey and new opportunities. IEEE Trans Eng Manag 2016;63(1):27–52. https://doi.org/10.1109/TEM.2015.2491283.
[13] Zheng P, Xu X, Yu S, Liu C. Personalized product configuration framework in an adaptable open architecture product platform. J Manuf Syst 2017;43:422–35. https://doi.org/10.1016/j.jmsy.2017.03.010.
[14] Zheng P, Chen CH, Shang S. Towards an automatic engineering change management in smart product-service systems – a DSM-based learning approach. Adv Eng Inf 2019;39(January):203–13. https://doi.org/10.1016/j.aei.2019.01.002.
[15] Zheng P, Lin T-J, Chen C-H, Xu X. A systematic design approach for service innovation of smart product-service systems. J Clean Prod 2018;201:657–67. https://doi.org/10.1016/j.jclepro.2018.08.101.
[16] Grieves M. Digital twin: manufacturing excellence through virtual factory replication. Whitepaper 2014. https://doi.org/10.5281/zenodo.1493930.
[17] Nanda J, Thevenot HJ, Simpson TW, Stone RB, Bohm M, Shooter SB. Product family design knowledge representation, aggregation, reuse, and analysis. Artif Intell Eng Des Anal Manuf 2007;21(2):173–92. https://doi.org/10.1017/S0890060407070217.
[18] Simpson TW, Jiao JR, Siddique Z, Hölttä-Otto K. Advances in product family and product platform design: methods & applications. 2014.
[19] Yao X, Moon SK, Bi G. A cost-driven design methodology for additive manufactured variable platforms in product families. J Mech Des 2016;138(April):1–12. https://doi.org/10.1115/1.4032504.
[20] Jiao J, Simpson TW, Siddique Z. Product family design and platform-based product development: a state-of-the-art review. J Intell Manuf 2007;18(1):5–29. https://doi.org/10.1007/s10845-007-0003-2.
[21] Ma J, Kim HM. Product family architecture design with predictive, data-driven product family design method. Res Eng Des 2016;27(1):5–21. https://doi.org/10.1007/s00163-015-0201-4.
[22] Takenaka T, Yamamoto Y, Fukuda K, Kimura A, Ueda K. Enhancing products and services using smart appliance networks. CIRP Ann - Manuf Technol 2016;65(1):397–400. https://doi.org/10.1016/j.cirp.2016.04.062.

[23] Zheng P, Xu X, Chen CH. A data-driven cyber-physical approach for personalised smart, connected product co-development in a cloud-based environment. J Intell Manuf 2018:1–16. https://doi.org/10.1007/s10845-018-1430-y.
[24] D'Souza B, Simpson TW. A genetic algorithm based method for product family design optimization. Eng Optim 2003;35(1):1–18. https://doi.org/10.1080/0305215031000069663.
[25] Qiao L, Efatmaneshnik M, Ryan M, Shoval S. Product modular analysis with design structure matrix using a hybrid approach based on MDS and clustering. J Eng Des 2017;28(6):433–56. https://doi.org/10.1080/09544828.2017.1325858.
[26] Wei W, Tian Z, Peng C, Liu A, Zhang Z. Product family flexibility design method based on hybrid adaptive ant colony algorithm. Soft Comput 2019;23(20):10509–20. https://doi.org/10.1007/s00500-018-3622-y.
[27] Savarino P, Abramovici M, Göbel JC, Gebus P. Design for reconfiguration as fundamental aspect of smart products. Procedia CIRP 2018;70:374–9. https://doi.org/10.1016/j.procir.2018.01.007.
[28] Wang D, Du G, Jiao RJ, Wu R, Yu J, Yang D. A Stackelberg game theoretic model for optimizing product family architecting with supply chain consideration. Int J Prod Econ 2016;172:1–18. https://doi.org/10.1016/j.ijpe.2015.11.001.
[29] Wu Z, Kwong CK, Lee CKM, Tang J. Joint decision of product configuration and remanufacturing for product family design. Int J Prod Res 2016;54(15):4689–702. https://doi.org/10.1080/00207543.2015.1109154.
[30] Tao F, Cheng J, Qi Q, Zhang M, Zhang H, Sui F. Digital twin-driven product design, manufacturing and service with big data. Int J Adv Manuf Technol 2018;94(9–12):3563–76. https://doi.org/10.1007/s00170-017-0233-1.
[31] Lim K, Zheng P, Chen C. A state-of-the-art survey of digital twin: techniques, engineering product lifecycle management and business innovation perspectives. J Intell Manuf 2019;(November). https://doi.org/10.1007/s10845-019-01512-w.
[32] Schleich B, Anwer N, Mathieu L, Wartzack S. Shaping the digital twin for design and production engineering. CIRP Ann - Manuf Technol 2017;66(1):141–4. https://doi.org/10.1016/j.cirp.2017.04.040.
[33] Biancolini ME, Cella U. Radial basis functions update of digital models on actual manufactured Shapes. J Comput Nonlinear Dynam 2018;14(2):021013. https://doi.org/10.1115/1.4041680.
[34] Schluse M, Priggemeyer M, Atorf L, Rossmann J. Experimentable digital twins-streamlining simulation-based systems engineering for industry 4.0. IEEE Trans Ind Inform 2018;14(4):1722–31. https://doi.org/10.1109/TII.2018.2804917.
[35] Damjanovic-Behrendt V, Behrendt W. An open source approach to the design and implementation of Digital Twins for Smart Manufacturing. Int J Comput Integrated Manuf 2019:1–19. https://doi.org/10.1080/0951192X.2019.1599436.
[36] Guo J, Zhao N, Sun L, Zhang S. Modular based flexible digital twin for factory design. J Ambient Intell Humaniz Comput 2018;10(3):1189–200. https://doi.org/10.1007/s12652-018-0953-6.
[37] Alam KM, El Saddik A. C2PS: A digital twin architecture reference model for the cloud-based cyber-physical systems. IEEE Access 2017;5:2050–62. https://doi.org/10.1109/ACCESS.2017.2657006.
[38] Zheng P, Xu X, Chen CH. A data-driven cyber-physical approach for personalised smart, connected product co-development in a cloud-based environment. J Intell Manuf 2018:1–16. https://doi.org/10.1007/s10845-018-1430-y.
[39] Bao J, Guo D, Li J, Zhang J. The modelling and operations for the digital twin in the context of manufacturing. Enterp Inf Syst 2018;13(4):534–56. https://doi.org/10.1080/17517575.2018.1526324.

[40] Ding K, Chan FTS, Zhang X, Zhou G, Zhang F. Digital Twin-based Cyber-Physical Production System for autonomous manufacturing in smart shop floors. Int J Prod Res 2019:1−20. https://doi.org/10.1080/00207543.2019.1566661.
[41] Liu C, Vengayil H, Zhong RY, Xu X. A systematic development method for cyber-physical machine tools. J Manuf Syst 2018;48:13−24. https://doi.org/10.1016/j.jmsy.2018.02.001.
[42] Lu Y, Xu X. Cloud-based manufacturing equipment and big data analytics to enable on-demand manufacturing services. Robot Comput Integr Manuf 2019;57(October):92−102. https://doi.org/10.1016/j.rcim.2018.11.006.
[43] Tan Y, Yang W, Yoshida K, Takakuwa S. Application of IoT-aided simulation to manufacturing systems in cyber-physical system. Machines 2019;7(1):2. https://doi.org/10.3390/machines7010002.
[44] Zhang H, Zhang G, Yan Q. Digital twin-driven cyber-physical production system towards smart shop-floor. J Ambient Intell Humaniz Comput 2018. https://doi.org/10.1007/s12652-018-1125-4.
[45] Guo F, Zou F, Liu J, Wang Z. Working mode in aircraft manufacturing based on digital coordination model. Int J Adv Manuf Technol 2018;98(5−8):1547−71. https://doi.org/10.1007/s00170-018-2048-0.
[46] Luo W, Hu T, Zhang C, Wei Y. Digital twin for CNC machine tool: modeling and using strategy. J Ambient Intell Humaniz Comput 2018;10(3):1129−40. https://doi.org/10.1007/s12652-018-0946-5.
[47] Sharif Ullah AMM. Modeling and simulation of complex manufacturing phenomena using sensor signals from the perspective of Industry 4.0. Adv Eng Inf 2019;39:1−13. https://doi.org/10.1016/j.aei.2018.11.003.
[48] Coronado PDU, Lynn R, Louhichi W, Parto M, Wescoat E, Kurfess T. Part data integration in the Shop Floor Digital Twin: Mobile and cloud technologies to enable a manufacturing execution system. J Manuf Syst 2018;48:25−33. https://doi.org/10.1016/j.jmsy.2018.02.002.
[49] Liu LL, Wan X, Gao Z, Li X, Feng B. Research on modelling and optimization of hot rolling scheduling. J Ambient Intell Humaniz Comput 2019;10(3):1201−16. https://doi.org/10.1007/s12652-018-0944-7.
[50] Liu J, et al. Dynamic evaluation method of machining process planning based on the digital twin-based process model. IEEE Access 2019;7. https://doi.org/10.1109/access.2019.2893309. 1−1.
[51] Moreno A, Velez G, Ardanza A, Barandiaran I, de Infante ÁR, Chopitea R. Virtualisation process of a sheet metal punching machine within the Industry 4.0 vision. Int J Interact Des Manuf 2017;11(2):365−73. https://doi.org/10.1007/s12008-016-0319-2.
[52] Park KT, Im SJ, Kang YS, Do Noh S, Kang YT, Yang SG. Service-oriented platform for smart operation of dyeing and finishing industry. Int J Comput Integr Manuf 2019;32(3):307−26. https://doi.org/10.1080/0951192X.2019.1572225.
[53] Söderberg R, Wärmefjord K, Madrid J, Lorin S, Forslund A, Lindkvist L. An information and simulation framework for increased quality in welded components. CIRP Ann 2018;67(1):165−8. https://doi.org/10.1016/j.cirp.2018.04.118.
[54] Tabar RS, Wärmefjord K, Söderberg R. A method for identification and sequence optimisation of geometry spot welds in a digital twin context. Proc Inst Mech Eng Part C J Mech Eng Sci 2019;233(16):5610−21. https://doi.org/10.1177/0954406219854466.
[55] Xu X. Machine Tool 4.0 for the new era of manufacturing. Int J Adv Manuf Technol 2017;92(5−8):1893−900. https://doi.org/10.1007/s00170-017-0300-7.
[56] Zhao R, et al. Digital twin-driven cyber-physical system for autonomously controlling of micro punching system. IEEE Access 2019;7:9459−69. https://doi.org/10.1109/ACCESS.2019.2891060.

[57] Leng J, et al. Makerchain: A blockchain with chemical signature for self-organizing process in social manufacturing. J Clean Prod 2019;234:767–78. https://doi.org/10.1016/j.jclepro.2019.06.265.
[58] Söderberg R, Wärmefjord K, Carlson JS, Lindkvist L. Toward a digital twin for real-time geometry assurance in individualized production. CIRP Ann Manuf Technol 2017;66(1):137–40. https://doi.org/10.1016/j.cirp.2017.04.038.
[59] Zhang H, Liu Q, Chen X, Zhang D, Leng J. A digital twin-based approach for designing and multi-objective optimization of hollow glass production line. IEEE Access 2017;5:26901–11. https://doi.org/10.1109/ACCESS.2017.2766453.
[60] Lu Y, Xu X. Resource virtualization: A core technology for developing cyber-physical production systems. J Manuf Syst 2018;47(February):128–40. https://doi.org/10.1016/j.jmsy.2018.05.003.
[61] Sierla S, Kyrki V, Aarnio P, Vyatkin V. Automatic assembly planning based on digital product descriptions. Comput Ind 2018;97:34–46. https://doi.org/10.1016/j.compind.2018.01.013.
[62] Bilberg A, Malik AA. Digital twin driven human–robot collaborative assembly. CIRP Ann. 2019;68(1):499–502. https://doi.org/10.1016/j.cirp.2019.04.011.
[63] Nikolakis N, Alexopoulos K, Xanthakis E, Chryssolouris G. The digital twin implementation for linking the virtual representation of human-based production tasks to their physical counterpart in the factory-floor. Int J Comput Integr Manuf 2019;32(1):1–12. https://doi.org/10.1080/0951192X.2018.1529430.
[64] Petković T, Puljiz D, Marković I, Hein B. Human intention estimation based on hidden Markov model motion validation for safe flexible robotized warehouses. Robot Comput Integr Manuf 2019;57:182–96. https://doi.org/10.1016/j.rcim.2018.11.004.
[65] Baruffaldi G, Accorsi R, Manzini R, Baruffaldi G. Warehouse management system customization and information availability in 3pl companies A decision-support tool. Ind Manag Data Syst 2019. https://doi.org/10.1108/IMDS-01-2018-0033.
[66] Bottani E, Cammardella A, Murino T, Vespoli S. From the Cyber-Physical System to the Digital Twin: the process development for behaviour modelling of a Cyber Guided Vehicle in M2M logic. In XXII Summer School "Francesco Turco. – Industrial Systems Engineering 2017:96–102.
[67] Defraeye T, et al. Digital twins probe into food cooling and biochemical quality changes for reducing losses in refrigerated supply chains. Resour Conserv Recycl 2019;149:778–94. https://doi.org/10.1016/j.resconrec.2019.06.002.
[68] Arafsha F, Laamarti F, El Saddik A. Cyber-physical system framework for measurement and analysis of physical activities. Electronics 2019;8(2):248. https://doi.org/10.3390/electronics8020248.
[69] Schneider GF, Wicaksono H, Ovtcharova J. Virtual engineering of cyber-physical automation systems: The case of control logic. Adv Eng Inf 2018;39:127–43, 2019, https://doi.org/10.1016/j.aei.2018.11.009.
[70] Haag S, Anderl R. Digital twin – Proof of concept. Manuf Lett 2018;15:64–6. https://doi.org/10.1016/j.mfglet.2018.02.006.
[71] Iglesias D, et al. Digital twin applications for the JET divertor. Fusion Eng Des 2017;125(October):71–6. https://doi.org/10.1016/j.fusengdes.2017.10.012.
[72] Tao F, Zhang M. Digital twin shop-floor: A new shop-floor paradigm towards smart manufacturing. IEEE Access 2017;5:20418–27. https://doi.org/10.1109/ACCESS.2017.2756069.
[73] Zhuang C, Liu J, Xiong H. Digital twin-based smart production management and control framework for the complex product assembly shop-floor. Int J Adv Manuf Technol 2018;96(1–4):1149–63. https://doi.org/10.1007/s00170-018-1617-6.
[74] General Electric. Power Digital Solutions. GE Digital Twin; 2016.

[75] He Y, Guo J, Zheng X. From surveillance to digital twin: Challenges and recent advances of signal processing for industrial internet of things. IEEE Signal Process Mag 2018;35(5):120–9. https://doi.org/10.1109/MSP.2018.2842228.
[76] Coraddu A, Oneto L, Baldi F, Cipollini F, Atlar M, Savio S. Data-driven ship digital twin for estimating the speed loss caused by the marine fouling. Ocean Eng 2019;186:106063. https://doi.org/10.1016/j.oceaneng.2019.05.045.
[77] Ferguson S, Bennett E, Ivashchenko A. Digital twin tackles design challenges. World Pumps 2017;2017(4):26–8. https://doi.org/10.1016/s0262-1762(17)30139-6.
[78] Kannan K, Arunachalam N. A digital twin for grinding wheel: An information sharing platform for sustainable grinding process. J Manuf Sci Eng 2019;141(2):021015. https://doi.org/10.1115/1.4042076.
[79] MacDonald C, Dion B, Davoudabadi M. Creating a digital twin for a pump. Ansys Advant. Issue 2017;1.
[80] Wang HK, Haynes R, Huang HZ, Dong L, Atluri SN. The use of high-performance fatigue mechanics and the extended Kalman/particle filters, for diagnostics and prognostics of aircraft structures C Comput Model. Eng Sci 2015;105(1):1–24.
[81] Xu Y, Sun Y, Liu X, Zheng Y. A digital-twin-assisted fault diagnosis using deep transfer learning. IEEE Access 2019;7. https://doi.org/10.1109/access.2018.2890566. 1–1.
[82] Lu Y, Min Q, Liu Z, Wang Y. An IoT-enabled simulation approach for process planning and analysis: a case from engine re-manufacturing industry. Int J Comput Integr Manuf 2019;32(4–5):413–29. https://doi.org/10.1080/0951192X.2019.1571237.
[83] Popa CL, Cotet CE, Popescu D, Solea MF, Şaşcîm SG, Dobrescu T. Material flow design and simulation for a glass panel recycling installation. Waste Manag Res 2018;36(7):653–60. https://doi.org/10.1177/0734242X18775487.
[84] Wang XV, Wang L. Digital twin-based WEEE recycling, recovery and remanufacturing in the background of Industry 4.0. Int J Prod Res 2019;57(12):3892–902. https://doi.org/10.1080/00207543.2018.1497819.
[84a] Tao F, Qi Q, Wang L, Nee AYC. Digital twins and cyber–physical systems toward smart manufacturing and industry 4.0: Correlation and comparison. Engineering 2019;5(4):653–61.
[85] Lim KYH, Zheng P, Chen C, Huang L. A digital twin-enhanced system for engineering product family design and optimization. Journal of Manufacturing Systems 2020;57(August):82–93. https://doi.org/10.1016/j.jmsy.2020.08.011.
[86] Cai P, Chandrasekaran I, Zheng J, Cai Y. Automatic path planning for dual-crane lifting in complex environments using a prioritized multiobjective PGA. IEEE Trans Ind Inform 2018;14(3):829–45. https://doi.org/10.1109/TII.2017.2715835.
[87] Wang Z, Chen CH, Zheng P, Li X, Khoo LP. A graph-based context-aware requirement elicitation approach in smart product-service systems. Int J Prod Res 2019:1–17. https://doi.org/10.1080/00207543.2019.1702227.
[88] David J, Lobov A, Lanz M. Attaining learning objectives by ontological reasoning using digital twins. Procedia Manuf 2019;31:349–55. https://doi.org/10.1016/j.promfg.2019.03.055.

CHAPTER 5

Digital twin-driven fault diagnosis service of rotating machinery

Jinjiang Wang[1], Yilin Li[1], Zuguang Huang[2] and Qianzhe Qiao[1]
[1]School of Safety and Ocean Engineering, China University of Petroleum, Beijing, China;
[2]Chinese Machine Quality Supervision and Inspection Center, Beijing, China

5.1 Introduction

With the advent of Industry 4.0, smart manufacturing becomes more digital and productive, requiring equipment to be more flexible to adapt to emergencies. For long-term and reliable operation of equipment in a full life cycle process, not only intrinsically safe designs but also continuous condition monitoring is mandatory. The machinery in smart manufacturing is equipped with a large number of sensing and communication devices. Intelligent sensing and information fusion are the basis of machine condition monitoring. A lot of research works have been conducted on the continuous signal analysis of equipment under multiple working conditions. Emerging computer technologies, such as big data, industrial Internet of things (IIOT), cloud computing, etc., have provided the foundation for applying data-driven methods to machine fault diagnosis with improved efficiency. However, data-driven methods rely heavily on statistical models to determine the existence of faults and predict future operating conditions of machinery. Due to the various failure modes and the complex relationship between faults and system response, data-driven methods may not be able to accurately detect faulty status, and even cause false alarms.

As the integration of cyber-physical systems (CPS), digital twin provides a new approach for smart manufacturing with advanced communication and data processing technologies. In the production process, the state information collected from dispersed devices with advanced sensing technologies, such as vibration, rotating speed, and operating efficiency, is transmitted to a cloud server through high-speed and low-latency communication networks. Then, advanced data processing technologies such as machine learning are used for data mining, intelligent data analysis

Digital Twin Driven Service
ISBN 978-0-323-91300-3
https://doi.org/10.1016/B978-0-323-91300-3.00004-8

and decision-making, and mixed reality technology is used to visualize data statistics and decision-making, intelligently detecting the faults of machines.

Since rotating machinery plays an important role in manufacturing and any unexpected fault may incur extra expenses and unplanned outage, fault diagnosis of rotating machinery is very important. Considering the advantages of digital twin, this chapter presents a digital twin-driven fault diagnosis method and demonstrates its application for rotating machinery fault diagnosis.

5.2 The related works

Over the past few decades, fault diagnosis of rotating machinery has been developed rapidly. Many data analysis technologies, such as machine learning, deep learning, transfer learning, and meta-learning, have been applied. In general, the fault diagnosis methods of rotating machinery are mainly divided into three types: fault diagnosis methods based on physical models, statistical models, and artificial intelligence (AI).

The physical model-based fault diagnosis method relies on the theory of physical dynamics to establish mathematical models corresponding to different faults to describe their deterioration, and predict the remaining service life based on the cumulative effect of damage (such as crack length). Luo et al. developed a physical model-based fault prediction method using interactive multiple models to analyze structural degradation under coupled effects of vibration damage [1,2]. Lei et al. analyzed the crack growth behavior of elastic elements in a nonlinear system under random loads and the slow system stiffness degradation caused by crack growth [3]. Shih et al. studied the edge crack propagation behavior of rectangular plates under the coupled effect of vibration damage [4]. Oppenheimer constructed a crack growth model of a motor shaft based on the Forman crack growth model and predicted its remaining service life [5].

The fault diagnosis method based on statistical models estimates the remaining life of a machine by establishing a statistical model based on mathematical statistics, and the prediction results are usually given in the form of a conditional probability density function based on observations. Commonly used methods include autoregressive model, Markov model, principal component analysis (PCA), and proportional hazard model, etc. Abouelseoud et al. used wavelet filtering and statistical methods to identify gear damage in the initial stage of degradation [6]. Wan et al. used binary wavelet packet transform to extract high-dimensional redundant features, and then used PCA to reduce their dimensionality. Experiments proved that the extracted features can identify the severity of gear crack damage more accurately under different working conditions [7]. Shakya et al. merged time domain, frequency domain, and time-frequency domain features from the monitoring signal of rolling bearings into a feature parameter through Mahalanobis distance and then used the Chebyshev's inequality method to distinguish them to identify the different stages of bearing deterioration [8].

The continuous development of AI technology provides new solutions to equipment fault diagnosis. At present, the most representative AI-based fault diagnosis methods are based on expert systems, machine learning, and deep learning. Expert systems technology enables computers to help analyze and solve complex coupling faults in the rotary equipment [9]. It has been widely used in many fields, such as automobiles, aircraft, turbine, boiler [10], turbomachinery [11], etc. Many typical machine learning methods have been applied in the fault diagnosis of rotating equipment, such as Radial Basis Function (RBF) neural networks [12], selforganizing mapSOM [13], multilayer perception machine [14], and back propagation neural network [15], etc. However, these methods have the disadvantages of unclear physical meaning and large amount of training data requirement, which impedes their applications in industry. Recurrent neural network (RNN) and its variants perform well in long-term time series prediction by considering the correlation between different time steps of time series. Malhi et al. [16] and Yam et al. [17] used RNN to predict the deterioration of rotating components on mechanical equipment. Elsheikh et al. used a two-way long short-term memory (LSTM) neural network to predict the remaining service life of aerospace engines with high accuracy [18]. Abdeljaber et al. established a structural damage detection system using a one-dimensional convolutional neural network (CNN) [19]. Taking advantage of the feature learning of CNN, Guo et al. directly imported the collected raw data into CNN to adaptively extract features for the construction of health indicators [20]. However, RNN and LSTM have many weights parameters, and are prone to overfitting when faced with small training samples. Aiming at this issue, support vector machine (SVM) shows unique characteristics in processing small samples and has been widely used for machinery fault diagnosis. Zhang et al. [21] conducted an in-depth study on the SVM algorithm and concluded that SVM is very suitable for fault pattern recognition of rotating machinery. Hu et al. proposed a new multiclass SVM algorithm and applied it to the fault diagnosis of diesel engines. The experimental results showed that the method has good classification accuracy and robustness [22].

Digital twin is an advanced data analysis technology that integrates multiple physical quantities, multiple temporal and spatial scales, and multiple probabilities and provides a new approach to troubleshooting rotating machinery. Digital twin connects the physical space and virtual space. The rotating machinery prediction model based on digital twin has five characteristics as follows:

(1) Combination of virtual and real: The digital twin aims to establish a two-way mapping between the physical entity and the virtual model, and it needs to realize the deep integration of virtual and real. On the one hand, the physical forms of the rotating machinery's deterioration can be displayed in the digital twin dynamically and in real time. On the other hand, digital twin can perform intelligent analysis and prediction based on sensor data, historical data, and real-time data, improve insight into the state of the rotating machinery, and accurately predict the life of the rotating machinery.

(2) Timeliness: The rotating machinery twins predict the wear situation based on the real physical model and are gradually improved by merging the real-time status data of the rotating machinery. Among them, the acquisition of the real physical model and the status of the rotating machinery needs to meet the timeliness in order to realize the dynamic and real-time monitoring of the wear status of the rotating machinery.

(3) Multidiscipline/multiphysics: The rotating machinery twin is an entity of digital mapping model based on physical characteristics. It needs to describe not only the geometric characteristics of the rotating machinery, but also the various physical characteristics of its physical entities, for example, structural mechanics, thermodynamics, stiffness, strength, etc.

(4) High-fidelity: Digital twins can describe the behavior and characteristics of different tools and should also be modeled using uniform standards, so as to perform high-performance simulations of the state and behavior of the rotating machinery robustly, efficiently, and accurately.

(5) Uncertainty: There are many uncertainties in digital twin modeling of the rotating machinery. The uncertainty is mainly due to a wide variety of causes for rotating machinery wear, and more training data are needed to make up for it to establish a robust digital twin model.

5.3 Digital twin-driven fault diagnosis framework

The proposed digital twin-driven fault diagnosis framework for rotating machinery is displayed in Fig. 5.1. This framework includes four components: the construction of digital twin models, machine interconnection and interoperability, integrative virtual and real data analytics, and applications. The first component aims to construct digital twin models which can cover massive physics information of rotating machinery. The interconnection and interoperability component focuses on the bidirectional operation and communication between the physical entity and the digital twin model. Integrative virtual and real data analytics processes the data from the physical entity and the digital twin model to construct advanced

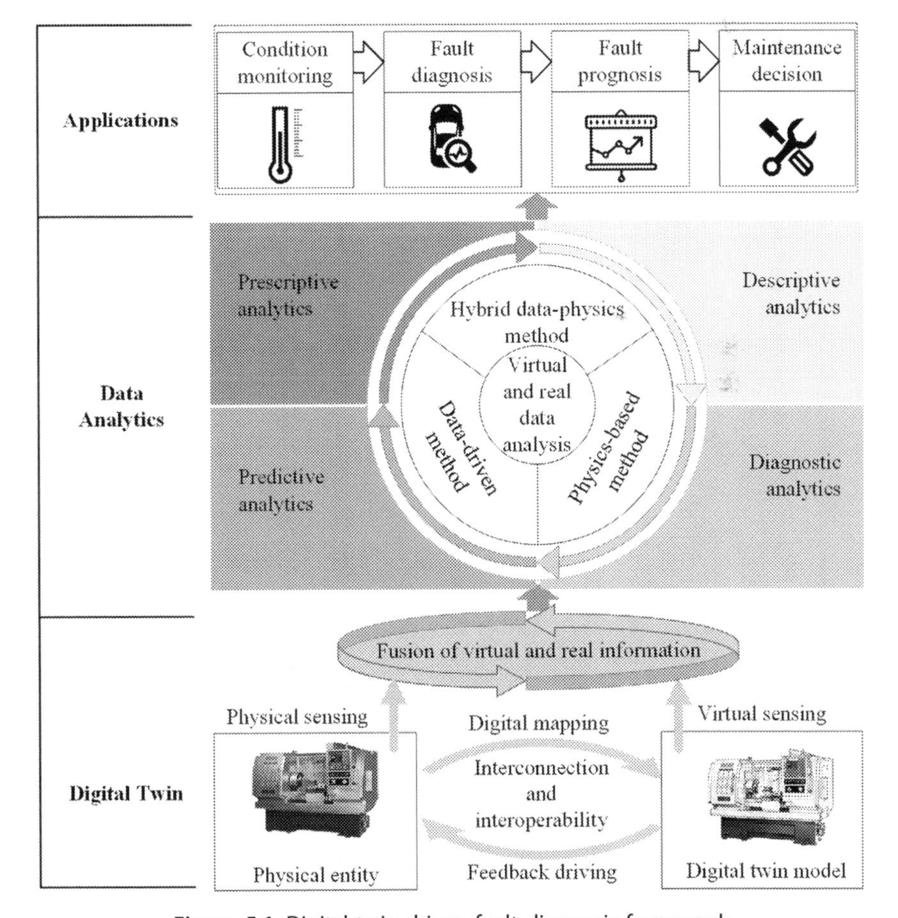

Figure 5.1 Digital twin-driven fault diagnosis framework.

data analysis systems with outstanding analytic abilities. These advanced data analysis systems are applied for condition monitoring, fault diagnosis, fault prognosis, and maintenance decision.

5.3.1 Construction of digital twin models

The construction of digital twin models aims to build a three-dimensional (3D) model to record the initial information of rotating machinery. Initial information denotes static parameters collected when rotating machinery is offline, such as the geometric parameters and structure parameters. These parameters are input into virtual space to construct the static digital twin model which cannot interact with the physical entity yet. In the next step, the correlation between the physical entity and the digital twin model is constructed.

5.3.2 Machinery interconnection and interoperability

Machinery interconnection and interoperability include digital mapping and feedback driving, as shown in Fig. 5.2. Digital mapping denotes using the information of a physical entity to update the digital twin model in virtual space. The Internet of Things (IoT) is used to collect data from rotating machinery. Data communication transfers the collected data into virtual space to update the virtual entity continuously, which realizes the synchronous movement of the physical entity and the digital twin model

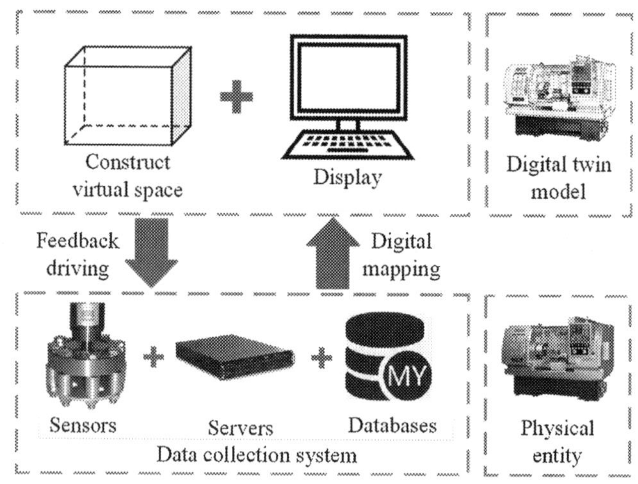

Figure 5.2 Interconnection and interoperability between physical entity and digital twin model.

and ensures the accuracy of the digital twin model. By operating the digital twin model in virtual space, users modify the operations of the physical entity, termed feedback driving. The digital mapping and feedback driving between the physical entity and the digital twin model is termed as interconnection and interoperability. The data from the physical entity and the digital twin model are termed as physical sensing and virtual sensing, respectively, which is fused for analysis and modeling in the next step.

5.3.3 Integrative virtual and real data analytics

Integrative virtual and real data analytics is the core of data analysis systems to explore virtual and real information of rotating machinery. Commonly used virtual and real data analysis methods include physics-based methods, data-driven methods, and hybrid data-physics methods. Fig. 5.3 displays a comparison of the three methods at analyzing the relationship or correlation between variables. Physics-based methods, such as empirical equations and multiphysics simulation, depend on physics mechanism. The relationship between physical variables is explored with massive physics information. Data-driven methods aim to mine hidden information in monitoring signals and build correlation between variables. As a classical method for feature mining, signal processing can extract fault information of rotating machinery, which contributes to the health monitoring of rotating machinery.

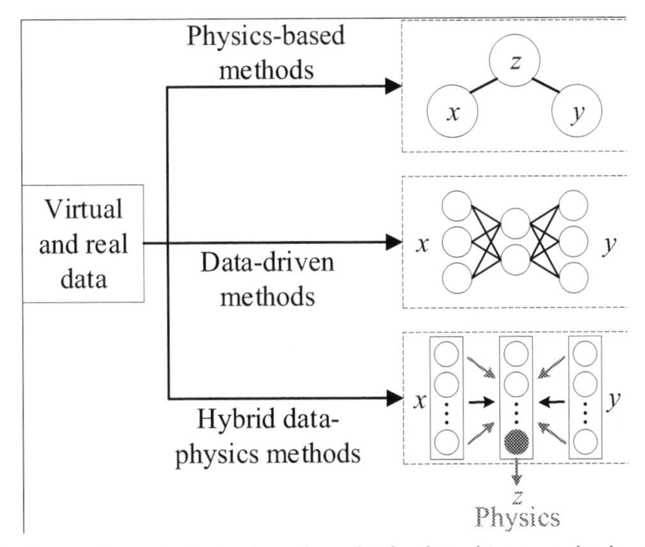

Figure 5.3 Comparison of physics-based methods, data-driven methods, and hybrid data-physics methods at coping with virtual and real data.

As a well-known technique of data-driven models, AI, such as machine learning and deep learning, constructs the mathematical correlation between variables without considering their physics characteristics. The emerging hybrid data-physics methods combine physics mechanisms with data-driven analysis to achieve interpretable and robust analysis, which improves the availability of data-driven methods.

5.3.4 Applications for fault diagnosis

Data analysis systems supported by virtual and real data analysis possess four strong analytic abilities, termed descriptive analytics, diagnostic analytics, predictive analytics, and prescriptive analytics. Descriptive analytics fuses the data from physical sensing and virtual sensing to extract the virtual and real features which represent the health condition of rotating machinery in a current or past time. Diagnostic analytics analyzes the virtual and real features to judge the fault types and search the fault reasons of rotating machinery, termed fault diagnosis. Predictive analytics supports the fault prediction of rotating machinery by estimating the deterioration of mechanical parts and analyzing their fault trends. Prescriptive analytics analyzes the results sourced from state monitoring, fault diagnosis, and fault prediction to determine the scheme for maintenance decisions.

5.4 Case studies

5.4.1 Digital twin for CNC interconnection and interoperability

This section sets up an interconnection and interoperability test bed as shown in Fig. 5.4 to illustrate the realization of the digital twin of an interconnected CNC equipment. The test bed is a simulation production line composed of multiple control systems, including a dual-channel truss loading and unloading test bench, a robot loading and unloading test bench, and a numerical control test bench. The interconnection and

(a) Truss loading and unloading bench

(c) Cross slide bench

(c) Robot loading and unloading bench

Figure 5.4 CNC interconnection and interoperability test bed.

interoperability among heterogeneous equipment are realized via the OLE for process control unified architecture (OPC-UA) communication protocol.

To construct the 3Dmodel, a laser scanner is used to scan the equipment to obtain 3D point cloud data. Combining with on-site inspection of the workshop, the 3D point cloud data is used to construct the digital models of the workshop and the equipment. The obtained point cloud data and the constructed digital twin models are shown in Fig. 5.5. Appropriate materials and textures are selected for rendering through photos taken on site. The rendering effect is shown in Fig. 5.6.

The digital twin model is driven real time by the Unity3D virtual reality engine. The Unity3D engine uses real-time operating data of equipment to drive the digital model to realize the synchronous movement of the digital model and the equipment. The framework of equipment data drive system is shown in Fig. 5.7, including a display interface layer, a logic transfer layer, and a data layer as the main structure. The display interface layer presents the production status of the equipment to users through 3D visualization and at the same time transmits users' instructions to the server through the logic layer. The data layer includes a server, a data acquisition system, and a database. The data acquisition system collects real-time status data of the equipment and stores it in the database. When the server receives users' instructions from the logic layer, it obtains the real-time data in the database as the source data for the data-driven engine to drive the digital model to be consistent with the real equipment in production process. Then, the logic

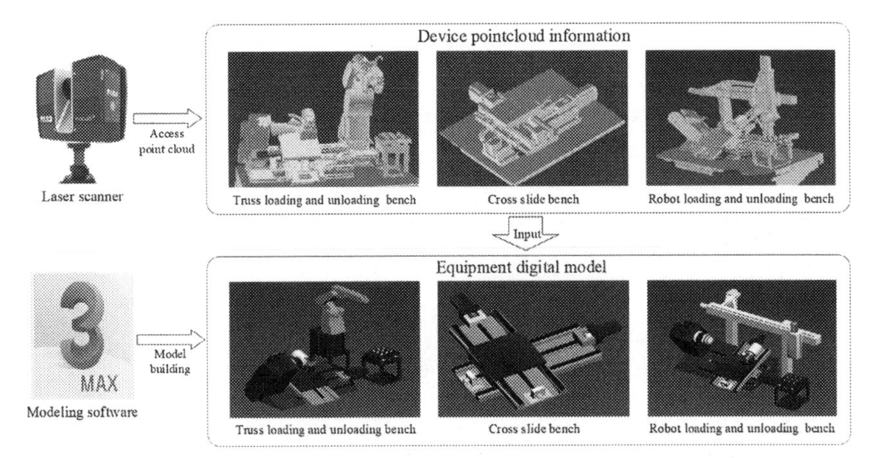

Figure 5.5 The obtained point cloud data and constructed twin model.

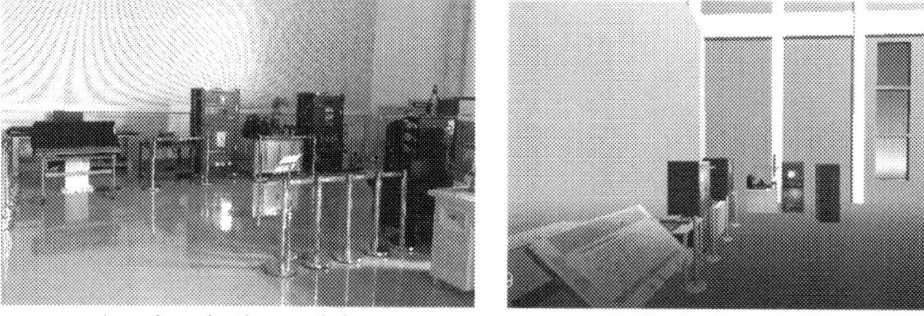

Figure 5.6 The digital twin workshop after rendering.

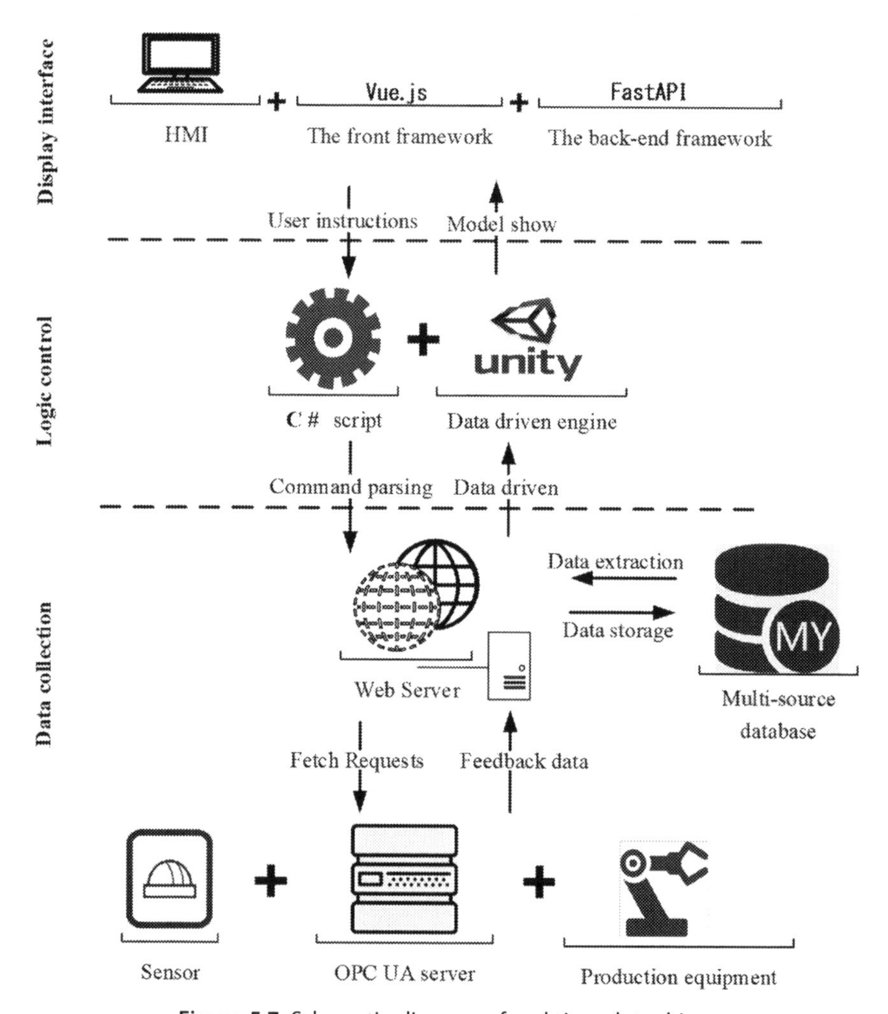

Figure 5.7 Schematic diagram of real-time data drive.

layer passes the data to the display interface layer to drive the digital workshop in real time.

After the OPC-UA server is installed in the actual production workshop, it first searches the data source of the equipment in the address space to obtain the component information of the equipment. Then, it gathers the data and stores it in the database after converting the format. Finally, the Unity data-driven engine is used to logicalize the data, and the logic control layer written in C# obtains the real-time status parameters of each device from the database, so as to interact with the digital model of the production equipment in the control layer to realize the highly realistic operation of the digital model and real-time monitoring of equipment operation status.

The visualization system for the interconnection and interoperability of numerical control equipment is established based on the Brower/Server framework. This system takes advantages of fast running speed, convenient operation and maintenance, and fast packaging of the FastAPI back-end framework and the Vue front-end framework, which greatly facilitates the maintenance and learning of users in a later period. The system function is shown Fig. 5.8, taking the robot's loading and unloading test bench as an example. The system continuously records the real-time status parameters of each operating cycle of the production equipment in the form of dynamic curves combined with digital coordinates, visualizing the changes and development trends of equipment operating data, and laying foundation for equipment optimization. At the same time, the equipment digital twin model is combined with the real-time monitoring of the production site to record the production process of the equipment. The working

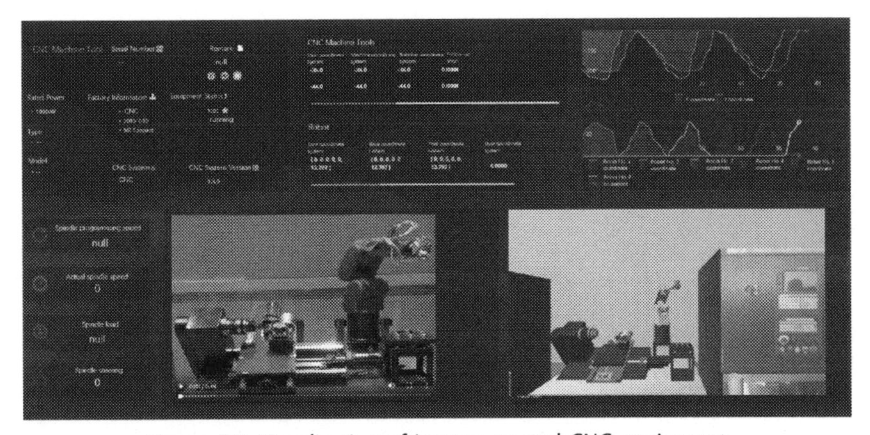

Figure 5.8 Visualization of interconnected CNC equipment.

process of the monitored equipment is displayed to users in an all-round and 3D manner. Compared with traditional methods using data to record the production process in a workshop, it greatly improves the traceability of the production accidents of the CNC equipment in the production workshop and the predictability of the development trend.

5.4.2 Digital twin for rotating machinery diagnosis

In order to demonstrate the feasibility of digital twin in the fault diagnosis of rotating machinery, an experimental study is conducted for the fault diagnosis of rotor unbalance. The model used in this section is a typical rotor-support system. It is composed of an elastic shaft with distributed mass in the axial direction, several rigid discs, and two supporting bearings at both ends of the shaft. A finite element model of the rotor-support system is shown in Fig. 5.9. The shaft is divided into 14 parts, and the node numbers are from1 to 15 from left to right. The nodes 2 and 14 are bearing supports. Disc 1 is located at node six and disc 2 is located at node 10. The maximum speed is set as 3000 rpm. The degree of unbalance in the equipment is affected by the manufacturing precision of the discs and accessories. The three-point method can determine the initial unbalance of the system of 2.7×10^{-5} kg m. In order to verify the effectiveness of the rotating machinery fault diagnosis method driven by the digital twin, the unbalance quantities of 5.5×10^{-5} kg m and 7.3×10^{-5} kg m are set on the disc 2, respectively. Finally, an electric eddy current sensor is used to collect the vertical and horizontal displacement vibrations at nodes three and 13. Four displacement sensors are mounted on the both ends of the rotor support in vertical and horizontal directions to collect vibration signals. The sampling frequency of the data acquisition system is set as 10 kHz.

A finite element model is used to obtain the dynamic response of the rotor system, so it is necessary to construct a high-precision finite element model of the rotor system. Among them, the vibration mode is an inherent

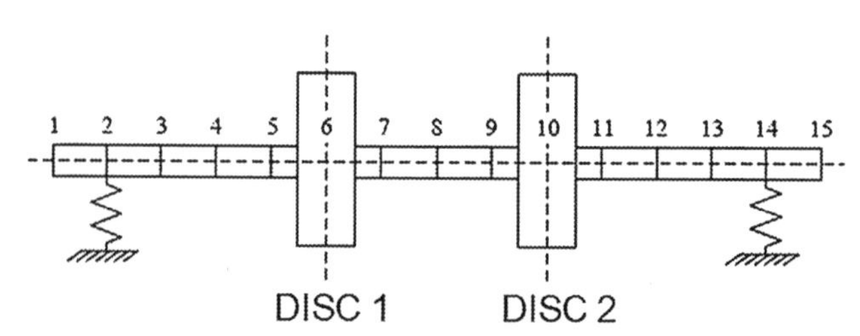

Figure 5.9 Finite element model of a rotor-support system.

property of the system, and its quantitative analysis directly affects the results of dynamic calculations. In a rotor system, the critical speed and frequency response are directly affected by the vibration mode. The vibration mode obtained from finite element analysis is called computational modal analysis, and the system modal parameter obtained by experiment is called experimental modal analysis. If the error between the system natural frequencies obtained through computational modal analysis and experimental modal analysis is small, the established finite element model meets the requirements of accuracy.

In the experimental modal analysis, spring ropes are used to suspend the rotor system to simulate free boundary conditions, and the transient response of the system is obtained by transient excitation. After data processing, the natural frequency and modal shape of the rotor system is obtained. The first four natural frequencies of the rotor system are presented in Table 5.1.

Then the finite element model of the rotor system is constructed. In the digital twin model of the rotor-support system, its dynamic response needs to be considered, including the motion equations of the elastic shaft, the rigid disc, and the bearing. The motion equation of the overall rotor system can be obtained by merging the motion equations of the three elements. The transfer matrix method and the finite element method are usually used to calculate the dynamic equations of the rotor-support system.

For a rotor system with N nodes connected by $N-1$ shaft segments, the displacement vector is assumed to be [23]:

$$\{u_0\} = \left[x_1, y_1, \theta_{y1}, -\theta_{x1}, x_2, y_2, \theta_{y2}, -\theta_{x2}, \ldots, x_N, y_N, \theta_{yN}, -\theta_{xN}\right] \quad (5.1)$$

Combining the dynamic equations of the elastic shaft, the rigid disc and the bearing unit, the dynamic equation of the rotor system can be obtained [23]:

$$M\ddot{u}_0 + (\omega G + C)\dot{u}_0 + Ku_0 = Q_0 \quad (5.2)$$

where M, G, C, K, Q_0, u_0 refer to the global mass matrix, the global gyroscopic matrix, the global damping matrix, the global stiffness matrix,

Table 5.1 The first four natural frequencies of the rotor system.

Mode order	1	2	3	4
Natural frequency (Hz)	34.7	101.2	351.0	535.5

the node force vector, and the node displacement vector, respectively. $\dot{u}_0$ is the first derivative of u_0. $\ddot{u}_0$ is the second derivative of u_0. For a rotor system without damping and external force, $Q_0 = 0$.

The critical speed is a natural frequency proportional to the angular velocity, defined as: $\lambda = \alpha\omega$. Substituting it into Eq. (5.2) in the case of $Q_0 = 0$ yields [23]:

$$\left(K - \lambda^2 \overline{M}\right)\phi = 0 \tag{5.3}$$

where ϕ is the mode shape, and λ is the natural frequency corresponding to the critical speed of the rotor system.

There is a centrifugal force due to unbalance in the rotor system. The relationship between centrifugal force and unbalance can be expressed as:

$$F = mR \cdot \omega^2 = Q\left(\frac{2\pi n}{60}\right)^2 (N) \tag{5.4}$$

where F is the centrifugal force, m is the rotor mass, R is the centrifugal radius, and ω is the angular velocity. Since the fault studied in this section is an unbalance fault, its equivalent load can be regarded as the centrifugal force of the discs. Q is the unbalance quantity, and n is the rotor speed.

Under unbalanced conditions, a new dynamic equation of the rotor system can be obtained by replacing the right side of Eq. (5.2) with $Q_0 + F$ [23].

In order to calculate the equivalent load of an unbalanced rotor system, it is necessary to obtain the vibration response of the rotor system with full degrees of freedom. The more degrees of freedom a system is divided into, the more accurate will be the calculated equivalent load. However, in an actual system, it is impossible to obtain vibration responses of all degrees of freedom through normal signal collection due to the limited number of installed sensors. The modal expansion method solves this problem. The principle of the method is to estimate the vibration response under full degrees of freedom by using the first several modes of a system and the measured vibration signals. The equivalent load of each node in the rotor system obtained by using the modal expansion method can be calculated as [23]:

$$\Delta Ft = \left\{\Phi\left[(C\Phi)^T (C\Phi)\right]^{-1} (C\Phi)^T\right\}\Delta F_{Mt} \tag{5.5}$$

where C is the measurable matrix, and Φ is a modal matrix composed of mode vectors.

Table 5.2 The parameters for constructing digital twin model of rotor system [23].

Parameter	Initial value
Length of shaft	0.56 m
Diameter of shaft	0.008 m
Disc radius	0.0305 m
Disc weight	0.806 kg
Initial unbalance mass	2.7×10^{-5} kg m
Young's modulus	2.1×10^{11} N/m^2
Poisson ratio	0.3
Density	7.8×10^3 kg/m^3
X/Y bearing stiffness of the drive end and free end of the rotor system	1.00×10^6 N/m
X/Y bearing damping at the drive end and free end of the rotor system	5×10^3 N s/m

The initialized parameters used to construct the rotor digital twin model are listed in Table 5.2.

Although the initial digital twin model of the rotor system can describe the dynamic rotor system, it still has a gap with the actual model, as shown in Fig. 5.10A. Therefore, it is proposed to construct a modified digital twin model based on the response surface. The idea of model modification based on response surface is as follows. Based on the multiparameter samples obtained from experiments, the response surface is constructed using the response surface reconstruction technology that can reflect the relationship between the modified parameters and the target variable, and optimize iteratively on the fitted response surface to reduce the difference between the model and the test model, so as to obtain the correction parameters.

Since the inherent characteristics of the rotor system under operating conditions are difficult to obtain, the characteristic variables such as the first-order critical speed and unbalance response under the operating conditions of the rotor are selected to update the model. According to the updated parameters and characteristic variables, the objective of the updated model is [23]:

$$R(p, u_m) = f_D(C, u_m) - f_M(C, u_m) \tag{5.6}$$

where f_D and f_M are the functions of updating parameters of the critical speed C and the amplitude unbalance u_m from the digital system and physical system, respectively. The error $R(p, u_m)$ is minimized for the best fitting to update the digital twin model at different levels of unbalance of the rotor system.

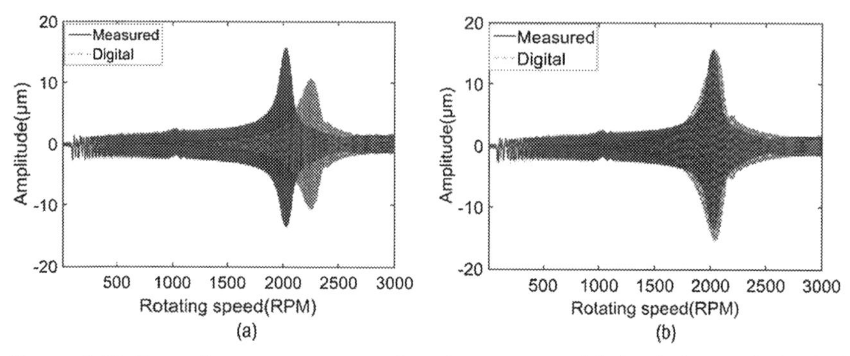

Figure 5.10 Dynamic response comparison between model simulation and physical system, (A) before model updating, (B) after model updating [23].

Using a multiparameter optimization scheme based on the PSO algorithm, the objective function gradually decreases and tends to zero through continuous optimization iterations. The optimized parameter results of the critical speed C and the amplitude unbalance u_m are shown in Table 5.3. The dynamic response of the updated digital twin model of the rotor system is shown in Fig. 5.10B, which proves that the updated digital twin model is a more realistic physical model. Additionally, two sets of unbalance testing are carried out to simulate the dynamic behavior of the rotor system, as shown in Fig. 5.11.

Unbalance faults are quite common in rotating systems. At present, signal processing techniques are mostly used for fault diagnosis for this problem. Fig. 5.12 shows the unbalance fault diagnosis using Fourier transform and wavelet transform. It can be seen from Fig. 5.12 that for different unbalance faults, the possible balance weight value can be displayed in time.

As mentioned above, the fault diagnosis method based on digital twin can effectively combine the mathematical models of the rotor system with

Table 5.3 The optimized parameters in the model updating process [23].

Index	Updated parameter	Before updating	After updating
1	Critical speed C	2256 r/min	2047 r/min
	Vibration amplitude u_m	0.955 μm	1.220 μm
2	Critical speed C	2047 r/min	2137 r/min
	Vibration amplitude u_m	1.220 μm	2.257 μm
3	Critical speed C	2137 r/min	2086 r/min
	Vibration amplitude u_m	2.257 μm	3.523 μm

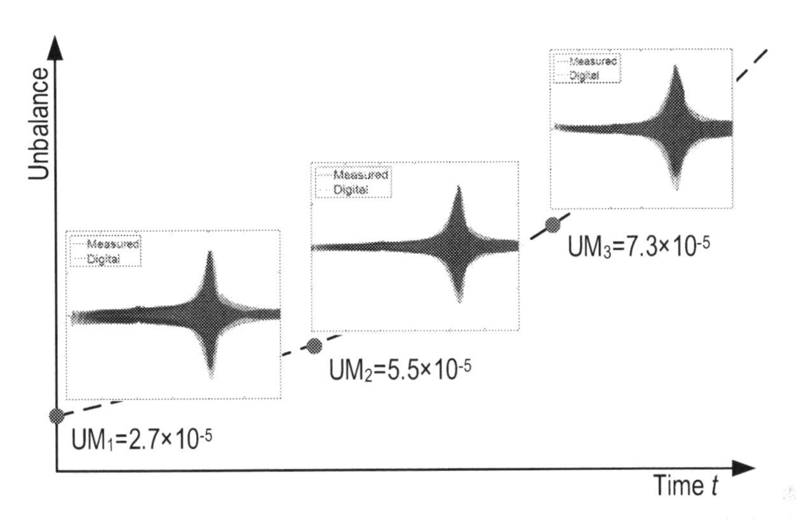

Figure 5.11 Dynamic response comparison between model simulation and physical system under different levels of unbalance [23].

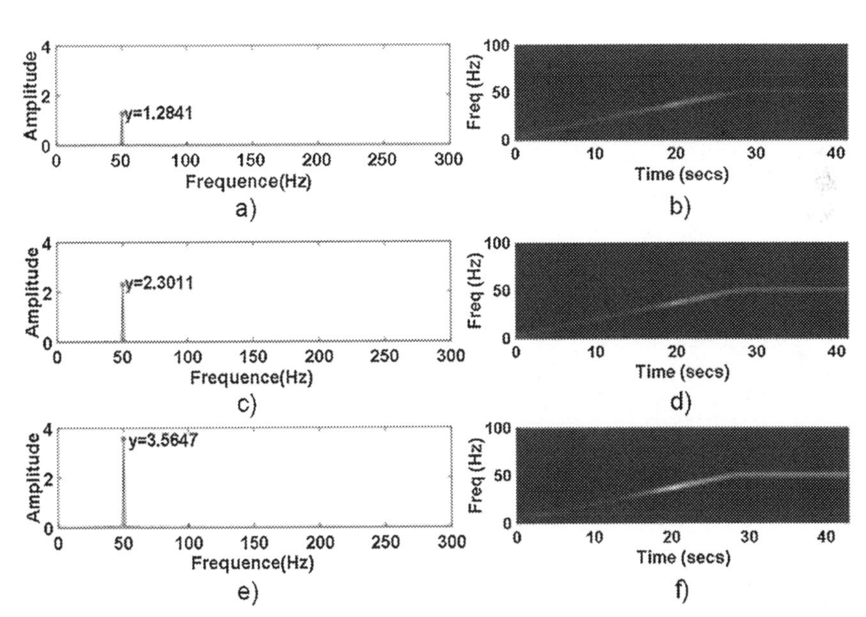

Figure 5.12 Spectrum analysis and time-frequency analysis results of rotor unbalance faults with unbalance mass: (A and B) 2.7×10^{-5} kg m, (C and D) 5.5×10^{-5} kg m, (E and F) 7.3×10^{-5} kg m [23].

the collected data, identify the failure parameters of the rotor system and distinguish quantitative diagnosis of the rotor system. Compared with the conventional methods, the fault diagnosis based on digital twin can realize the quantification of the unbalance and the positioning of the fault diagnosis, which is useful for accurate diagnosis and maintenance of rotating machines.

5.5 Summary

To analyze the complex fault modes and the relationship between faults and system response, this study presents a novel framework termed digital twin-driven fault diagnosis for rotating machinery. It illustrates the integration of information from a physical entity and a digital twin model for state monitoring, fault diagnosis, fault prediction, and maintenance decision. Depending on the proposed framework, digital twin models which include geometric and physical characteristics of rotary machinery are constructed. Experiments demonstrate that this model can realize the synchronous movement of the digital models and the physical entities and accurately analyze the fault response of rotating machinery. Fault diagnosis based on digital twin, as a new concept, is seldom studied at the theoretical level. Future work includes the following aspects:

(1) High-precision modeling and real-time analysis of complex equipment.

In the experimental analysis, the models and data are all derived from a simplified rotating system. If they are applied in practical production, the modeling and analysis capabilities should be improved. The method of simplifying the modeling and improving the analysis ability needs to be further studied.

(2) The generalization problem of the quantitative diagnosis of the rotating system.

The quantitative diagnosis of the rotating system studied is only for the fault type that can equate the system state to the external load. However, a real rotating system is usually coupled when the system fault is in operation when it is difficult to use mathematical models to perform the description. Therefore, integrating the physical model and data-driven model in the digital twin model will contribute to understanding the operation status of the rotating system, which is more conducive to the quantitative diagnosis of general faults in a rotating system.

References

[1] Luo J, Namburu M, Pattipati K, et al. Model-based prognostic techniques. Autotestcon; 2003. p. 330−40.
[2] Luo J, Bixby A, Pattipati K, et al. An interacting multiple model approach to model-based prognostics. In: Systems, man and cybernetics, IEEE international conference on. IEEE; 2003.
[3] Lei Y, Zhu WQ. Fatigue crack growth in degrading elastic components of nonlinear structural systems under random loading. Int J Solid Struct 2000;37(4):649−67.
[4] Shih Y, Wu G. Effect of vibration on fatigue crack growth of an edge crack for a rectangular plate. Int J Fatig 2002;24(5):557−66.
[5] Oppenheimer C, Loparo K. Physically based diagnosis and prognosis of cracked rotor shafts. Proc SPIE Comp Syst Diag Prog Health Manag 2002;4733:122−32.
[6] Abouel-seoud SA, Elmorsy MS, Dyab ES. Robust prognostics concept for gearbox with artificially induced gear crack utilizing acoustic emission. Energy Environ Res 2011;1(1):81−93.
[7] Wan X, Wang D, Tse PW, et al. A critical study of different dimensionality reduction methods for gear crack degradation assessment under different operating conditions. Measurement 2016;78:138−50.
[8] Shakya P, Kulkarni MS, Darpe AK. A novel methodology for online detection of bearing health status for naturally progressing defect. J Sound Vib 2014;333(21):5614−29.
[9] Alty J, Coombs MJ. Expert systems: concepts and examples. National computing centre publications; 1984.
[10] He Q, Zhao X, Du D. A novel expert system of fault diagnosis based on vibration for rotating machinery. J Measure Eng 2013;1(4):219−27.
[11] Siu C, Shen Q, Milne R. A Fuzzy expert system for turbomachinery diagnosis. In: IEEE international conference on Fuzzy systems; 1997.
[12] Cao Z, Shen J. Research on sensor fault diagnosis method based on RBF time series predictor. Trans Microsyst Technol 2010;29(05):63−65+69.
[13] Xu Z, Wang H, Xu C, et al. Analog circuit fault diagnosis method based on preferred wavelet basis and Fuzzy SOM. Measure Contr Technol 2016;35(11):5−8+13.
[14] Souahlia S, Bacha K, Chaari A. MLP neural network-based decision for power transformers fault diagnosis using an improved combination of Rogers and Doernenburg ratios DGA. Int J Electr Power Energy Syst 2012;43(1):1346−53.
[15] Han B. Analog circuit fault diagnosis method based on BP neural network. In: 2012 4th electronic system-integration technology conference; 2012.
[16] Malhi A, Yan R, Gao RX. Prognosis of defect propagation based on recurrent neural networks. IEEE Trans Instrum Measur 2011;60(3):703−11.
[17] Yam R, Tse PW, Li L, et al. Intelligent predictive decision support system for condition-based maintenance. Int J Adv Manuf Technol 2001;17(5):383−91.
[18] Elsheikh A, Yacout S, Ouali MS. Bidirectional handshaking LSTM for remaining useful life prediction. Neurocomputing 2019;323(5):148−56.
[19] Abdeljaber O, Avci O, Kiranyaz S, et al. Real-time vibration-based structural damage detection using one-dimensional convolutional neural networks. J Sound Vib 2017;388(3):154−70.
[20] Guo L, Lei Y, Li N, et al. Deep convolution feature learning for health indicator construction of bearings. Prognostics and System Health Management Conference; 2017.
[21] Zhang Z, Fei Y, Song L, et al. Research on multi-classification algorithm based on Fuzzy support vector machine. J Comput Appl 2008;28(7):1681−3.

[22] Hu Z, Cai Y, Ye L, et al. Data fusion for fault diagnosis using Dempster-Shafer theory based multi-class SVMs. In: Advances in natural computation, first international conference; 2005.
[23] Wang J, Ye L, Gao RX, et al. Digital Twin for rotating machinery fault diagnosis in smart manufacturing. Int J Prod Res 2018;57:3920–34.

CHAPTER 6

Digital twin-driven energy-efficient assessment service

Wenjun Xu, Zhenrui Ji, Yongli Ma, Yanping Ma and Zude Zhou
School of Information Engineering, Wuhan University of Technology, Wuhan, Hubei, China

6.1 Introduction

Social, environmental, and economic aspects are considered as the three bottom lines of sustainability by almost all literature [1]. The aim of sustainable development is to achieve a balance between social, environmental, and economic dimensions. Manufacturing is the main pillar of human society, which consumes a lot of energy, materials, and other resources to provide products for human beings and will play an essential role in sustainable development. In the manufacturing industry, according to the definition of The U.S. Department of Commerce's International Trade Administration, sustainable manufacturing is as "the creation of manufactured products that use processes that minimize negative environmental impacts; are safe for employees, communities, and consumers; and are economically sound" [2].

Sustainable manufacturing reduces the impact on the environment and energy consumption during the production process to achieve economic benefits while taking into account social and environmental benefits. According to the statistics, the manufacturing industry accounts almost 90% energy consumption of the whole industrial sector [3]. Enormous energy is being consumed worldwide with an estimated annual increase of 1.4%, and energy-related carbon emissions tend to reach 36 billion metric tons in 2020 [4]. With ever-increasing energy consumption, the issue of improving energy efficiency and reducing resource consumption has attracted much attention. To achieve this, energy-efficient assessment is one of the prerequisites of the solution. This can assess the sustainability of the manufacturing system through the quantitative indicators by providing support for optimizing the operation of the production process.

Energy-efficient assessment can reflect the efficiency and quality characteristics of energy utilization of products, and it assesses the output per unit of energy, which is a more scientific method to assess the energy

Digital Twin Driven Service
ISBN 978-0-323-91300-3
https://doi.org/10.1016/B978-0-323-91300-3.00007-3

performance of products [5]. In tradition, data-driven energy-efficient assessment can obtain more accurate results but is highly data-intensive, time-consuming, and requires high hardware cost. Moreover, with a wide variety of manufacturing equipment, systematic structures, data sources, and data formats, it is challenging to use universal information models and management models to standardize the management of manufacturing information. It is desirable to provide accurate and reasonable assessment as the decision-making basis for the overall optimization of system performance.

The idea of digital twin is to create a digital counterpart to mirror the characteristics and behaviors of its corresponding object in the physical world, which can realize the information mapping and interconnection between information space and physical space [6]. Digital twin can also minimize the delay of data collection in the process of manufacturing system operation and maintenance and modeling and provide the sufficient high-quality data which are fundamental to realizing real-time, high transparency, strong interoperability, and rapid response of the cyber-physical manufacturing system [7]. Powered by digital twin, the more efficient approach for energy-efficient assessment could be achieved. Digital twin-driven energy-efficient assessment, on the one hand, can integrate the information from the overall physical manufacturing resources into the cyberspace to solve the information island problem, which helps to improve the comprehensiveness of assessment; on the other hand, could also utilize the data generated from the simulation in cyberspace, which is high-fidelity mapping to the physical world, to obtain the predictive assessment results without any real-world tests which are normally time-consuming and costly.

Moreover, as socialization and servitization are becoming the development trend of manufacturing, service is playing an increasingly significant role in manufacturing. Manufacturing service has the characteristics of on-demand use, dynamic reconfiguration, and cross-platform features [8], which provide manufacturing with the possibilities of resource sharing and multiparty collaboration. With the concept of Everything-as-a-Service, the full potential of digital twin could be unleashed by service which could share the components and functionalities of digital twin and offer access to the third-part services [9].

In view of the power of the aforementioned digital twin and service, this chapter presents a digital twin-driven energy-efficient assessment method for quantifying sustainable manufacturing capabilities and provides solution for encapsulating services and sharing assessment functions. The following first introduces the energy efficiency assessment indicators, assessment model, and assessment approaches for the problem of quantifying manufacturing capability; then, after the introduction of how to apply digital twin to achieve more efficient and accurate energy-efficient assessment, a framework of digital twin-driven energy-efficient assessment is presented for manufacturing services; finally, the feasibility verification of the above-mentioned methods is carried out through two cases: a digital twin-driven assessment service platform, and an industrial robots (IRs) energy-efficient assessment service.

6.2 Energy-efficient assessment for manufacturing capability

The concept of manufacturing capability was first proposed by Skinner in 1969, and it is composed of cost, quality, production time, and the relationships between them [10]. With the rise of sustainable manufacturing, some researchers have introduced energy efficiency into the assessment of manufacturing systems, evaluating the manufacturing capability to make manufacturing decision from the economic, environmental, and social aspects [11,12]. Improving energy efficiency in manufacturing systems can effectively mitigate the impact on the environment [13]. Accurate and reasonable energy-efficient assessment for manufacturing capacity is one of the ways to realize energy-efficient operations of the manufacturing system. Therefore, energy-efficient assessment for manufacturing capability has become a worldwide issue in the manufacturing industry. Energy-efficient assessment helps to improve the environmental performance of manufacturing processes and makes energy efficiency a key success factor for sustainable production, which is a crucial link to increase productivity and reduce energy consumption.

6.2.1 Energy-efficient assessment indicators

In the entire process of energy efficiency assessment for equipment or system operation, clarifying equipment assessment goals and building a comprehensive assessment indicators system are the basic guarantees for a smooth assessment process. The status of the target equipment should be fully analyzed in the process of defining assessment indicators. This section will explain the relevant principles and the connotation of the energy efficiency assessment indicators system.

6.2.1.1 Principles for assessment indicator selection

The indicator selection for energy-efficient assessment is the first step of the entire assessment process, and it is also the key step to ensure the acquisition of the valid and reasonable assessment results. In the indicators construction, it is essential that the constructed indicators can fully reflect the characteristics of the assessment target, which would make the final assessment result more objective and reasonable. Combined with the relevant design principles and selection methods of assessment indicators [14], the following six basic principles have been considered for the assessment indicators construction.

A. **Scientific.** The establishment of indicators needs be based on relevant scientific theories. The setting of indicators should avoid redundancy and be objective and reasonable.
B. **Integrity.** In the process of the assessment indicators selection, it is inadvisable to consider the decision-making problem from separate indicators. The appropriate method is to consider the relationship between the indicators from the perspective of the system as a whole.
C. **Independence.** Assessment indicators at the same unit (e.g., different manufacturing equipment) or system (e.g., production line) level should be as independent as possible.
D. **Dynamic.** The dynamics of the operating state of the manufacturing system is constantly changing during the manufacturing process, and this must be fully considered during the process of assessment indicators selection.
E. **Hierarchical.** Assessment indicators need be graded and refined step by step, so as to establish a clear hierarchical structure of the indicators and clarify the weighted impact of indicators at different levels on the energy-efficient capability of the entire system.
F. **Feasibility.** The first is that the selection of indicators should be concise and complete; the second is to select indicators that reflect the evaluation characteristics as much as possible, so that the final assessment results are easily understandable; the third is to be measurable, which means that the calculation methods involved in the assessment process should not be too complicated to solve, and the required data should be accessible as well.

6.2.1.2 Framework of assessment indicators

According to existing literature, the form and granularity of the manufacturing capability can be divided into three levels, namely, the manufacturing unit level, the production line level, and the workshop level [15]. Manufacturing capability of unit level mainly describes the manufacturing capabilities of a single device. Different manufacturing unit-level equipment manufacturing capabilities form production line-level manufacturing capabilities in the form of combination and coordination to meet the needs of complex manufacturing tasks. At present, most researchers mainly focus on the energy-efficient assessment of manufacturing capability at the manufacturing unit level and the production line level.

The unit-level framework of assessment indicators for manufacturing capability is shown in Fig. 6.1, and the production line-level framework of assessment indicators for manufacturing capability is shown in Fig. 6.2. In

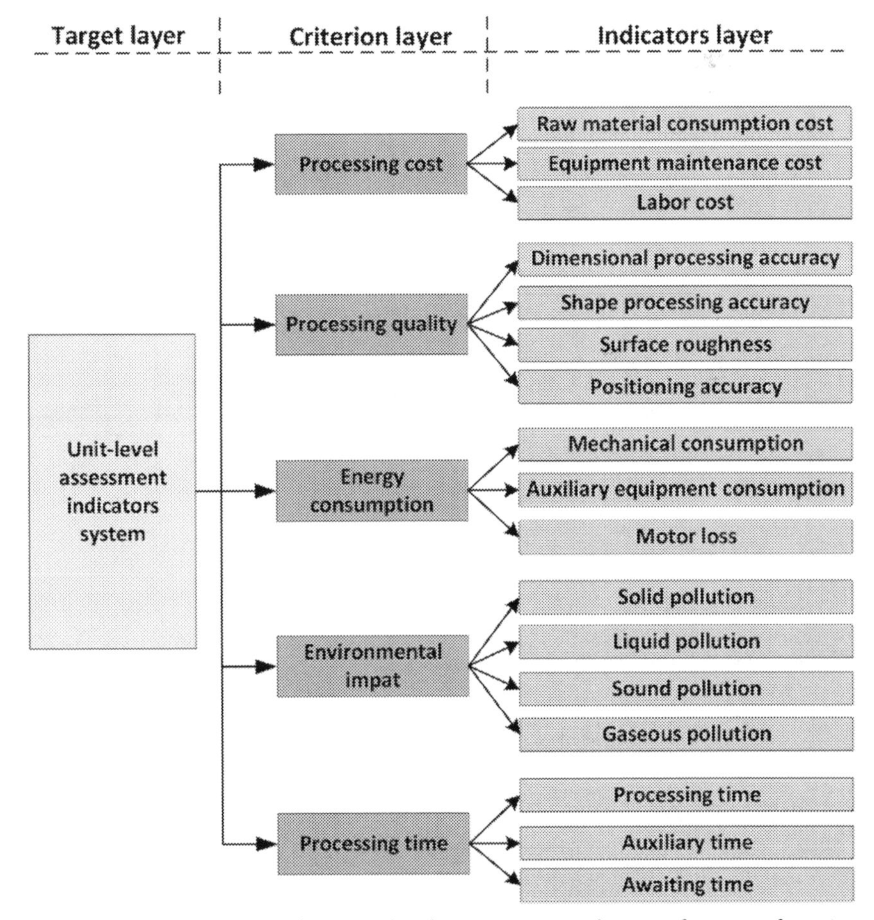

Figure 6.1 The unit-level framework of assessment indicators for manufacturing capabilities [12].

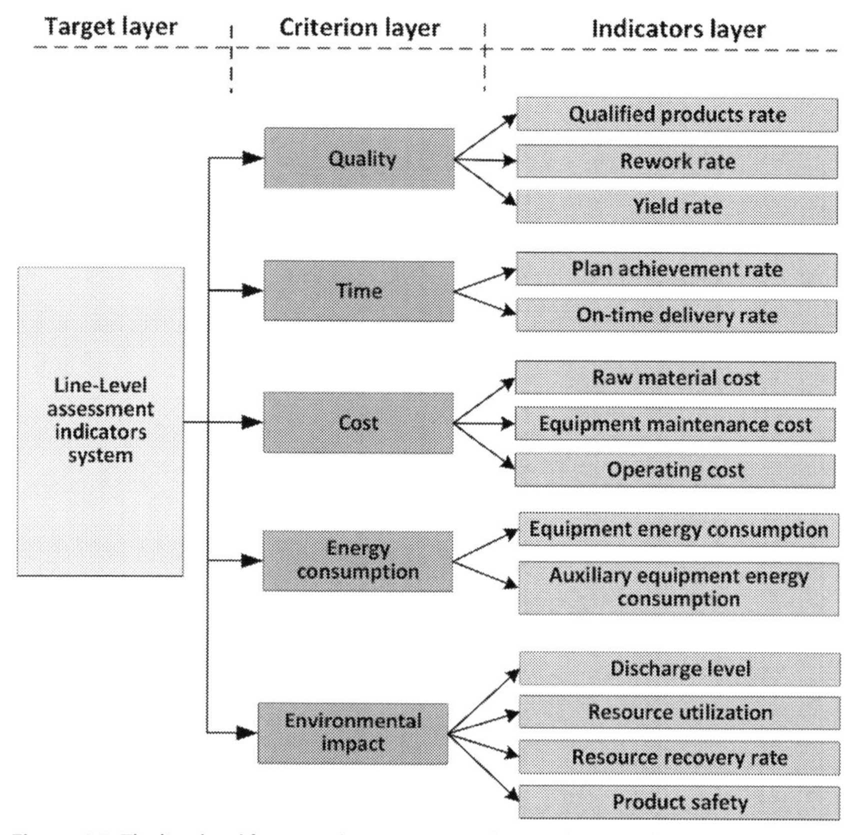

Figure 6.2 The line-level framework assessment indicators for manufacturing capabilities.

the indicators for manufacturing capacity assessment at the unit level, there are five criteria for assessment, namely, processing cost, processing quality, energy consumption, environmental impact, and processing time. The following indicators are generated on the basis of the corresponding criterion. Similarly, in the indicators for manufacturing capacity assessment at the line level, there are five criteria for assessment, namely, quality, time, cost, energy consumption, and environmental impact, followed the corresponding indicators.

In general, the manufacturing unit and production line are not separated, but related to each other, which makes the capability assessment of the manufacturing system particularly complicated. For the robot manufacturing unit, its performance might be dynamic to some extent when completing production tasks in the complex and changeable workshop environment. Different units cooperate with each other to complete the production tasks, which leads to the coupling relationship between different manufacturing unit-level attributes. At the same time, the production line is composed of multiple manufacturing units. When some attributes of the manufacturing unit are affected by external interference, the relevant attributes of the production line then fluctuate accordingly.

6.2.1.3 Extensible dynamic multidimensional assessment indicator framework

The traditional assessment indicators cannot satisfy the needs of newly joined equipment types in the manufacturing system due to its poor scalability. Due to the different production processes involved in various types of equipment, the indicators used to characterize their manufacturing capability are different as well. Therefore, it is not advisable to use unified and fixed indicators to cover all types of equipment that may enter the evaluation system in the future. With regard to this, an extensible multidimensional assessment indicators framework was proposed [16]. The presented extensible indicators framework can be dynamically expanded in the assessment process. From the overall structure, it contains multiple subindicators, and each subindicator is established for a specific type of equipment. Although they are all established based on five indicators criteria such as processing time, processing quality, processing cost, environmental impact, and energy consumption, the following refinement indicators under the corresponding criterions are different from each other due to the differences in processing technology. In addition, each subindicator module contains a multidimensional indicator structure formed by the above five-indicators criteria. For the subindicators which corresponding criterion has been determined, the multidimensional indicators under the different criteria have been selected and determined; on the other hand, the indicators of the subindicators to be expanded present a kind of null status. When an external individual expands it, the initial state null value of the subindicators will be covered by the that of the newly joined equipment of the new type.

In terms of expansion, when a new type of equipment is added to the assessment framework or the user proposes some personalized requirements for improving the original indicators framework, the multidimensional indicators framework only needs to create a new subindicator or to change the indicators vector dimension to which the corresponding indicator to be expanded belongs in the existing subindicators, which would not affect the other subindicators. Therefore, the extensible dynamic multidimensional assessment framework can effectively improve the situation of the poor scalability of the traditional assessment indicators framework.

6.2.2 Energy-efficient assessment model

6.2.2.1 Dynamic assessment model

Since the performance of manufacturing equipment varies with the age and wear of equipment, the indicator values reflecting the equipment manufacturing capability also change dynamically. Abnormal mutations of data may occur due to some reasons, leading to the distortion of the assessment results. Therefore, it is necessary to introduce time dimension factors to dynamically assess equipment manufacturing capabilities, which means that the energy-efficient assessment for the manufacturing capability is then a multidimensional decision-making issue of three dimensions with indicators space, object space, and time space.

According to the connotation of manufacturing capability and the demands of energy-efficient manufacturing, a set of indicators is established including quality, time, cost, resource consumption and environmental impact, and a new dynamic assessment model is presented, as shown in Fig. 6.3. It can be described that $A = \{s_1, s_2, \ldots, s_n\}$ represents manufacturing equipment objects to be assessed, $X = \{x_1, x_2, \ldots, x_m\}$ represents the m selected indicators, and $x_{ij}(t_k)$ $(i = 1, 2, \ldots, n, j = 1, 2, \ldots, m)$

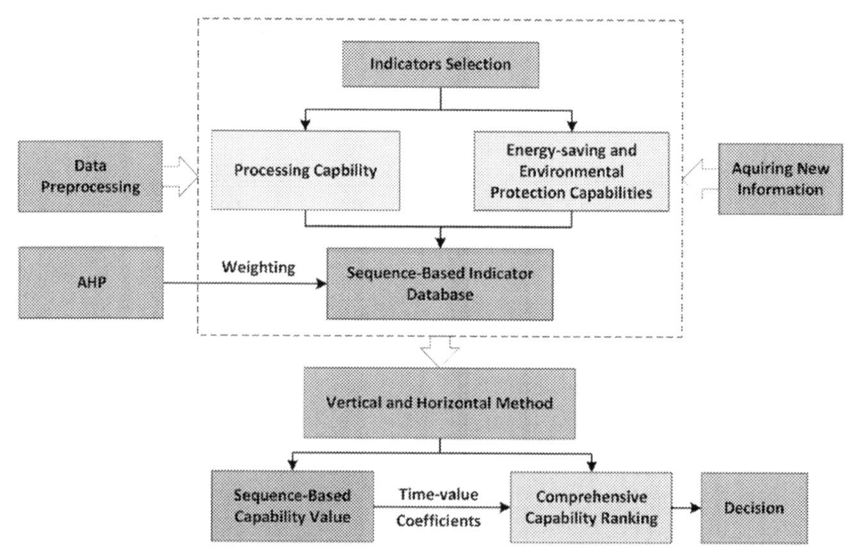

Figure 6.3 The dynamic assessment model [11].

indicates the j-th attribute value of the equipment s_i at time $t_k(k = 1, 2, ..., N)$. First, the subjective weight vector $w' = (w'_1, w'_2, ..., w'_m)$ can be obtained using the AHP method. Then the weighted timing database can be obtained from Eq. (6.1)

$$x'_{ij}(t_k) = w'_j x_{ij}(t_k), i = 1, 2, ..., n; \ j = 1, 2, ..., m \tag{6.1}$$

Subsequently, the subjective dynamic weight vector $w = (w_1, w_2, ..., w_m)$ can be obtained by the so-called "vertical and horizontal" method. The assessment result $y_i(t_k)$ of equipment s_i at time $t_k(k = 1, 2, ..., N)$, which can be shown in Eq. (6.2).

$$y_i(t_k) = \sum_{j=1}^{m} w_j x'_{ij}(t_k), \ \ k = 1, 2, ..., N; \ \ i = 1, 2, ..., n \tag{6.2}$$

Finally, $v = (v_1, v_2, ..., v_N)$ indicates the time weight vector. The comprehensive capability assessment value of manufacturing equipment s_i in the entire assessment time and space can be obtained by the following function:

$$y_i = \sum_{k=1}^{N} y_i(t_k) v_k, i = 1, 2, ..., n \tag{6.3}$$

6.2.2.2 Dynamic assessment model based on correlation model

Most of the traditional assessment models focused on the evaluation of manufacturing capabilities of manufacturing unit equipment or combination, ignoring the manufacturing capability correlation of individual and combined equipment. In order to solve this problem, Xie et al. [16] proposed a dynamic assessment model for the sustainable manufacturing capability of industrial cloud robots (ICRs) based on the correlation model. The correlation model included service correlation established through service describing and potential data correlation obtained through data mining. Based on it, the dynamic assessment model is established from processing capability, energy-saving capability, and environmental protection capability aspects.

In this assessment model, the evaluation algorithm uses the indicators values considering the service correlation of ICRs as input. Then, an improved ANP algorithm that considers the interaction relationships of indicators and enhances the objectivity of the ANP algorithm was used to obtain the weights. The indicators network and the judgment matrix were established by the objective data correlation in the improved ANP method. The time weighting factor was introduced to reflect the importance of manufacturing data in different time periods. Finally, combining objective weights in time scale and subjective weights, the final dynamic weights can be obtained to make a dynamic comprehensive assessment in quantifying manufacturing capability.

6.2.2.3 Concurrent assessment model

As for the robotic manufacturing system, the assessment model needs to have the characteristics of a multilevel system, including the workshop level, the production line level, and the manufacturing unit level. The system levels are interrelated and affect each other. The interaction between manufacturing units affects the operating performance and the scheduling of the production line, which will also put requirements of forward response for the operation of the manufacturing unit. At this stage, the traditional assessment method of the energy efficiency of the robotic manufacturing system mostly focuses on a single level, ignoring the close correlation between the evaluation indicators of the various levels of the manufacturing system. It would result in the deviations in the overall performance evaluation of the robotic manufacturing system and unable to provide accurate and reasonable decision-making basis for the overall optimization of system performance.

To solve this problem, Wang et al. [12] proposed an energy-efficient concurrent assessment model of the industrial robotic manufacturing system based on association rules, as shown in Fig. 6.4. The concurrent assessment model aims to reduce energy consumption and ensure production performance. A unit-level and production-line-level assessment indicators system for the robot manufacturing system was established. In order to eliminate errors in the process of subjective assessment (e.g., in model weight valuing) as much as possible and to obey objective rule (e.g., the coupling relationship between different units), based on a large amount of statistical data, a multilayer association rule mining method, namely, the positive—negative association rule mining algorithm with the interesting multiple level minimum supports [17], was used to obtain the association rules of assessment indicators of the production line level and unit level. Then the more objective weights of the selected assessment indicators were obtained based on the mined association rules. The correlation between indicators is changing constantly as the environment and time changed, resulting in dynamic adjustment of weights. Using the objective

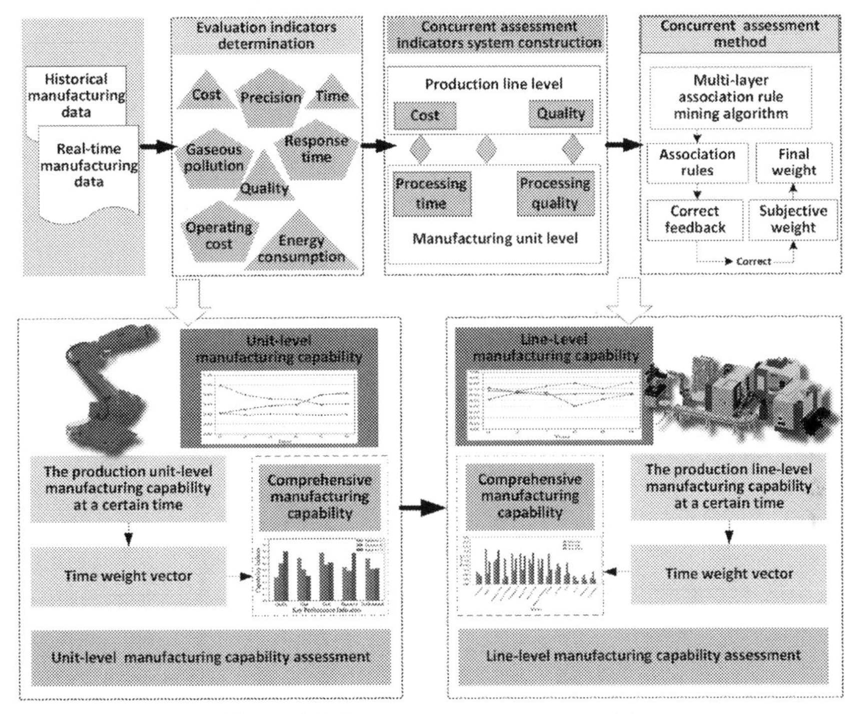

Figure 6.4 Concurrent assessment model.

information of the indicators to modify the subjective weight of the indicators and performing a comprehensive subjective and objective evaluation, the assessment result of the manufacturing capability of the manufacturing unit and the production line at a certain time was obtained. Finally, the linear weighted model was used to extend the assessment result at a certain time to the entire time sequence and assign weight coefficients to the time to realize dynamic concurrent assessment.

6.2.3 Dynamic energy-efficient assessment approach

The energy-efficient assessment for manufacturing capacity of the manufacturing system belongs to a multiattribute comprehensive assessment problem, that is, a certain target is used to measure multiple essential attributes of the assessment object, and through certain assessment criteria and methods to convert these attributes into objective and quantitative values. The multiattribute comprehensive assessment methods are usually divided into three categories: subjective assessment method, objective assessment method, and combination weighting method. Based on these assessment methods, considering the dynamics of the assessment process, the time factors are also introduced into the assessment process, which make the assessment process a multidimensional issue of three dimensions with indicators space, object space, and time space.

6.2.3.1 Subjective assessment method

The subjective assessment method mainly relies on expert experience. Experts determine the relative importance of assessment indicators through their knowledge and then obtain indicator weights. It is subjective and may cause errors due to different decision makers. Common subjective evaluation methods include analytic hierarchy process (AHP) [18] and analytic network process (ANP) [19], fuzzy comprehensive evaluation method [20]. The AHP method was widely used in the field of IRs. Taking energy consumption and flexibility as the standard, the AHP method was used to evaluate the manufacturing capability of IRs for milling, and the most suitable and effective robot was selected for users from various general robot types [21]. Tan et al. [22] studied the performance assessment and analysis of the human-robot collaboration system based on the AHP, with productivity, quality, human fatigue, and safety as the evaluation requirements, so as to select a suitable robot to complete the collaboration task.

6.2.3.2 Objective assessment method

Objective assessment method is based on objective data under the selected indicators and uses related data models to obtain the weights, usually including entropy weight method [23], gray correlation analysis method [24], ideal point method (TOPSIS method) [25], etc. The objective assessment method is in a manner that evaluates the manufacturing capability from the objective criteria under the support by the actual measured data, so it would reduce errors caused by the subjective assessment method that might introduce the bias or prejudice of evaluator. Therefore, it was widely used by many scholars. Wang et al. [26] proposed an evaluation decision method based on cloud model and TODIM method when dealing with the problem of robot selection with fuzzy language information, in which the entropy weight method was used to obtain the weight coefficient of the evaluation index. Princely et al. [27] used the TOPSIS method to solve the multicriteria optimization problem of robot deburring parameters in the finishing process. However, the preferences of decision makers cannot be well represented by the objective assessment methods.

6.2.3.3 Combination weighting method

Combination weighting method is a weighting method that integrates subjective and objective weighting results. It can take into account subjectivity and experts' preference for indicators, while avoiding subjective arbitrariness in indicator weighting, thereby avoiding the limitation of single weighting method. In order to choose suitable energy-efficient plan for enterprises, an AHP-fuzzy method was proposed to comprehensively evaluate the selected qualitative and quantitative indicators, and to give suggestions for the future energy-efficient development of enterprises [28]. The gray relational analysis method and the AHP were combined to comprehensively evaluate the human-robot collaboration capabilities during the disassembly process [29]. Rao et al. [30] proposed a subjective and objective comprehensive multiattribute decision-making method to complete the selection of the robot, which considers the objective weight of the importance of the attributes and the subjective preference of the decision maker to determine the comprehensive weight of the attributes. However, most of the multiindicator comprehensive assessment methods are static evaluation methods, that is, the assessment process just evaluates the assessed object at a certain time, without considering the sequential feature of the data.

6.2.3.4 Dynamic assessment approach

The state of the assessed object is usually under uncertain dynamics, which may cause the fluctuation of the assessment results. Thus, the changing state of the assessed object over time should be considered. For the problem of dynamic assessment, considering the change of equipment manufacturing capability in the time dimension, a dynamic assessment method based on the entropy weight method of time series data and gray correlation for weighting was proposed to comprehensively evaluate and monitor the performance changes during processing [11]. Liu et al. [31] analyzed the factors that affect equipment capabilities and used AHP to obtain multi-attribute index weights. And then considering the time dimension of equipment manufacturing tasks, the time entropy was combined to complete the assessment of equipment manufacturing capability in the sequential time. Xie et al. [16] proposed a dynamic assessment method of energy-efficient manufacturing capability for ICRs based on the correlation model. The time weighting factor was introduced to reflect the role of data correlation in different time periods. As reported in the above literature, the above methods are of high fitness for the assessed objects with uncertain dynamic states to resist the fluctuation of the actual measured data, by which the more comprehensive and accurate assessment can be achieved.

6.3 Digital twin-driven assessment service

With the development of service-oriented [32], intelligent, and sustainable manufacturing, reducing the energy consumption of manufacturing equipment in the production process has attracted more experts and scholars. As a key link to improving the performance of the manufacturing system effectively, the energy-efficient assessment research is currently focused on a single level, ignoring the relationship and influence of the assessment indicators at each level of the manufacturing system [12]. There are few studies on energy-efficient assessment based on virtual model data and cyber-physical fusion data. Digital twin technology realizes the information mapping and interconnection between information space and physical space [6]. In this section, based on the five-dimension digital twin model [33], a digital twin-driven manufacturing system energy-efficient assessment framework is established, and the energy-efficient assessment is further abstracted and encapsulated as a service to provide users, production lines, and enterprises with customized, efficient, and reliable services.

6.3.1 Framework of digital twin-driven assessment

In terms of the manufacturing mode, sustainable manufacturing pays more attention to energy consumption in various processing links such as cutting, casting, welding, spraying, etc., including energy-efficient modeling and evaluation under different equipment, different process routes, and different production environments. In sustainable manufacturing, the energy consumption of a single processing equipment in continuous manufacturing directly affects the energy-efficient of the production process, so its efficiency mainly depends on whether the processing equipment used is energy-saving. On manufacturing and services, product manufacturing aims to study the energy consumption status of finished products in the entire life cycle of product design, manufacturing, use, and recycling, and to improve the overall energy-efficient of the manufacturing system through simulation and prediction [34]. In the manufacturing process, the effective assessment of the operating energy of the manufacturing system is of a great significance to improving the efficient operation of manufacturing equipment and realizing the sustainable manufacturing. However, most of the existing manufacturing enterprise production mode is the large quantities of flow shop, many varieties of medium and small batch production, and single-piece production. There are many types of manufacturing equipment, different structures, diverse manufacturing data sources, and different formats, which makes it difficult to use common information model and management mode to manage manufacturing information uniformly. Therefore, it is easy to lead to the emergence of *information islands*, which cannot provide accurate and reasonable decision basis for the overall optimization of system performance.

The cyber-physical system realizes the function of monitoring the physical world and interacting with the computer system by integrating computing resources with physical processes [28,35]. The digital twin technology provides a solution for achieving the information mapping and interconnection between the information space and the physical space and has been widely focused by the academia and industry community [36,37]. The digital twin used in the manufacturing process takes the coupling of the manufacturing system and its equivalent system as the basis for optimization. By simulating the design, development, manufacturing, and use process of the physical product in the virtual space before the physical product is processed, not only can it save a lot of time and economic costs, improve the safety of the manufacturing process [38,39], but also minimize the delay of data collection. This has impact on the process of manufacturing system operation and maintenance and modeling and ensures that sufficient high-quality data are provided to provide a data foundation for the realization of high real-time, high transparency, strong interoperability, and rapid response of the cyber-physical manufacturing system [7].

It can be seen that the research on the digital twin-driven energy-efficient assessment of manufacturing system solves the complex issue in establishing a unified energy-efficient assessment model for different production modes and manufacturing equipment of the manufacturing system on the one hand, and on the other hand, it also realizes the information fusion and interconnection between the physical space and the information space of the robot manufacturing system.

Considering the power of digital twin mentioned above, a framework of digital twin-driven energy-efficient assessment is presented, as shown in Fig. 6.5, which includes the physical manufacturing system, the virtual manufacturing system, the manufacturing system database, and the energy-efficient assessment cloud platform.

In the physical manufacturing system with processing workshop production line, features are extracted from the physical equipment resources

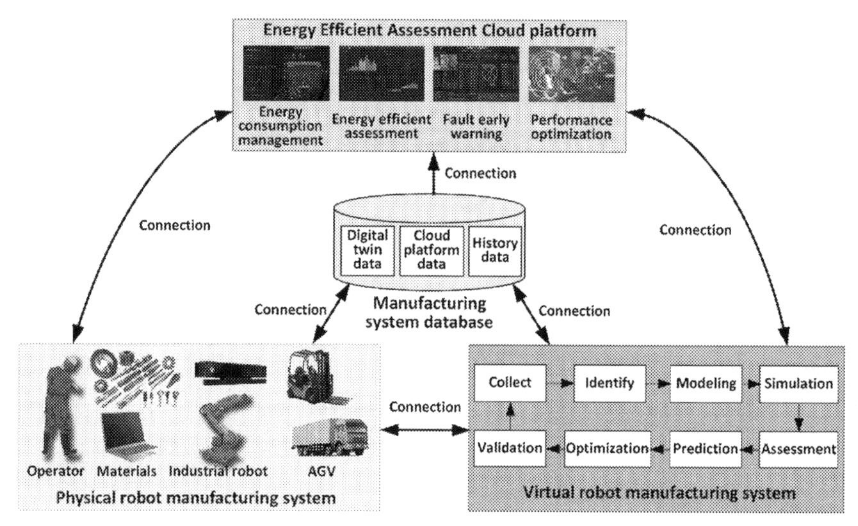

Figure 6.5 The framework of digital twin-driven energy-efficient assessment.

such as IRs and sensors, production and processing energy consumption such as robot assembly, disassembly and transportation, as well as manufacturing environment with real-time intelligent perception. Real-time interaction is conducted with virtual system, cloud platform, and database.

In the virtual manufacturing system, the physical space equipment entity is used as the object to map the virtual model with high fidelity, and the robot equipment model is simulated, evaluated, optimized, and predicted in real time in the virtual system according to user needs.

In the manufacturing system database, the data are divided into three parts: twin data, cloud platform data, and historical data. The digital twin data includes the data set mapped to the virtual model, and various model data generated by the virtual manufacturing system for model simulation, energy-efficient assessment, and prediction optimization. The digital twin data are the cornerstone that drives the execution of digital twin and to a certain extent supports the iterative optimization of the model. The cloud platform data mainly refer to user data, historical data, the index system of energy-efficient assessment, and other data. Historical data provide a data basis for mining the correlation between the index systems of energy-efficient assessment.

In the energy-efficient assessment cloud platform, it provides users with personalized services such as energy consumption management visualization, energy-efficient assessment, failure warning, optimization and prediction, and provides a reference basis for energy-efficient assessment for manufacturing companies and users in the selection of manufacturing solutions, and promotes enterprise intelligence, green development.

Following Tao et al. [40], we built a digital twin-driven energy-efficient assessment model of the manufacturing system based on the five-dimensional model [41] integrates the physical space and the virtual space through the analysis of the twin data to achieve a comprehensive, unified, and efficient energy-efficient assessment of the manufacturing system. The model [42] is shown as Eq. (6.4).

$$M_{DT-En} = (P_{En}, V_{En}, D_{En}, S_{En}, C_{En}) \tag{6.4}$$

where P_{En} represents the physical entity of the energy-efficient assessment equipment, V_{En} represents the virtual model of the energy-efficient assessment equipment, D_{En} represents all the data in the physical space and the virtual space and the integration between them, S_{En} represents the energy-efficient assessment service of the physical manufacturing system and the virtual manufacturing system, and C_{En} represents the interconnection between P_{En}, V_{En}, D_{En} and S_{En}.

P_{En} is the sum of physical elements related to energy efficiency in the production process of the robot manufacturing system, such as manufacturing equipment, raw materials, operators, and other elements. On the one hand, various types of energy consumption data of physical equipment are sensed, collected, and transmitted to the virtual space in real time through industrial Internet devices such as Radio Frequency Identification (RFID), sensors, and Programmable Logic Controller (PLC). On the other hand, compared to the one-way non—closed-loop energy-efficient assessment and optimization control, P_{En} quickly and flexibly respond to the energy-efficient optimization and control commands issued by the virtual space, and realize the virtual-real two-way closed-loop energy-efficient assessment and iterative optimization.

V_{En} is a multiscale, high-fidelity real-time interactive mapping of P_{En} from the perspective of energy efficiency, including physical model $G_{V_{En}}$, geometric model $P_{V_{En}}$, energy-efficient behavior model $B_{V_{En}}$, and energy-efficient rule model $R_{V_{En}}$, as shown in Eq. (6.5).

$$V_{En} = (G_{V_{En}}, P_{V_{En}}, B_{V_{En}}, R_{V_{En}}) \tag{6.5}$$

In Eq. (6.5), $G_{V_{En}}$ is a three-dimensional virtual model describing the geometric parameters and relationships of $P_{V_{En}}$ and has a one-to-one mapping relationship with $P_{V_{En}}$. $P_{V_{En}}$ adds information such as gravity, friction, and other physical properties and constraints to $G_{V_{En}}$. $B_{V_{En}}$ includes material energy consumption assessment model, IR energy consumption monitoring model, product processing energy-efficient assessment model, etc. $R_{V_{En}}$ includes energy-efficient association rule mining model, energy-efficient optimization method, energy-efficient early warning, and feedback, etc.

D_{En} is the data energy source for the digital twin-driven energy-efficient assessment of the robotic manufacturing system, expressed as Eq. (6.6).

$$D_{En} = (D_P, D_V, D_S, D_K, D_F) \tag{6.6}$$

Among them, D_P refers to the energy-efficient data related to the physical entity elements such as raw materials and IRs in the manufacturing system, including real-time energy-efficient data such as manufacturing unit energy-efficient data, production line energy-efficient data, and workshop energy-efficient data. D_V represents energy-efficient data related to V_{En}, including manufacturing unit energy-efficient simulation data, production line energy-efficient simulation data, workshop energy-efficient evaluation and forecast data, etc. D_S stands for energy-efficient data related to S_{En}, which mainly includes energy-efficient real-time monitoring service data, energy-efficient assessment and analysis service data, system energy-efficient prediction, and optimization service data, etc. D_K stands for knowledge data in the field of energy efficiency, including energy model data, energy-efficient assessment rule data, intelligent algorithm optimization data, and energy-efficient assessment expert knowledge. D_F is the energy efficiency-related fusion data after data screening, preprocessing, correlation, and integration of D_P, D_V, D_S, and D_K, including information and physical fusion data obtained by fusing physical real-time energy-efficient data, historical energy-efficient data, etc.

S_{En} refers to providing services such as real-time monitoring of energy efficiency, energy-efficient assessment and analysis, failure prediction, and green manufacturing program optimization for the energy-efficient assessment process of the robot manufacturing system with physical entities and virtual models in the virtual and real space as objects, driven by twin data.

C_{En} is based on intelligent perception and interactive devices, and two-way communication and pairwise interaction according to different interfaces and protocols among P_{En}, V_{En}, D_{En} and S_{En}, as shown in Eq. (6.7).

$$C_{En} = \left(P_{En}\text{-}V_{En}, P_{En}\text{-}D_{En}, P_{En}\text{-}S_{En}, V_{En}\text{-}D_{En}, V_{En}\text{-}S_{En}, D_{En}\text{-}S_{En}\right) \tag{6.7}$$

where the operator "_" represents the data exchange between two involved elements.

6.3.2 Energy-efficient assessment as services

As one of the key enabling technologies for intelligent manufacturing, cloud computing has the advantage of providing high-availability, reliability, and scalability computing services to distributed requests as needed. The concept of cloud manufacturing (CMfg) is an example of the combination of cloud computing and traditional manufacturing [43]. The core concept of CMfg is that different manufacturing companies provide on-demand, flexible, and configurable cloud services in the form of modular and loosely coupled software components for users distributed in different geographical locations. Distributed manufacturing resources are encapsulated into cloud services and centralized management is performed on the cloud service platform. User requests and calls for various services involved in the product life cycle on demand. Service search, intelligent matching, recommendation, and invocation are executed by the cloud service platform [43]. CMfg uses a service-oriented architecture (SOA) to define common interfaces and protocols, so that components do not depend on the development language, hardware resources, and software systems that implement services, ensuring that services interact in a unified and standard way. SOA helps to sort out the way components are developed and deployed within an enterprise architecture, improving the speed, reliability, and reusability of business systems to help enterprises more confidently face the rapidly changing market environment and business needs. Service-oriented manufacturing provides a business model and implementation strategy for CMfg through the hierarchical design of the cloud platform covering the physical resource layer, infrastructure layer, global service layer, and application layer and realizes the sustainability of manufacturing through energy-efficient methods, such as improving real-time response, environmental friendliness, cost-effectiveness, etc. [44].

Energy efficiency assessment is an important link to realize sustainable manufacturing. The traditional full life cycle assessment of product energy consumption mainly focuses on the energy consumption and emissions in the production process of manufacturing products and pays less attention to the energy consumption in the product use stage. For the service-oriented CMfg model, the reduction in the total evaluation of service energy consumption throughout the life cycle is a useful service for energy conservation and emission reduction [42]. Xiang et al. [45] proposed a comprehensive evaluation framework for energy consumption of CMfg resource services based on the Internet of Things (IoT) and a CMfg service energy consumption calculation model. The calculation method of comprehensive energy consumption for manufacturing services is given. The key technologies involved in each level of the energy consumption assessment framework are analyzed, and a CMfg resource access and information management system is preliminarily designed.

Digital twin provides an effective way for the cyber-physical integration of manufacturing. The combination of intelligent manufacturing services and digital twin fundamentally changes various services in the full life cycle of product manufacturing and helps to realize intelligent manufacturing through the two-way connection between virtual and physical manufacturing. Zhang et al. [46] proposed an equipment energy consumption management system based on a digital twin workshop, including physical equipment, virtual equipment, energy consumption management services, and various data. The application of the system in energy consumption monitoring, energy analysis, and energy optimization was discussed.

To realize the servitization of energy-efficient assessment, following Rao et al. [30], a framework of digital twin-driven energy-efficient assessment for manufacturing services is presented, as shown in Fig. 6.6, which includes four parts: physical equipment layer, virtual service layer, service application layer, and manufacturing database.

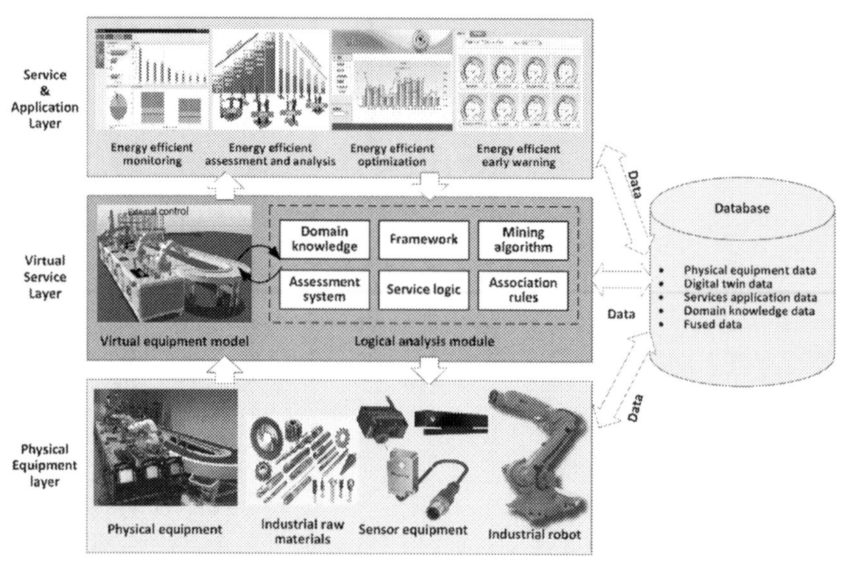

Figure 6.6 The framework of digital twin-driven energy-efficient assessment for services.

The physical equipment layer is the target of energy-efficient assessment. Different sensors and RFID are used for intelligent perception and information collection. Energy consumption and energy efficiency-related data are the foundation of the digital twin-driven energy-efficient assessment system for manufacturing services. The physical properties of IRs include motion speed, acceleration, friction coefficient, rotation angle, etc. The material status is divided into raw materials, semifinished products, and finished products.

The virtual service layer is composed of virtual equipment model and logic analysis module. The virtual equipment model is a high-fidelity reproduction of the physical equipment model. The logic analysis module integrates multisource data (physical equipment data, twin data, expert knowledge, etc.) to analyze the logic and rules of the current virtual equipment model, mine the association relationship, and optimize the energy efficiency of the current manufacturing scheme based on the energy-efficient assessment system. On the one hand, the virtual equipment layer maintains real-time interaction and information transmission with physical equipment through the data collected by sensors. On the other hand, it interacts with the expert knowledge, association rules, and mining algorithms in the logic analysis module to assess and analyze the energy efficiency of the current operating scheme. On the basis of current data combined with historical data, the energy efficiency of the virtual equipment model is further optimized to guide physical manufacturing and realize energy-efficient and sustainable manufacturing.

The service application layer is a collection of energy-efficient assessment service applications encapsulated in the top layer of the manufacturing system, including energy-efficient monitoring, energy-efficient assessment and analysis, energy-efficient optimization, and energy-efficient early warning services. Different users formulate personalized service schemes and combinations according to their own needs to realize the many-to-one and many-to-many combination of "centralized use of decentralized resources" and "centralized resource decentralized services" [47].

The manufacturing database includes physical equipment data, twin data, service application data, expert knowledge data, and fusion data. The physical equipment data are collected by sensors, mainly including operation states and working conditions. Twin data refer to the data generated during the simulation verification process of the virtual equipment model, which mainly includes IR parameters, simulation results, assessment, and analysis data, etc. Service application data mainly include manufacturing scheme data, model optimization data, etc. Expert knowledge data are obtained from existing energy-efficient assessment-related information systems or databases and used for mining association rules among energy-efficient assessment models. Fusion data refer to the comparison, association, and combination of two or more data among the above four types of data.

The digital twin-driven energy-efficient assessment of manufacturing services enables the interactive mapping of the complex physical environment and the virtual model to realize the integration of the physical world and the information world. Using interactive feedback data, the energy efficiency of the manufacturing process is dynamically reflected in the information space with data stream in real time. Meanwhile, the combination of feedback data and historical data enables continuous optimization of the energy-efficient assessment model, therefore building an agile, reliable, transparent, and sustainable manufacturing system.

6.4 Case studies

In this section, two cases are presented to demonstrate the feasibility of the presented methods. To begin with, a digital twin-driven assessment service platform is introduced, which can encapsulate the assessment function driven by digital twin as services to make assessment more accurate and accessible. Moreover, an energy-efficient assessment service for IRs is presented, which introduces the digital twin model to make a more accurate energy consumption prediction for the energy-efficient assessment.

6.4.1 Digital twin-driven assessment service platform

In this part, focusing on manufacturing capability assessment, authors develop a digital twin-driven service platform. As shown in Fig. 6.7, the framework of the developed platform consists of four parts: physical production line, digital twin model, digital twin data, and assessment service platform. A detailed description is as follows.

- Physical production line, the assessment object, is equipped multisource sensors, for example, RFID, smart meters for current and voltage, etc., so that its running state can be monitored in real time. Moreover, its controller is servicized which make itself can be manipulated by both the local and remote operators.
- Digital twin model mirrors the running state of the physical production line and updates itself according to the feedback sensory data from the

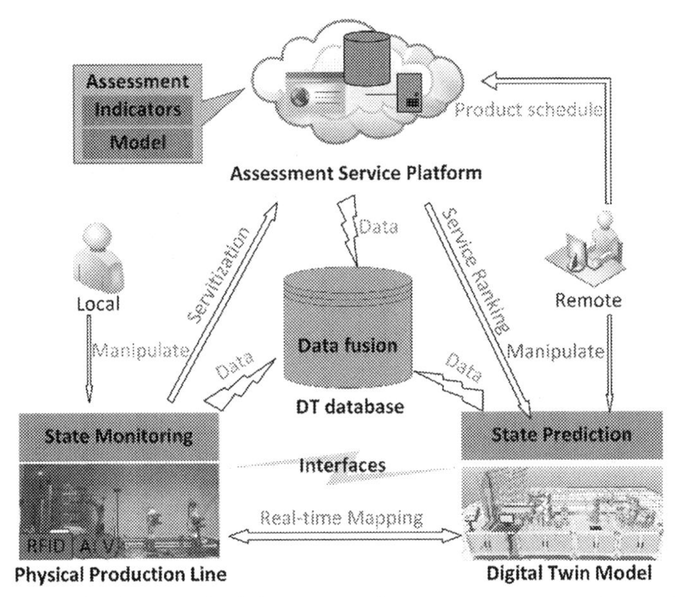

Figure 6.7 The framework of digital twin-driven assessment service platform.

physical world. The digital twin model can also be controlled by the assessment service platform when it acts as a simulation platform (service provider) for the decision-making process of assessment. The state of the digital twin model can be monitored by the remote operators through Web streaming.

- Digital twin data are collected in the manufacturing process from different parts of the system, such as the real-time data captured by RFID, smart electricity meters, etc. in the physical space, various parameters, and attributes of models in the digital space, property, and functions of the assessment service platform. It is then stored in the database after data fusion and can be accessed by the other parts of the system.
- Assessment service platform consists of indicators and models for manufacturing capability assessment which can calculate the corresponding capability value according to the assessment request (production scheduling alternatives) sent by the service requester and give the ranking result to the service requester through the visual feedback (3D model animation, data charts, text, etc.) in digital twin model.

Based on the above framework, authors developed a platform to provide the assessment service for the production line-level and manufacturing

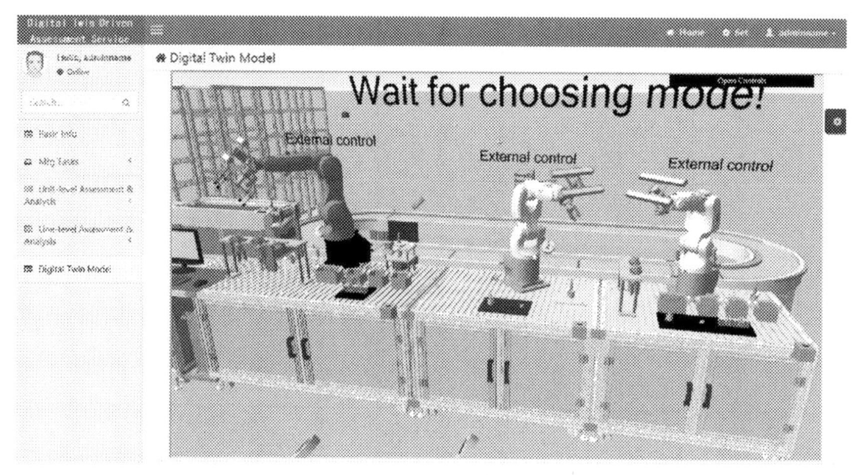

Figure 6.8 The digital twin-driven simulation.

cell-level manufacturing capabilities of the robot manufacturing system, thereby improving the management and decision-making efficiency of workshop production tasks, and improving the efficiency of energy utilization of production process.

This service platform is oriented as a robotic production line with two KUKA KR3 R540, one KUKA KR6 R700, and the counterpart digital robot models. The production tasks-related information is stored in the database (SQL Server/MySQL) in ontology form. Demo 3D-based digital IRs reads the tasks-related data ontology, performs simulation, and controls the digital models to implement the corresponding production tasks, as shown in Fig. 6.8. After that, energy-efficient assessment can be computed by the assessment service module which is shown in Fig. 6.9, as well as the analysis according to the assessment result can be realized by the analysis service module, as shown in Fig. 6.10. According to the assessment and analysis, the best energy-efficient production schedule corresponding to the given production tasks can be obtained.

6.4.2 Energy-efficient assessment service for industrial robots

In this part, one considers the IRs as the assessment object of energy efficiency. The framework of energy-efficient assessment service for IRs is shown in Fig. 6.11, which consists of four parts: the physics-based energy model of IRs, the virtual robot model, the unified digitized description model, and the digital twin data [48].

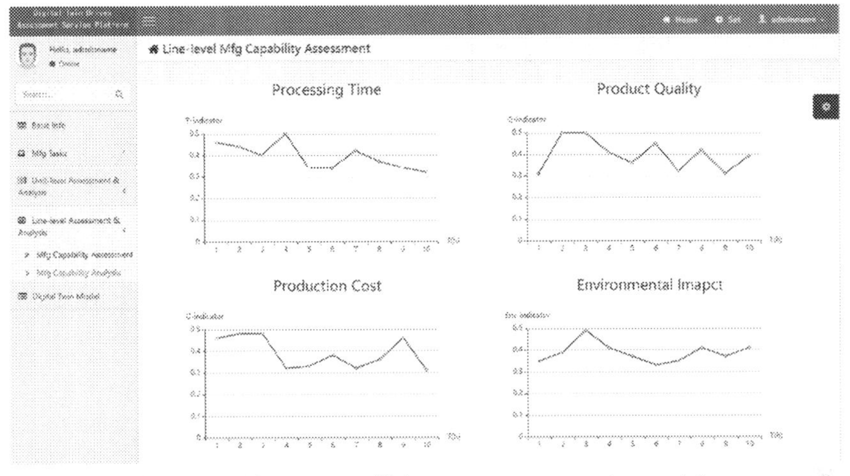

Figure 6.9 The energy-efficient assessment service module.

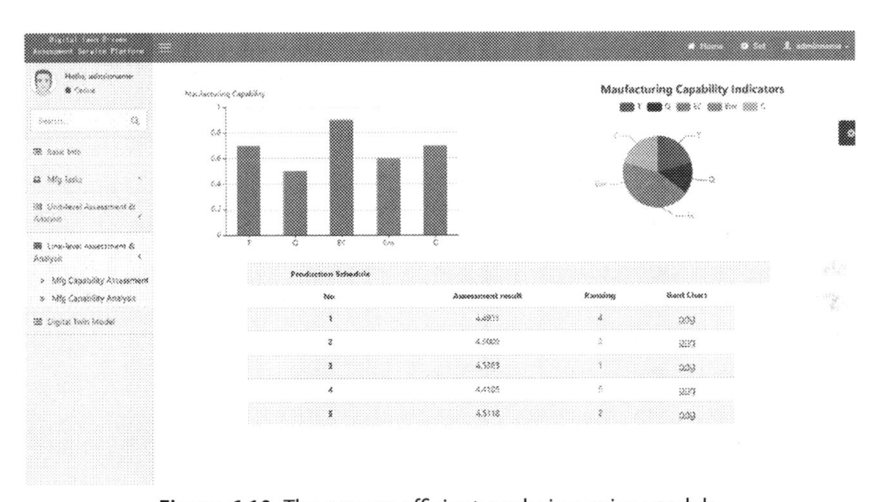

Figure 6.10 The energy-efficient analysis service module.

6.4.2.1 Physics-based energy modeling

The physics-based energy model aims to analyze and model the energy consumption law of IRs using the input of the perception data collected from the physical space. The workflow of the physics-based energy modeling of IRs is shown in Fig. 6.12 [48].

Using the Denavit-Hartenberg notation, robotic kinematics and dynamics, and geometric information as prior knowledge, a linear model of

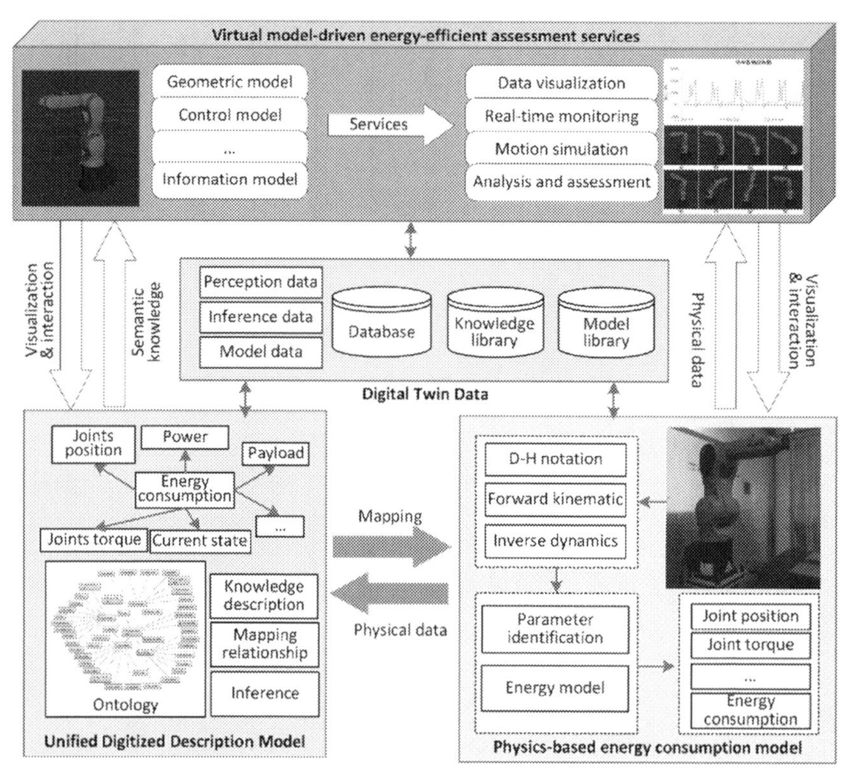

Figure 6.11 The framework of energy-efficient assessment service for industrial robots [50].

torque is developed. Using that model, the data for dynamic parameters identification can be acquired with the preprocessed data of the design of excitation trajectory. Then the identification process can be completed by the identification algorithms, like Gbest-guided artificial bee colony algorithm [49]. After the correctness validation and iterative optimization, the energy model is obtained.

6.4.2.2 Virtual industrial robot modeling

Virtual IR model aims to realize visual monitoring and high-fidelity simulation assessment of the energy consumption of the physical IRs in cyberspace and to provide support for further optimization of energy efficiency. In this setting, in order to achieve a correct mapping of robotic energy consumption from physical space to cyberspace, authors modeled a virtual IR from multiple dimensions including geometry, behavior, control, information, and rule.

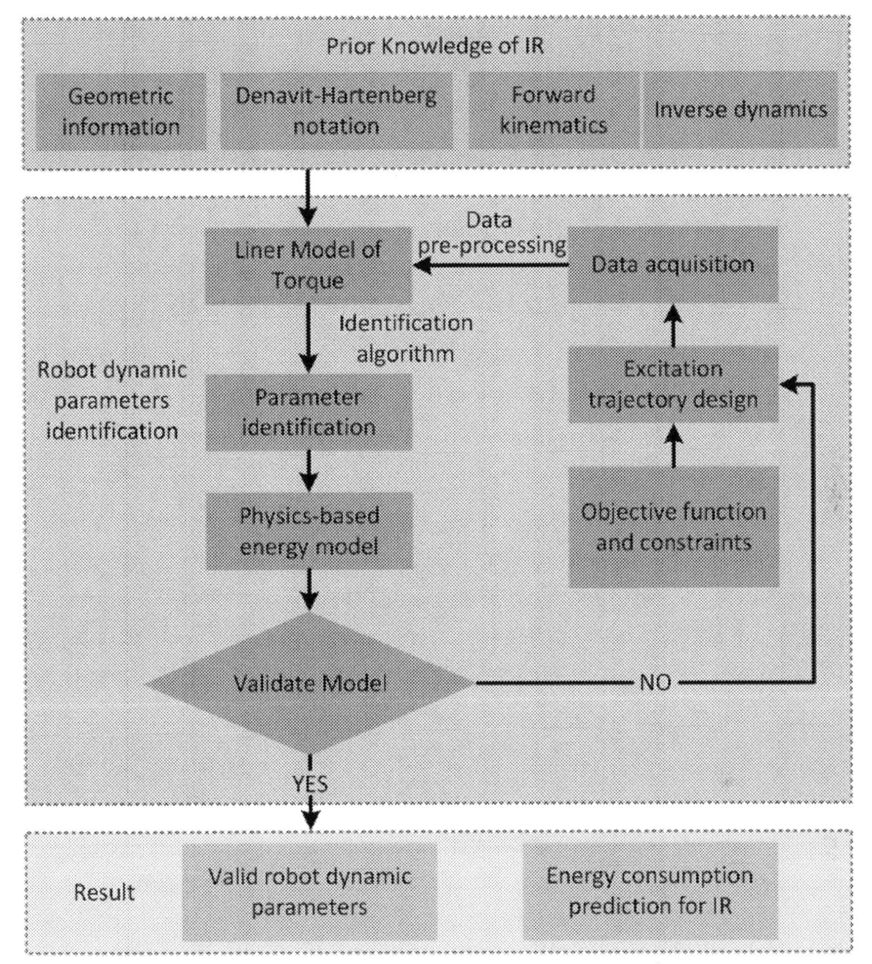

Figure 6.12 The workflow of the physics-based energy modeling of IRs [48].

As shown in Fig. 6.13, the geometric model is a prerequisite for the visualization of virtual IRs, which is the representation of the physical IR's appearance attributes, rigid body structure, specifications, and dimensions in the virtual space. The behavior model is a description of the dynamic operating behaviors of IRs driven by task-level commands, such as the transporting behavior, obstacle avoidance behavior, and collaborative behavior, by which the virtual IRs can be given the behavior characteristics and response mechanisms corresponding to the physical ones. The control model is built on the basis of the geometric model, so it can enable the virtual robot produce dynamic behavior and execute under the control

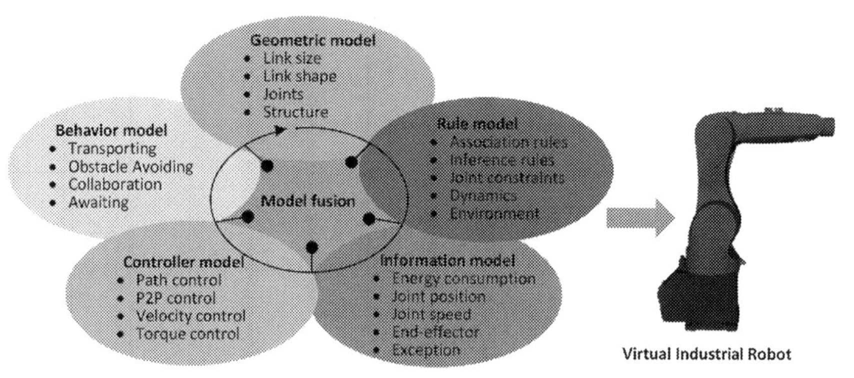

Figure 6.13 Multidimensional model of virtual industrial robot [50].

mechanism corresponding to the physical IR. The rule model refers to the description of the rules and constraints of IRs on the aforementioned geometric, behavior, and control model. The information models the association and combination of the IRs' information of energy consumption, joint position, joint speed, end-effector position, etc., which can drive the function of information visualization. The above five models are then integrated and fused from both structure and function to form a comprehensive model of IR energy consumption, which enable the real-time monitoring and high-confidence simulation of the IR energy-efficient status.

6.4.2.3 Unified digitized description model

In order to ensure the correct mapping between the virtual robot and the corresponding physical model, the unified digitized description model is established to provide the knowledge supported by the encapsulation of related ontology attributes and mapping relationships in the field of IR energy consumption.

Ontology is the formal specification of conceptual models which has good machine-understanding capability to enable the interaction and knowledge-transmission easily. In this setting, authors utilized Web Ontology Language to build the ontology model for the unified digitized description of IRs. The IR energy consumption ontology (*IR_EnergyConsumption*), given in Eq. (6.8), is composed of the robot basic information ontology (*IR_information*), the robot technical parameters ontology (*IR_TechParam*), the energy consumption process ontology (*EC_Process*), and the robot running tasks ontology (*IR_RunTask*). Each ontology has its own instance. The example of energy consumption

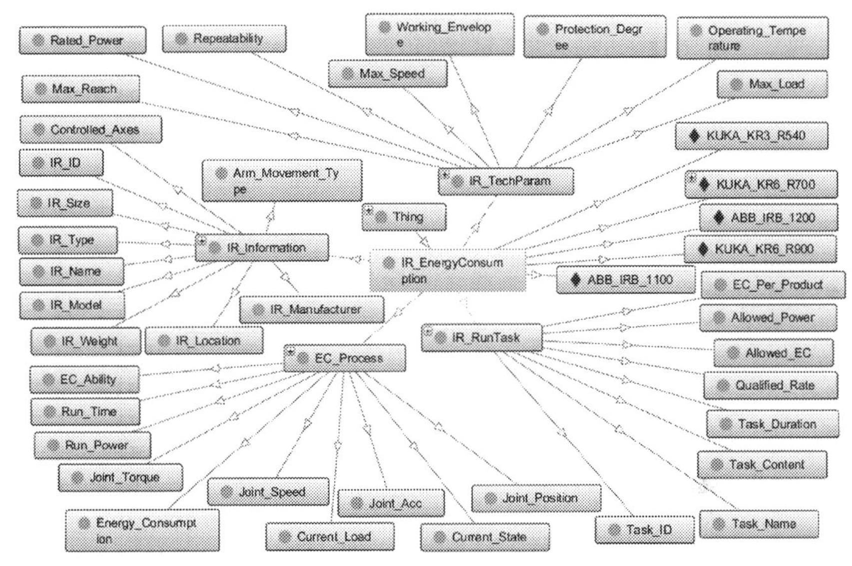

Figure 6.14 Ontology of industrial robot energy consumption.

ontology of robots including KUKA_KR6_R700, ABB_IRB _1200, etc. is shown in Fig. 6.14. As shown in the example, the IR energy consumption ontology and four subontologies are related through their instances.

$$IR\_EnergyConsumption = (IR\_Information, IR\_TechParam, EC\_Process, IR\_RunTask) \tag{6.8}$$

After developing the above ontology, the mapping relationship between the ontology attributes and real-time data collected from physical world is then established. A knowledge library is developed to ensure the interaction between the virtual model and respective physical energy model. This interaction process is reflected in the simultaneous update of the instance of ontology according to the real-time changeable physical data. The update approach is based on Jena, a free and open source Java framework for building semantic applications.

6.4.2.4 Implementation

In this part, authors integrated the above subsystems to a whole energy-efficient assessment service system for an IR, KUKA KR6 R700 sixx. As shown in Fig. 6.15, the users can specify the robot's initial and target configurations through the digital twin robot built in ROS, and then the digital twin robot can perform a simulation-based energy consumption

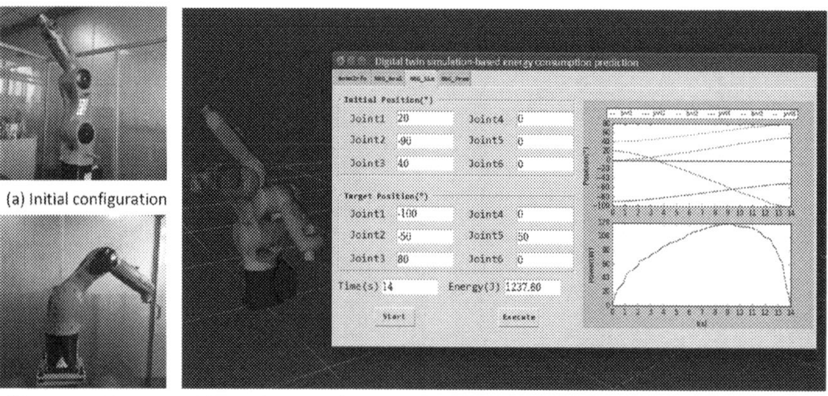

(a) Initial configuration

(b) Target configuration (c) Digital twin simulation-based energy consumption prediction

Figure 6.15 Energy-efficient assessment service system for KUKA KR6 [50].

prediction, which can provide support to the energy-efficient assessment. The simulation results can be displayed to users with the 3D virtual models and the curve of predicted power. After the energy consumption optimization, the valid path planning will be executed on the physical robot.

6.5 Summary

Sustainable manufacturing is a much pursued topic in industry and academia, which reflects the common vision of the human society. Improving the utility of energy consumption can make a great contribution for sustainable manufacturing. To achieve that, quantifying the energy efficiency of production process is essential. Energy-efficient assessment, which can indicate the sustainable capability of manufacturing systems, could provide the comprehensive performance indices to support the optimization of production operations.

In this chapter, a digital twin-driven energy-efficient assessment service framework is presented, which combines the latest technological achievements of cyber-physical systems, industrial IoT, cloud service technologies, etc. The key technologies of energy-efficient assessment including the selection of assessment indicators, the establishment of assessment indicators, and the dynamic assessment approach were further discussed. Following that, a digital twin-driven assessment service framework is developed, which integrates the advantages of the digital twin and cloud service technology to endow the assessment system with the characteristics of reconfigurability, servitization, and resources sharing. Finally, two case studies, which are the digital twin-driven assessment service platform, and IR energy-efficient assessment service, were presented to demonstrate the feasibility and effectiveness of the presented assessment approaches for sustainable manufacturing capability and to verify the functionalities of the digital twin-driven service framework.

The presented framework depicts the potential application of digital twin and service technology in the energy-efficient assessment of manufacturing system. There are still many challenges, for example, digital twin modeling should ensure the accuracy of energy consumption models, behavior models, rule models, etc. of each element of the manufacturing system. Besides, the service quality of the assessment platform is highly affected by the delay of multisource data collection, the time consumption of data processing and analysis, the capacity of service, etc. However, with the continuous improvement of the digital and information technologies, the above problems will be solved gradually by combining multidisciplinary and multidomain knowledge, which could then fully unleash the potential of digital twin and service technology and contributes toward sustainable manufacturing.

References

[1] Garetti M, Taisch M. Sustainable manufacturing: trends and research challenges. Prod Plann Control 2012;23(2–3):83–104.
[2] USDOC. How does commerce define sustainable manufacturing? 2011. Available from: http://www.trade.gov/competitiveness/sustainablemanufacturing/how_doc_defines_SM.asp.
[3] Schipper M. Energy-related carbon dioxide emissions in US manufacturing. Energy Information Administration; 2006. p. 8–10.
[4] U.S. Energy Information Administration. International energy outlook. 2016. Available from: http://www.eia.gov/pressroom/presentations/sieminski_05112016.pdf.
[5] Zhang C, Su B, Zhou K, Yang S. Analysis of electricity consumption in China (1990–2016) using index decomposition and decoupling approach. J Clean Prod 2019;209:224–35.
[6] Tao F, Cheng Y, Cheng J, Zhang M, Xu W, Qi Q. Theories and technologies for cyber-physical fusion in digital twin shop-floor. Computer Integrated Manufacturing Systems; 2017.
[7] Uhlemann TH-J, Lehmann C, Steinhilper R. The digital twin: realizing the cyber-physical production system for industry 4.0. Procedia CIRP 2017;61:335–40.
[8] Lightfoot H, Baines T, Smart P. The servitization of manufacturing: a systematic literature review of interdependent trends. Int J Oper Prod Manag 2013;33(11–12):1408–34.
[9] Qi Q, Tao F, Zuo Y, Zhao D. Digital twin service towards smart manufacturing. Procedia CIRP 2018;72:237–42.
[10] Skinner W. Manufacturing-missing link in corporate strategy. 1969.
[11] Xie L, Jiang X, Xu W, Wei Q, Li R, Zhou Z. Dynamic assessment of sustainable manufacturing capability for CNC machining systems in cloud manufacturing. In: IFIP

international conference on advances in production management systems; 2015. p. 396–403.
[12] Wang S, Zhang X, Xu W, Liu A, Zhou Z. Energy-efficient concurrent assessment of industrial robot operation based on association rules in manufacturing. In: 2018 IEEE 15th international conference on networking, sensing and control (ICNSC); 2018. p. 1–6.
[13] Yingjie Z. Energy efficiency techniques in machining process: a review. Int J Adv Manuf Technol 2014;71(5–8):1123–32.
[14] Li Z, Ping B, Zhong-yu W. Assessment of accident emergency plan based on analytic hierarchy process and fuzzy comprehensive evaluation. J Safety Sci Technol 2015;11(9):126–31.
[15] Xu W, Yu J, Zhou Z, Xie Y, Pham DT, Ji C. Dynamic modeling of manufacturing equipment capability using condition information in cloud manufacturing. J Manuf Sci Eng 2015;137(4).
[16] Xie X, Zhang X, Xu W, Liu J, Yu Q, Zhou Z. Dynamic assessment of sustainable manufacturing capability for industrial cloud robotics based on correlation model. In: 49th international conference on computers and industrial engineering(CIE 2019), Beijing, China; 2019.
[17] Swesi IMAO, Bakar AA, Kadir ASA. Mining positive and Negative Association Rules from interesting frequent and infrequent itemsets. In: 2012 9th international conference on fuzzy systems and knowledge discovery; 2012. p. 650–5.
[18] Huang D. A supplier selection method based on AHP. J Phys Conf 2019;1176:042055.
[19] Saaty TL. Fundamentals of the analytic network process—dependence and feedback in decision-making with a single network. J Syst Sci Syst Eng 2004;13(2):129–57.
[20] Chen L, Yang C, Huang W, Sun Z. Selecting Training method of a rehabilitation robot Based on fuzzy comprehensive evaluation. In: 5th international conference on mechanical engineering, materials and energy (5th ICMEME2016); 2016.
[21] Breaz RE, Bologa O, Racz SG. Selecting industrial robots for milling applications using AHP. Procedia Comput Sci 2017;122:346–53.
[22] Tan J, Arai T. Analytic evaluation of human-robot system for collaboration in cellular manufacturing system. In: 2010 IEEE/ASME international conference on advanced intelligent mechatronics; 2010. p. 515–20.
[23] Li Y, Dong S. Study on supplier selection of manufacturing in lean closed-loop supply chain. In: Proceedings of the ninth international conference on management science and engineering management; 2015. p. 275–82.
[24] Li J, Dai W. Multi-objective decision algorithm based on adaptive genetic algorithm and grey relation degree. In: 2006 9th international conference on control, automation, robotics and vision; 2006. p. 1–5.
[25] Rashid T, Beg I, Husnine SM. Robot selection by using generalized interval-valued fuzzy numbers with TOPSIS. Appl Soft Comput 2014;21:462–8.
[26] Wang J, Miao Z, Cui F, Liu H. Robot evaluation and selection with entropy-based combination weighting and cloud TODIM approach. Entropy 2018;20(5):349.
[27] Princely FL, Senthil P, Selvaraj T. Application of TOPSIS method for optimization of process parameters in robotic deburring. Mater Today Proceed 2020;27:2137–41.
[28] Wan K, Man K, Hughes D. Towards a unified framework for cyber-physical systems (cps). In: 2010 first ACIS international symposium on cryptography, and network security, data mining and knowledge discovery, E-commerce and its applications, and embedded systems; 2010. p. 292–5.
[29] Cheng H, Xu W, Ai Q, Liu Q, Zhou Z, Pham DT. Manufacturing capability assessment for human-robot collaborative disassembly based on multi-data fusion. Proced Manuf 2017;10:26–36.
[30] Rao RV, Patel BK, Parnichkun M. Industrial robot selection using a novel decision making method considering objective and subjective preferences. Robot Autonom Syst 2011;59(6):367–75.

[31] Liu H, Xin S, Xu W, Zhao Y. Dynamic comprehensive evaluation of manufacturing capability for a job shop. In: International conference in swarm intelligence; 2013. p. 360–7.
[32] Tao F, Zhang M. Digital twin shop-floor: a new shop-floor paradigm towards smart manufacturing. IEEE Access 2017;5:20418–27.
[33] Tao F, Liu W, Zhang M, Hu T, Qi Q, Zhang H, Sui F, Wang T, Xu H, Huang Z, Ma X, Zhang L, Cheng J, Yao N, Yi W, Zhu K, Zhang X, Meng F, Jin X, Liu Z, He L, Cheng H, Zhou E, Li Y, Lv Q, Luo Y. Five-dimension digital twin model and its ten applications. Comput Integr Manuf Syst 2019;25(1):1–18.
[34] Tuo J, Liu F, Liu P, Zhang H, Cai W. Energy efficiency evaluation for machining systems through virtual part. Energy 2018;159:172–83.
[35] Zhu K, Zhang Y. A cyber-physical production system framework of smart CNC machining monitoring system. IEEE ASME Trans Mechatron 2018;23(6):2579–86.
[36] Leng J, Zhang H, Yan D, Liu Q, Chen X, Zhang D. Digital twin-driven manufacturing cyber-physical system for parallel controlling of smart workshop. J Ambient Intell Human Comput 2019;10(3):1155–66.
[37] Bauernhansl T, WGP-Standpunkt Industrie 4.0. 2016: WGP, Wissenschaftliche Gesellschaft für Produktionstechnik.
[38] Ferguson S, Bennett E, Ivashchenko A. Digital twin tackles design challenges. World Pumps 2017;2017(4):26–8.
[39] Moreno A, Velez G, Ardanza A, Barandiaran I, Infante AR, Chopitea R. Virtualisation process of a sheet metal punching machine within the Industry 4.0 vision. Int J Interact Des Manuf 2017;11(2):365–73.
[40] Tao F, Liu W, Liu J, Liu X, Liu Q, Qu T, et al. Digital twin and its potential application exploration. Comput Integr Manuf Syst 2018;24(1):1–18.
[41] Grieves M. Digital twin: manufacturing excellence through virtual factory replication. White Paper 2014;1:1–7.
[42] Xiang F, Huang Y, Zhang Z, Jiang G, Zuo Y, Tao F. New paradigm of green manufacturing for product life cycle based on digital twin. Comput Integr Manuf Syst 2019;25(06):1505–14.
[43] Xu X. From cloud computing to cloud manufacturing. Robot Comput Integrated Manuf 2012;28(1):75–86.
[44] Xu W, Yao B, Fang V, Xu W, Liu Q, Zhou Z. Service-oriented sustainable manufacturing: framework and methodologies. In: Proceedings of the 2014 international conference on innovative design and manufacturing (ICIDM); 2014. p. 305–10.
[45] Xiang F, Hu Y-F, Tao F, Zhang L. Energy consumption evaluation and application of cloud manufacturing resource service. Comput Integr Manuf Syst 2012;18(9):2109–16.
[46] Zhang M, Zuo Y, Tao F. Equipment energy consumption management in digital twin shop-floor: a framework and potential applications. In: 2018 IEEE 15th international conference on networking, sensing and control (ICNSC); 2018. p. 1–5.
[47] Li B, Zhang L, Wang S, Tao F, Cao J, Jiang X, Song X, Chai X. Cloud manufacturing: a new service-oriented networked manufacturing model. Comput Integr Manuf Syst 2010;16(1):1–7.
[48] Yan K, Xu W, Yao B, Zhou Z, Pham DT. Digital twin-based energy modeling of industrial robots. In: Asian simulation conference; 2018. p. 333–48.
[49] Zhu G, Kwong S. Gbest-guided artificial bee colony algorithm for numerical function optimization. Appl Math Comput 2010;217(7):3166–73.
[50] Yan K. Research on digitized modeling of energy consumption for industrial robots driven by physics-based model. Master thesis. Wuhan University of Technology; 2019.

CHAPTER 7

Digital twin-driven cutting tool service

Huibin Sun and Yuanpu Yao
School of Mechanical Engineering, Northwestern Polytechnical University, Xi'an, Shaanxi, China

7.1 Introduction

Cutting tools are of great importance to machining quality, cost, efficiency, and sustainability. Worn cutting tools deteriorate surface integrity and lead to rejected parts. Although cutting tools account for about 4% of the machining cost [1], their worn states can lead to 10%–40% of machine tool downtime, and indirectly affect up to 30% of total machining cost [2]. Moreover, many rare elements (such as tungsten, chromium, molybdenum, vanadium, cobalt, etc.) require high energy in cutting tool production [3]. Consequently, the more durable and efficient a cutting tool is, the more cost, time, resources, and energy can be saved.

Unfortunately, to avoid poor surface quality, cutting tools are normally underused, resulting in huge waste. Studies showed that only 50%–80% of cutting tool life was rationally used [4]. American companies believed that up to 30% cutting tool life were wasted [2]. European and Japanese manufacturers used no more than 70% of cutting tool lives [5]. Inevitably, underuse will lead to cutting tool life reduction, increase machining cost, and cutting tool waste. Precise usage of cutting tools can improve economic, environmental, and social benefits greatly. However, it is a pertinent issue for cutting tool consumers. Although a cutting tool has potential for prolonged utilization, to unleash this fully is very difficult. The lack of reliable prediction has rendered cutting tool utilization improvement with much difficulty.

Traditionally, cutting tool manufacturers only market physical cutting tools. Customers make their own decision of cutting tools selection and replacement. To provide competitive edge for the cutting tools, cutting tool manufacturers strive to develop innovative industry solutions for different application areas such as aerospace, automotive, wind power, etc.

Nowadays, cutting tool manufacturers are extending their business chains. They find that consumers need cutting tool services rather than

Digital Twin Driven Service
ISBN 978-0-323-91300-3
https://doi.org/10.1016/B978-0-323-91300-3.00010-3

physical cutting tools alone. Therefore, they integrate intangible cutting services with physical cutting tools to satisfy consumers' machining requirements [6]. Consumers only pay for cutting tool services without owning the physical cutting tools. Cutting tool manufacturers are in charge of monitoring cutting tool conditions and ensuring the cutting tool services [7]. Moreover, third-party cutting tool service providers are also available. They have professional knowledge of cutting tools and can provide excellent expert cutting tool services [8]. Obviously, cutting tool services help cutting tool service consumers, cutting tool manufacturers, and the third-party cutting tool service providers to realize a win-win situation [9].

However, implementing cutting tool services is still a pending issue. Prior to its success, some significant problems must be solved. For example, both service quality and profit depend on the accuracy of tool condition estimation. Professional knowledge and specific skills are needed to evaluate cutting tool conditions precisely and in a timely manner. Then, reliable cutting tool selection and replacement decisions should be made. In fact, the dynamic and time-varying cutting tool conditions could not be known exactly. This phenomenon leads to the lack of reliable support for cutting tool selection and replacement.

Recently, digital twin (DT) is attracting attention from both academia and industry [10,11]. DT provides a unique means to achieve the cyber-physical integration, which is a notion embraced by increasing number of enterprises [12]. DT means an organic whole of physical entity as well as its digitized model, which mutually communicate, promote, and coevolve with each other through bidirectional interactions [13]. Through various digitization technologies, the entities, behaviors, and relations in the physical world are digitized holistically to create high-fidelity virtual models [14,15]. Such virtual models depend on real-world data from the physical world to formulate their real-time parameters, boundary conditions, and dynamics, leading to a more representative reflection of the corresponding physical entities [16,17]. Based on physical-virtual convergence, DT has been considered as an emerging technology to improve the accuracy and efficiency of prognostics and health management of complex equipment [18].

Therefore, DT is a service-oriented technology, which can provide a unique way to implement cutting tool services. Together with the sensory data acquisition, big data analytics, as well as artificial intelligence (AI), DT can be used for cutting tool condition monitoring and prediction [19,20]. Through the assessment of ongoing states, and the prediction of future trends, DT can provide more comprehensive supports for decision-making of cutting tool selection and replacement. Due to the strategic benefits, DT can be used to enable cutting tool services; however, related researches are very limited.

This chapter focuses on DT-driven cutting tool service. Its concept, framework, and function are addressed in detail. A digital model for cutting tools in a machining process is proposed to enable DT-driven cutting tool services. DT-driven cutting tool condition monitoring (TCM) service, tool condition forecast (TCF) service, remaining useful life (RUL) prediction service, cutting tool selection decision-making service are also discussed in detail. A prototype is developed to illustrate and validate the model. Then, some intelligent, proactive, and predictive cutting tool services can be enabled to realize the win-win situation discussed above.

7.2 Related works

As an integration of a product and related services [21], an industrial product service system (iPSS) accelerated the resource and capability sharing among industries [22]. In the past few years, many iPSSs have been studied and developed, including the machine tool iPSS (mt-iPSS) [23], the warehouse product service system (wPSS) [24], turning process iPSS (TP-iPSS) [25], etc. Recently, data monitoring, storage, and processing trigger new and intelligent services [26]. Then, in-process machining condition monitoring (MCM) has shown great potential for iPSS [26]. For example, DMG MORI Messenger provides machine tool monitoring services on various mobile devices [27]. SKF @ptitude Connect enables online bearing monitoring and maintenance decision-making [28].

Nowadays, the concept of cutting tool service is becoming more promising [29], because consumers can obtain more professional cutting tool services with lower costs. Traditional cutting tool providers are extending their business chains by integrating intangible cutting services with physical cutting tools [7]. They create valuable profit sources by providing cutting tool services. For example, the COMET ToolScope system [30] aims to provide some process and cutting tool monitoring services, including cutting breakage, collision, uneven running, vibrations, etc. As a third-party, Tool Consulting & Management Group provides cutting tool management services for Shanghai General Motors [31].

Although the concept of cutting tool service has brought a win-win business model for cutting tool service consumers, cutting tool manufacturers, and the third-party cutting tool service providers, the study with regard to its implementation is insufficient. Especially, the cutting tool service quality and profit depend on the accuracy of TCM, TCF, and RUL prediction. To enable successful cutting tool services, DT of a cutting tool was introduced as a digital replica of a physical tool [32]. It collected data and monitored the cutting tool and allows for a better understanding of the machining process, and better prediction of the behavior and results [33]. Modeling, application, and service strategy have also been studied to provide guidance for the development of DT for cutting tools [34]. Although DT has bridged the concept of cutting tool service and its implementations, more detailed investigations are still pending.

7.3 Framework of digital twin-driven cutting tool service

As an integration of industrial products and services, DT-driven cutting tool services involve both hardware and software [7]. As Fig. 7.1 shows, the framework of DT-driven cutting tool service is composed of four layers as follows.

(1) Physical basis layer provides fundamental hardware, including the machine tool, cutting tool, clamps, etc. In-process signals and raw data are collected using sensors, computers, and data acquisitions. Normally,

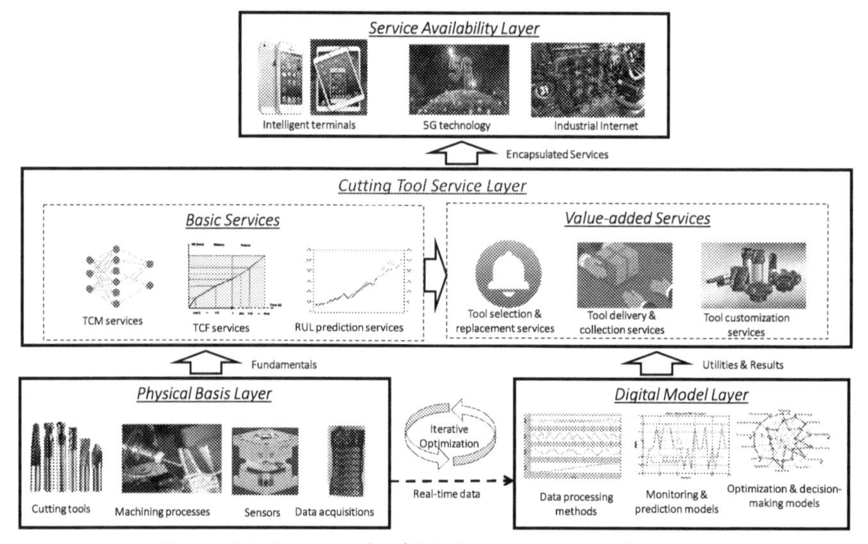

Figure 7.1 Framework of DT-driven cutting tool service.

cutting force, vibration, and AE signals are external signals. Instantaneous spindle speeds, driving speeds, tool center point position, etc. can also be acquired from CNC systems.

(2) Digital model layer is composed of various submodels. Using data acquired from physical basis layer, these submodels extract features and make judgments or decisions. Here, the optimization and decision-making models are based on the cutting tool condition monitoring and prediction models. The precision and reliability of these models are improved by iterative optimization based on physical basis layer.

(3) Cutting tool service layer encapsulates utilities and results of digital model layer as basic services and value-added services. Basic services include TCM service, TCF service, and cutting tool RUL prediction service. Based on basic services, value-added services include cutting tool selection/replacement service, cutting tool delivery/collection service, cutting tool customization service, etc.

(4) Service availability layer makes encapsulated intangible services available ubiquitously based on the mobile computing technology as follows.

- Time-dependent services are available at a fixed frequency. For example, TCM service can be provided every 5 min.
- Event-dependent services are available when some predefined events occur. For example, service consumers are informed when the cutter wear condition is 0.1, 0.2, or 0.3 mm.
- Alarm services are triggered by some emergent events, such as rapid cutter wear, tool broken, tool collision, etc.

To balance the requirement of private protection and service sharing, the service accession control policy may apply. In addition, offline services are delivered in the real world.

In summary, DT-driven cutting tool services are listed in Table 7.1.

7.4 Enabling digital twin-driven cutting tool service

7.4.1 Digital model for cutting tools in machining process

To enable cutting tool services, a digital model for cutting tools should be built. The model includes multitype, multitime scale, multigranularity factors, behaviors, rules, and knowledge. It involves preprocess, in-process, and after-process time scales. Both macroscopic and microcosmic factors are included. Its main functions are detection, judgment, prediction, and decision-making. Due to the complexity, various submodel should be built

Table 7.1 Cutting tool service catalog.

Service type	Service	Online/ offline	Regularity	Availability
Basic service	TCM service	Online	Time-dependent, event-dependent, emergent alarm.	Ubiquitous
	TCF service	Online	Time-dependent, event-dependent, emergent alarm.	Ubiquitous
	RUL prediction service	Online	Time-dependent, event-dependent, emergent alarm.	Ubiquitous
	Cutting tool selection service	Online	Time-dependent, event-dependent, emergent alarm.	Ubiquitous
	Cutting tool replacement service	Online	Time-dependent, event-dependent, emergent alarm.	Ubiquitous
Value-added services	Cutting tool warehouse allocation service	Offline	Time-dependent	Location-based
	Demand prediction service	Online	Time-dependent.	Ubiquitous
	Delivery/collection service	Offline	Time-dependent, event-dependent.	Location-based

for different levels and application fields [35]. As shown in Fig. 7.2, a digital model for cutting tools includes multiple submodels as follows.

(1) The design model describes the ideal length, radius, flank angle, etc. It enables machining simulation and initial tool usage decision-making.

(2) The geometric model describes a cutting tool's actual length, radius, flank angle, etc. It can be used in length compensation, radius compensation, runout compensation.

(3) The task model describes a cutting tool's states in every cutting task, including cutting parameters (spindle speed, feed rate, cutting depth, cutting width, etc.), machine tool, workpiece, material properties, machining features, cutter path, cutting length, overhanging length, etc. They can be used in runout evaluation, cutting parameter optimization, and cutting tool selection.

(4) The kinematics model reflects a cutting tool's physical behavior in the machining process. Based on cutter workpiece engagement,

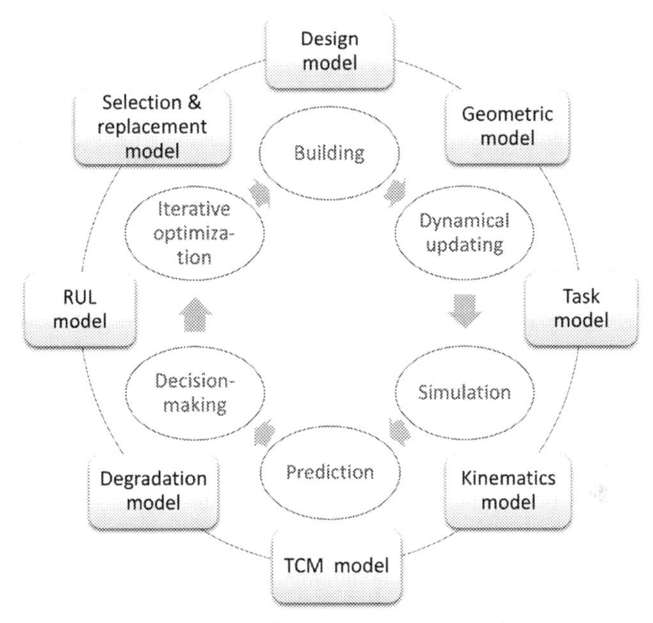

Figure 7.2 Digital model for cutting tools in machining process.

undeformed cutting chip, instantaneous cutting force, and machining heat generation are calculated and simulated. They can be used to optimize cutter path and cutter pose, and generate control commands.

(5) The TCM model evaluates tool wear condition based on acquired signals, such as current, cutting force, vibration, AE, etc. The TCF model estimates tool conditions in the nearest future based on historical tool condition data.

(6) The degradation model describes a cutting tool's condition changing path. Both mathematical models and AI-based models can be used to reflect the path. The degradation model supports decision-making about cutting tool selection or replacement.

(7) The RUL model estimates the cutting tool's RUL considering tool wear, machining accuracy, or surface roughness. Rather than the time dimension, cutting tool RUL can also be measured by cutting length or material removal volume (MRV).

(8) The cutting tool selection and replacement model makes decisions about cutting tool usage. Then, cutting tool utilization can be improved to reduce cost, improve efficiency, and sustainability. On the other hand, machining parameters can also be optimized to extend cutting tool lives.

Although these submodels have different functions, they have the similar work flow as follows.

(1) Model building. Define factors, behaviors, and rules, build a unique instance for a single cutting tool.
(2) Dynamical updating. Acquire on-site data using sensors or data interfaces, update the model.
(3) Calculation or simulation. Calculate, simulate, or monitoring cutting tool status, using historical data and real-time data.
(4) Judgment and prediction. Estimate cutting tool conditions, forecast future tool conditions, or predict cutting tool RUL.
(5) Decision-making. Generate cutting tool compensation command, and cutting tool selection or replacement advice.
(6) Iterative optimization. Improve the model's precision by optimizing or refining the parameters.

The interaction between a cutting tool and its digital model is vital to successful DT-driven cutting tool services. Using sensors and data interfaces, the digital model is aware of the variables and states of the physical cutting tool. Based on predefined rules, algorithms, and constraints, the digital model simulates, forecasts, or predicts cutting tool conditions. Then, cutting tool selection and replacement decisions can be made to optimize the physical machining process. The interaction between a cutting tool and its digital model varies with different submodels and machining stages. For example, prior to the machining process, the actual cutter length and radius are measured to update the geometric model. The geometric model generates cutting tool compensation commands, which are sent to CNC system for controlling. During the machining process, the digital model acquires physical signals, such as current, cutting force, vibration, AE, etc. It evaluates tool conditions using mathematical models or AI-based models. Future tool conditions or cutting tool RUL can also be estimated. Considering machining accuracy, surface roughness, and machining cost, the digital model makes cutting tool selection or replacement advice for machining process optimization (shown in Fig. 7.3).

7.4.2 Digital twin-driven cutting tool condition monitoring service

TCM is a basic DT-driven cutting tool service. An accurate and timely TCM service can not only improve machining efficiency but also reduce machining cost. However, due to the difficulty of recognizing tool condition directly, the indirect TCM services are enabled. A typical indirect TCM service procedure is listed as follows [35].

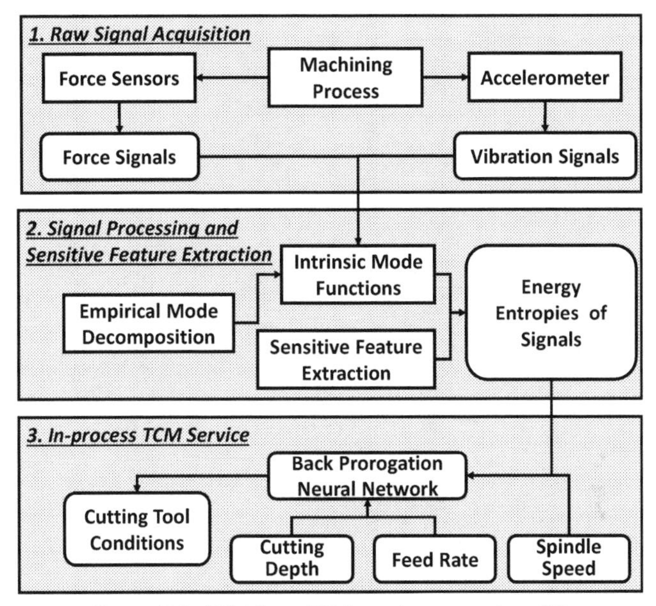

Figure 7.3 DT-driven TCM service procedure [7].

(1) Raw signal acquisition. Multiple physical signals in the machining process are acquired via various sensors, such as force sensors, AE sensors, accelerometers, etc.

(2) Signal processing and sensitive feature extraction. Acquired signals are processed in the time domain, frequency domain, and time-frequency domain. Fourier transform and wavelet transform may be used. Many sensitive features are extracted for further decision-making.

(3) In-process TCM service. On the basis of AI, such as back prorogation neural network, support vector machine, etc., an estimation about tool wear condition can be made.

In order to decompose the nonlinear and nonstationary signals, some adaptive signal decomposition methods can be used. For example, empirical model decomposition (EMD) is used to decompose the original signal into a set of orthogonal intrinsic mode functions. Due to EMD's strong adaptability in uncovering signals' local characteristics, high precision in both time and frequency domains may be obtained at the same time. Some Singular Value Entropies (SVEs) are calculated and extracted as sensitive features. As Fig. 7.4 shows, a back-propagation neural network (BPNN) model is built as the measurement model to map the relationship between SVEs and cutting tool wear conditions. At time t, SVEs of X force, Y force,

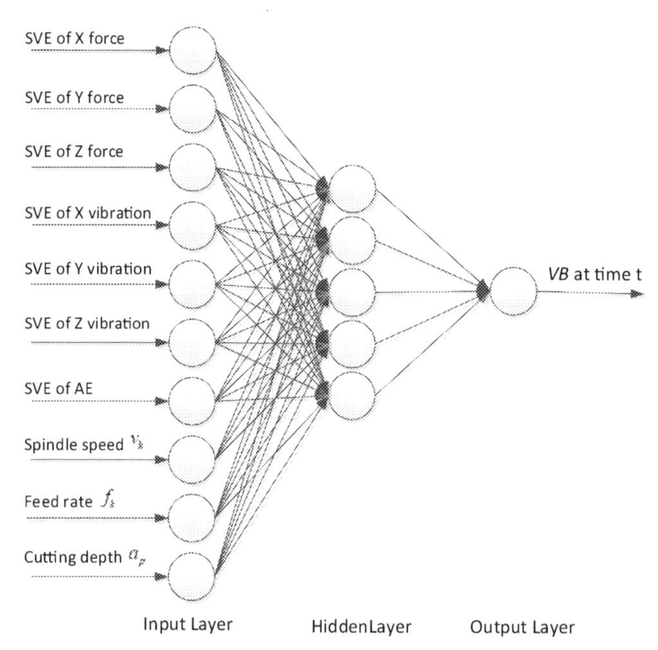

Figure 7.4 BPNN-based TCM service model [36].

Z force, X vibration, Y vibration, Z vibration, and AE are input parameters of the BPNN model, together with feed rate f_k, spindle speed v_k, and cutting depth a_p. The flank wear value (*VB*) is the output. Trained BPNN model is used to implement in-process TCM service.

In the past few years, deep learning-based TCM service models have been widely investigated. An in-process TCM service model based on residual neural network (ResNet) [37] is shown in Fig. 7.5. At time t, its inputs are seven-dimensional time series signal segments as $X_t = [x_{t1}, x_{t2}, x_{t3}, x_{t4}, x_{t5}, x_{t6}, x_{t7}]$, where x_{t1}, x_{t2}, x_{t3}, x_{t4}, x_{t5}, x_{t6}, and x_{t7} refer to X force, Y force, Z force, X vibration, Y vibration, Z vibration, and AE signals, respectively. A deterministic model, G, aims to learn a mapping from raw signal segments to a single output $\widetilde{y}_t$, which is *VB* value at time t. Formally, there are:

$$\widetilde{y}_t = G(X_t) = G(x_{t1}, x_{t2}, x_{t3}, x_{t4}, x_{t5}, x_{t6}, x_{t7}) \tag{7.1}$$

The ResNet-based in-process TCM service model includes a post-activation residual block and many preactivation residual blocks. The postactivation residual block starts with a convolutional layer (Conv). A batch normalization layer (BN) is used to accelerate the calculation and

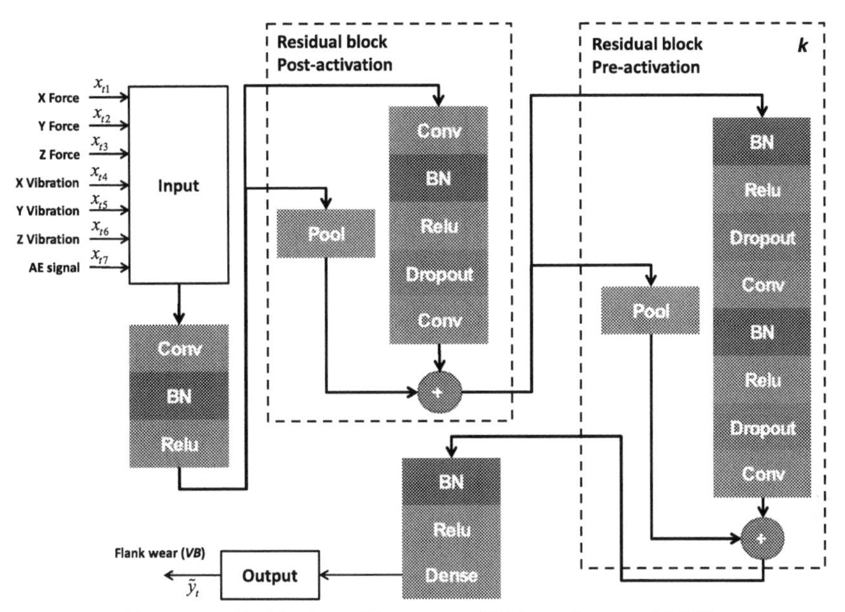

Figure 7.5 ResNet-based in-process TCM service model [37].

optimize the network. A rectifier linear unit (ReLu) is adopted to manipulate preactivation architecture. A dropout layer is added to diminish over-fitting in the training process and to improve the generalization ability. In order to reduce computation time and eliminate redundant features, a max-out pooling (Pool) layer is included in the shortcut of residual learning framework (RLF). Then, the postactivation residual block ends with a convolutional layer. The structure of convolutional layers and max-out pooling layers use classical parameters. The kernel size is 3 and the stride is 2. In every block, kernel numbers of two convolutional kernels are 64 and 128, respectively. A preactivation residual block starts with a BN, followed by Relu, Dropout, Conv in sequence. This structure repeats in every preactivation residual block. Finally, a fully connected layer (Dense) is used to produce a *VB* value.

An adaptive feature extraction mode is implemented based on the convolutional layer. In the convolutional process, the signal segments do convolutional operations with kernels of various sizes. During these steps, some specific patterns are emphasized, and noise is automatically filtered. Based on the excellent selectivity, the convolution kernel enables adaptive filtering. Parameters of the convolution kernel can be optimized in the training process. The trained kernels are able to fully extract features from the raw signals. The feature extraction process is nearly the same as the manual process.

Extracted features are selected based on multilayer perception (MLP), which is a feedforward neural network that simulates how signals pass two neurons. MLP is made up of many neurons in different layers. The output of a layer is the input of next layer. Assisted by the back-propagation (BP) algorithm, tool wear condition information could pass through these layers. Then, the low-level features are assembled into the high-level features. The high-level features approximate tool wear conditions, as no obvious clue could be found from the low-level features.

Gradient always vanishes when the multiplication of two gradients becomes smaller. Then, information produced by low-level layers may be wiped out, although tool wear condition always exists in minor clues. Based on RLF, the gradient vanishing phenomenon is prevented. Then, ResNet and RLF are merged to leverage their advantages. Moreover, over fitting is avoided by using BN, ReLu, and dropout layers.

7.4.3 Digital twin-driven cutting tool condition forecasting service

DT-driven TCF service focuses on the time-varying cutting tool wear curves during the machining processes. It estimates multiple *VB* values in the nearest future by using several sequential *VB* values measured in the latest past. Then, how to capture and continue data dependencies in historical *VB* values becomes the most significant issue. Although the standard recurrent neural network can retain the recent memories of input patterns, it has difficulty to model time series data when long time lag is present. To deal with this problem, a long short-term memory (LSTM) network is used [37]. By capturing long-range dependencies in time series data, the LSTM network contributes to accuracy improvement of the in-process TCF service. In Fig. 7.6, several sequential *VB* values measured recently, denoted by $y_{t-n+1}, y_{t-n+2}, \dots, y_t$ in sequence, are inputs of the LSTM network. Here, variable n stands for the number of historical data. The LSTM network outputs multiple *VB* values in the nearest future, denoted by $y_{t+1}, y_{t+2}, \dots, y_{t+m}$. Variable m stands for the number of forecast data. A deterministic model, H, aims to learn a mapping from the input values to the output values. Formally, there are:

$$\left(y_{t+1}, y_{t+2}, \dots, y_{t+m}\right) = H\left(y_{t-n+1}, y_{t-n+2}, \dots, y_t\right) \tag{7.2}$$

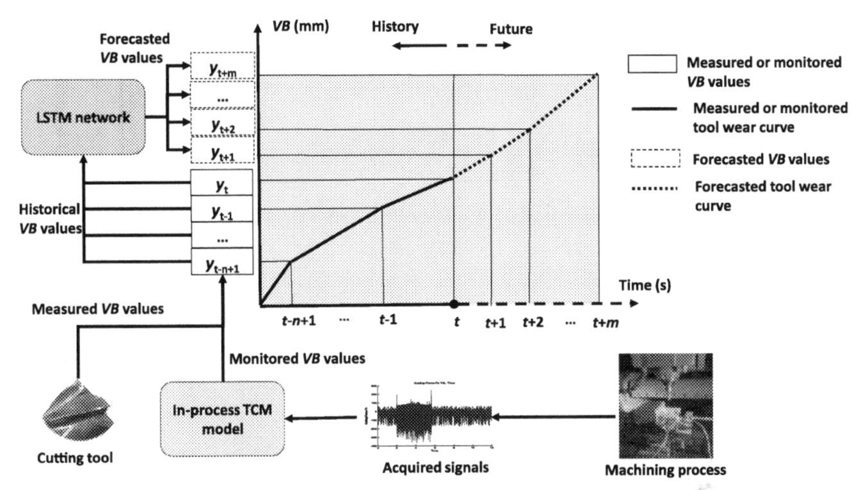

Figure 7.6 DT-driven in-process TCF service [37].

Normally, it is very difficult to measure *VB* values directly during the machining process. Under this case, the *VB* values outputted by the DT-driven in-process TCM service models (discussed in Section 7.4.2) could be used, without stopping the machining process.

As Fig. 7.7 shows, the LSTM network is logically composed of an encoder, a decoder, and a context tensor [38]. The encoder accepts n historical and sequential *VB* values and outputs the context tensor. The

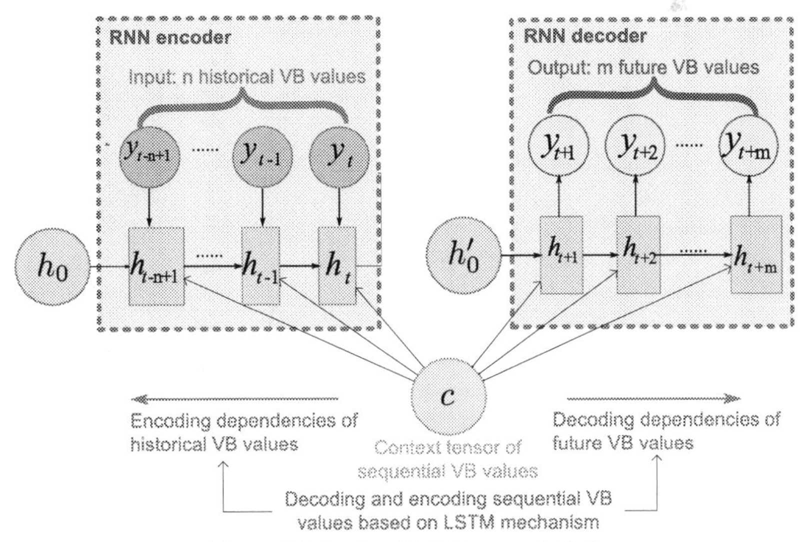

Figure 7.7 Logic of LSTM network [37].

decoder accepts the context tensor and estimates m future and sequential VB values. Both the encoder and the decoder use the same LSTM unit [39].

The LSTM network is shown in Fig. 7.8. Numbers and variables in brackets mean dimensions. The duplicated vector layer is used to extend 2D vectors to 3D vectors. The former two LSTM units are encoders, and the latter two LSTM units are decoders. Driven by the LSTM mechanism, the threshold of an LSTM unit is adjusted to save or delete time series information. Then, the network can use or ignore historical VB values in forecasting future tool conditions. By modifying n and m, future VB values in different time ranges can be forecasted. In addition, the LSTM network also contributes to back propagate error prevention from gradient vanishing or exploding problems.

If a new time range is used, it is time-consuming to train a new LSTM network. If the input VB values are insufficient, the LSTM network could work using its own outputs. On the basis of this so-called selfreferring mode, the LSTM network can extend its time range. However, if the thresholds are not updated accordingly, small errors will accumulate to great errors.

As shown in Fig. 7.9, LSTM unit t includes two states, the long-term state c_t and the short-term state h_t. The forget gate F_t, the input gate I_t,

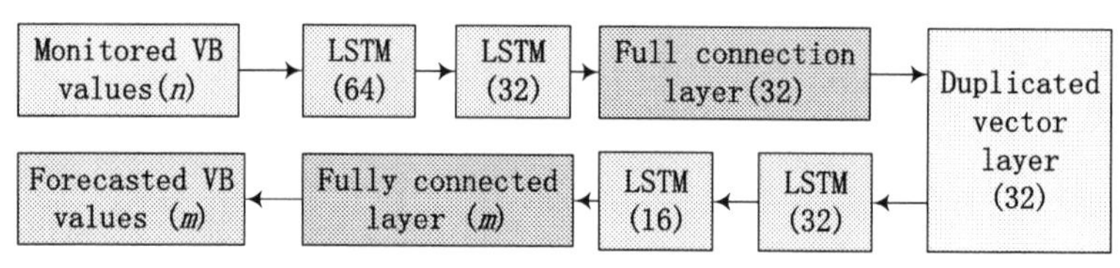

Figure 7.8 Structure of LSTM network [37].

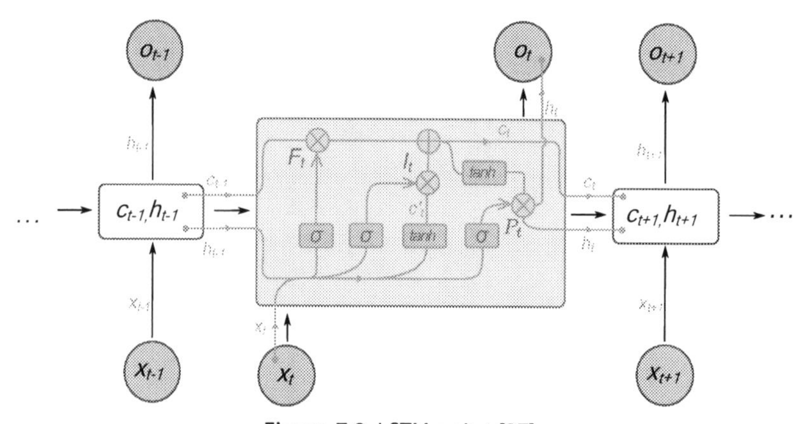

Figure 7.9 LSTM unit t [37].

and the output gate P_t are added to regulate the unit states. The forget gate F_t deletes the information from the previous long-term state c_{t-1}. The output gate P_t controls the formation of the current short-term state h_t using the information from the long-term state c_t. Its value is in the range [0, 1]. Values 1 and 0, respectively, stand for completely saving or removing historical information.

7.4.4 Digital twin-driven cutting tool remaining useful life prediction service

DT-driven cutting tool RUL prediction service estimates a specific cutting tool's RUL. It is essential for cutting tool utilization evaluation and replacement decision-making. The life of a cutting tool can be evaluated by a variety of indicators, such as cutting distance, surface roughness, tool wear, chip shape, etc. Tool wear is considered to be the prime criterion, due to its good measurability and availability. Especially, *VB* is the most widely used indicator. It is well known that *VB* is closely related to machining quality. *VB* develops with rapid cutting force increase, machining accuracy, and surface roughness deterioration. Without loss of generality, *VB* is considered to be the primary criterion of cutting tool failure. According to the nonlinear Wiener process model, cutting tool wear is regarded as a stochastic process $\{X(t), t \geq 0\}$ [40] and defined by

$$X(t) = X(0) + \int_0^t \mu(t;\theta)dt + \sigma_B B(t) \tag{7.3}$$

where $X(t)$ is *VB* value at time t, and $X(0)$ is the initial *VB* value. $B(t)$ stands for the standard Brownian motion (BM) as $B(t) \sim \mathrm{N}(0, \mathrm{t})$. The drift $\mu(t;\theta)$ is a nonlinear function of time t, which means the wear trend. $\theta = (a, b)$ is a parameter vector. Here, parameter a is a random variable representing individual characteristics of a single cutting tool. It obeys a normal distribution as $a \sim \mathrm{N}\left(\mu_a, \sigma_a^2\right)$. Parameter b is a constant for a batch of cutting tools under the same machining condition (MC). The random variable σ_B^2 stands for infinitesimal variance of *VB* values. Obviously, if $\mu(t;\theta)$ is a constant as $\mu(t;\theta) = \mu$, Eq. (7.3) describes a linear cutting tool wear process.

In general, cutting tool life ends when its *VB* value exceeds the predefined threshold w. The life of a cutting tool (T) is defined as

$$T = \inf\{t: X(t) \geq w | t > 0\} \tag{7.4}$$

RUL of a cutting tool at time t_k is denoted by T_k. According to first hitting time (FHT), T_k is the duration between time t_k and the time that VB value exceeds w for the first time.

$$T_k = \inf\{t: X(t + t_k) \geq w | X(t_k) < w, t > 0\} \tag{7.5}$$

In fact, although a batch of cutting tool lives can vary greatly, a typical statistical distribution characteristic still exists. This phenomenon can be formulated by the probability density function (PDF), denoted by $f_T(t)$. Regarding a cutting tool's RUL at time t_k, the statistical distribution characteristics also exists and is denoted by $f_{T_k}(t)$ as

$$f_{T_k}(t) = f_T(t + t_k)/F_T(t) \tag{7.6}$$

Here, $F_T(t) = 1 - \int_0^t f_T(t)dt$ is the cumulative distribution function (CDF).

As shown in Fig. 7.10, RUL regarding state $X(t_k)$ can also be estimated with quantized PDF and CDF. Under a specific confidence level α, the upper bound and the lower bound can be resolved.

Similarly, wear curves, RULs, PDFs, and CDFs under various MCs (shown in Fig. 7.11) can also be modeled or calculated. When more data are used, the RUL prediction service becomes more reliable.

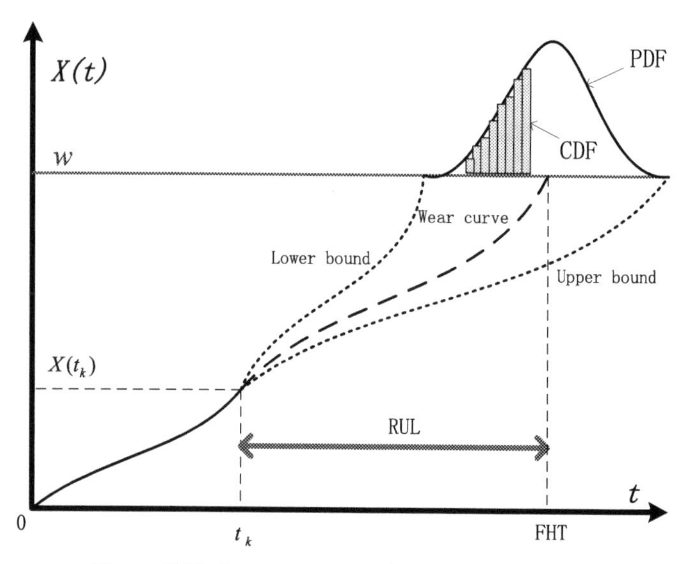

Figure 7.10 A wear curve under a certain MC [41].

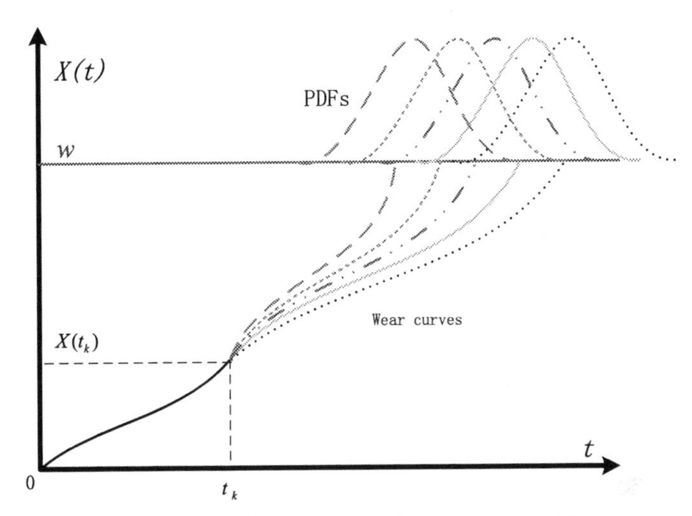

Figure 7.11 Wear curves under various MCs [41].

In order to estimate cutting tool RUL, the calculation of $f_T(t)$ becomes a vital issue.

The exponential function has been widely used to model equipment degradation process. Without loss of generality, it is also used here to build the cutting tool wear and RUL prediction model. If $\mu(t;\theta) = abt^{b-1}$ is true, Eq. (7.3) can be instantiated as

$$X(t) = X(0) + at^b + \sigma_B B(t) \tag{7.7}$$

In order to derive $f_T(t)$, the stochastic cutting tool wear process $\{X(t), t \geq 0\}$ is transformed into a standard BM, which has an explicit form of the PDF of FHT. The original Kolmogorov formula of the diffusion process is transformed into the standard BM's Kolmogorov formula. Formulating PDF of $X(t)$ crossing w for the first time is equivalent to calculating PDF of FHT for a standard BM $\{B(t), t \geq 0\}$ crossing a time-varying boundary $S_B(t)$.

According to the total probability formula, PDF of FHT is formulated as

$$f_{T|a}(t|a) \cong \frac{w - ab^t(1-b)}{\sigma_B\sqrt{2\pi t^3}} \exp\left(-\frac{\left(w - ab^t\right)^2}{2\sigma_B^2 t}\right) \tag{7.8}$$

If $b = 1$ is true, Eq. (7.8) is an inverse Gaussian distribution.

Because $a \sim \mathrm{N}\left(\mu_a, \sigma_a^2\right)$ is true, PDF of FHT can be formulated as

$$f_{T|a}(t|a) \cong \frac{1}{\sqrt{2\pi t^3\left(\sigma_a^2 t^{2b-1} + \sigma_B^2\right)}} \exp\left\{ -\frac{\left(w - \mu_a t^b\right)^2}{2t\left(\sigma_a^2 t^{2b-1} + \sigma_B^2\right)} \right\} \times \left(w - \left(t^b - bt^b\right) \frac{w\sigma_a^2 t^{b-1} + \mu_a \sigma_B^2}{\sigma_a^2 t^{2b-1} + \sigma_B^2} \right) \tag{7.9}$$

If $l_k = t - t_k \geq 0$ and $X(l_k) = X(l_k + t_k) - X(t_k)$ are true, RUL of a cutting tool at time t_k equals the duration between time t_k and FHT of the stochastic process $\{X(l_k), l_k \geq 0\}$ crossing the threshold $w_k = w - X(t_k)$. Then, Eq. (7.5) can be transformed into Eq. (7.10) as

$$L_k = \inf\{l_k : Y(l_k) \geq w_k | Y(0) = 0, l_k \geq 0\} \tag{7.10}$$

PDF of FHT can be formulated as Eq. (7.11) according to Eq. (7.6).

$$f_{L_k|a}(l_k|a) \cong \frac{1}{\sqrt{2\pi l_k^2\left(\sigma_a^2 \eta(l_k)^2 + \sigma_B^2 l_k\right)}} \times \exp\left(-\frac{\left(w_k - \mu_a \eta(l_k)^2\right)}{2\left(\sigma_a^2 \eta(l_k)^2 + \sigma_B^2 l_k\right)} \right) \left(w_k - \left(\eta(l_k) - bl_k(l_k + t_k)^{b-1}\right) \times \frac{\sigma_a^2 \eta(l_k) w_k - \mu_a \sigma_B^2 l_k}{\sigma_a^2 \eta(l_k)^2 + \sigma_B^2 l_k} \right) \tag{7.11}$$

with $\eta(l_k) = (l_k + t_k)^b - t_k^b$.

Prior to Eq. (7.11), some parameters should be estimated in advance based on historical data from a batch of the same cutting tools under the same MC. The unknown parameters are included in a vector as $\Theta = (\mu_{ak}, \sigma_{ak}, \sigma_B, b)'$ maximum likelihood estimation (MLE).

Then, $\widehat{b}$ and $\widehat{\sigma}_B$ can be estimated by maximizing the above profile log-likelihood function using a two-dimension searching algorithm. Based on estimated $\widehat{b}$ and $\widehat{\sigma}_B$, $\widehat{\mu}_a$ and $\widehat{\sigma}_a^2$ can be estimated.

Even under the same MC, a batch of cutting tool wear curves can vary greatly. Each cutting tool has its unique time-varying and dynamic wear curve. In order to predict a specific cutting tool's RUL more accurately, parameters estimated above should be updated on the basis of Bayesian model and real-time wear data. Then, the parameter vector is updated as $\widehat{\Theta} = \left(\widehat{\mu}_{ak}, \widehat{\sigma}_{ak}, \widehat{\sigma}_B, \widehat{b}\right)'$. By substituting it into Eq. (7.11), PDF of a cutting tool's RUL at time t_k is

$$f_{L_k|\hat{\Theta}}(l_k|\widehat{\Theta}) \cong \frac{1}{\sqrt{2\pi l_k^2\left(\widehat{\sigma}_{ak}^2\eta(l_k)^2 + \widehat{\sigma}_B^2 l_k\right)}} \times \exp\left(-\frac{\left(w_k - \widehat{\mu}_{ak}\eta(l_k)\right)^2}{2\left(\widehat{\sigma}_{ak}^2\eta(l_k)^2 + \widehat{\sigma}_B^2 l_k\right)}\right)$$
$$\times \left(\begin{array}{c} w_k - \left(\eta(l_k) - \widehat{b}l_k(l_k + t_k)^{\hat{b}-1}\right) \\ \times \dfrac{\widehat{\sigma}_{ak}^2\eta(l_k)^2 w_k + \widehat{\mu}_{ak}\sigma_B^2 l_k}{\widehat{\sigma}_{ak}^2\eta(l_k)^2 + \widehat{\sigma}_B^2 l_k} \end{array}\right) \tag{7.12}$$

with $\eta(l_k) = (l_k + t_k)^{\hat{b}} - t_k^{\hat{b}}$.

In addition to the variability of tool wear itself during machining process, variability also exists in the measurement. Inevitably, errors are included in the measured VB values. In order to improve the accuracy and reliability, these errors must be considered in the cutting tool RUL prediction model.

When variability is considered, the measured VB value at time t is denoted by $Y(t)$ and modeled as follows.

$$Y(t) = X(t) + \varepsilon \tag{7.13}$$

where ε is the variability error. It is a normally distributed random variable as $\varepsilon \sim N(0, \sigma_\varepsilon^2)$. Parameter σ_ε^2 is the variance to ensure the independency of variability errors at different times. Then, the unknown parameters vector discussed above is updated as $\Theta = (\mu_{ak}, \sigma_{ak}, \sigma_B, \sigma_\varepsilon, b)'$.

7.4.5 Digital twin-driven cutting tool selection service

DT-driven cutting tool selection service aims to improve cutting tool utilization. In order to prolong the use of every cutting tool at controllable risk, a cutting tool should be matched with a suitable machining task according to its specific RUL prediction, rather than a unified one [41]. Machining parameters can also be optimized based on RUL prediction under various MCs. The benefits and risks are balanced by using a confidence level.

(1) Cutting tool selection service based on a specific RUL prediction

Prior to a machining task, a correct cutting tool should be selected according to its specific RUL. Some cutting tools of the predefined types with different conditions and RULs are the candidates. A reliable decision could be made based on cutting tool RUL prediction. In order to maximize the use of every cutting tool at controllable risk, the confidence level α is used. To finish machining task j, CDF of cutting tool i regarding CT_j, denoted by $P_{T_{k,i}}(CT_j)$, should satisfy the following constraint.

$$P_{T_{k,i}}(CT_j) = \int_{t_k}^{t_k+CT_j} f_{T_{k,i}}(t)dt \leq 1 - \alpha \tag{7.14}$$

Here, CT_j is time duration of task i. $T_{k,i}$ is RUL of cutting tool i at time t_k. The difference between CT_j and $T_{k,i}$, denoted by $D_{i,j}$, stands for the margin. Then, the risk of a cutting tool selection service is calculable. RUL waste of a cutting tool is also measurable and controllable. By minimizing $D_{i,j}$ and satisfying Eq. (7.14), every cutting tool is matched with a right task. Interrelationship among α, CT_j and $D_{i,j}$ is shown in Fig. 7.12. Obviously, PDF affects the margin greatly. The benefit and risk can be balanced by adjusting the confidence level α.

(2) Machining parameter optimization service considering various MCs

Cutting tool wear curves vary greatly under different MCs. Machining parameters could be optimized to extend a cutting tool life. Without loss of generality, some factors, such as machine tool, workpiece material, and cutting fluid, are regarded as constants. Feed speed (f/mm/min), cutting width (a_e/mm), cutting depth (a_p/mm), etc. are considered as key machining parameters in this work. Here, an MC means a combination of them. As Fig. 7.13 shows, at time

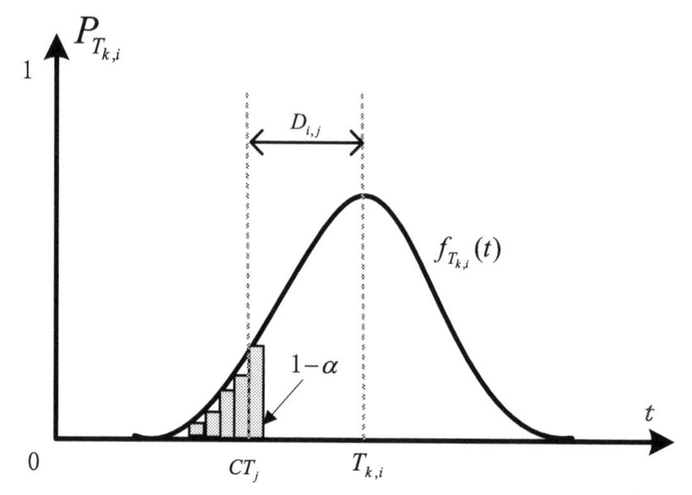

Figure 7.12 Pretask RUL prediction and probability [41].

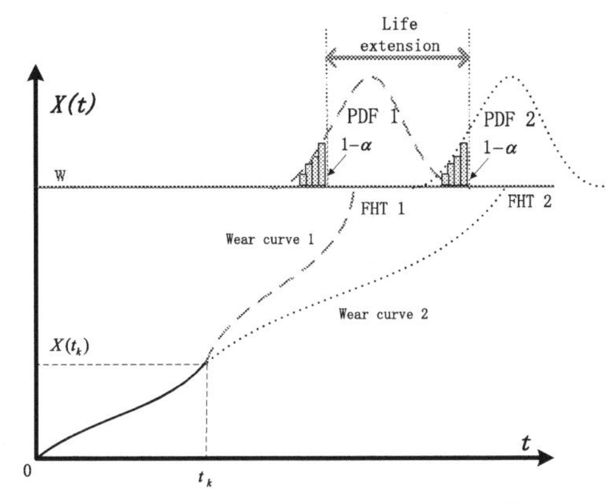

Figure 7.13 Cutting tool wear curves under different MCs [41].

t_k, 2 MCs result in different cutting tool wear curves, FHTs, RULs, and PDFs. Compared with MC 1, cutting tool life could be extended greatly by MC 2. Then, wear curve 2 is better than wear curve 1 under the same confidence level α. However, cutting tool life extension may lead to lower machining efficiency. To improve the benefit systematically, an MC should be evaluated according to the sustainability assessment metrics.

(3) Sustainability of DT-driven cutting tool selection service

According to the triple bottom line, sustainability of DT-driven cutting tool selection service can be assessed in economic, environmental, and social dimensions.

The economic sustainability is assessed by cutting tool life and cost. DT-driven cutting tool selection service can maximize the use of every cutting tool. Due to tool life extension, fewer cutting tools are needed to fulfill the same machining time requirement, and cutting tool cost could be reduced. Moreover, cutting tool life extension leads to fewer cutting tool replacements, which means more machining time. The equipment utilization ratio can also be increased.

The environmental sustainability is assessed by the ratio of a cutting tool's RUL to MRV. If MRV of cutting tool i is improved by machining parameter optimization, average carbon emission could be decreased. Therefore, environmental sustainability could be improved by enhancing cutting tool durability and capability.

Regarding the social sustainability, traditional operators select cutting tools based on experience at their own risks. It is quite conservative and arbitrary to some extent. Based on a specific RUL prediction, DT-driven cutting tool selection service comes with a quantized confidence level α, which makes the cutting tool usage decision-making more reliable. Machining quality could be guaranteed at controllable risk without artificial factors. Operators are free from cutting tool selection decision-making. Moreover, cutting tool life extension reduces cutting tool replacement frequency, which also decreases operator workload. Then, nonproductive time can be reduced, and cutting tools can be used precisely to avoid unexpected downtime and scrapped components.

7.5 Case study

According to above discussions, some DT-driven cutting tool services are implemented as follows.

(1) DT-driven in-process TCM service

IEEE PHM 2010 challenge dataset [42] is used. The DT-driven in-process TCM service is implemented on the basis of ResNet. Based on Keras framework, all algorithms are implemented with TensorFlow. The model with 20 residual blocks is selected due to its outstanding performance. According to the summary given by Keras, the average absolute training errors and validation error are 2 and 2.6 μm respectively. The trained model estimates tool wear conditions within less than 15 ms. As shown in Fig. 7.14, monitored *VB* values fit nicely with measured tool wear curve.

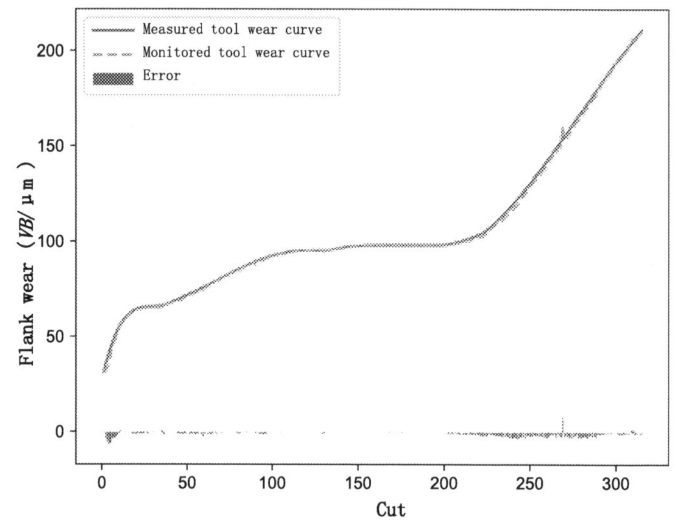

Figure 7.14 ResNet-based TCM service [37].

(2) DT-driven TCF service

Based on Keras framework, DT-driven TCF service is also implemented with TensorFlow. In the training process, the inputs are several measured time series *VB* values. The outputs are multiple time series *VB* values. Based on Keras framework, TensorBoard is used to monitor the training process. The time consumption of a training iteration is less than 1s. Although the trained LSTM networks have no knowledge of the testing datasets, the forecast accuracy is good enough.

As shown in Fig. 7.15, both smaller forecast ranges and larger historical ranges lead to smaller errors. Short-term forecast is more accurate than long-term forecast, because more input means greater reliability. However, when the historical range increases, the marginal benefits are reduced.

By integrating the ResNet-based in-process TCM service with the LSTM network, forthcoming *VB* values could be forecasted when the machining process is progressing. As shown in Fig. 7.16, the forecasted *VB* curves are not always increasing. The LSTM network is not as accurate as it works independently. Both median-based correction and mean-based correction contribute to accuracy improvement of forecasted tool wear curves. By removing singular points or scatter points, fluctuations in the forecasted cutting tool wear curves are reduced.

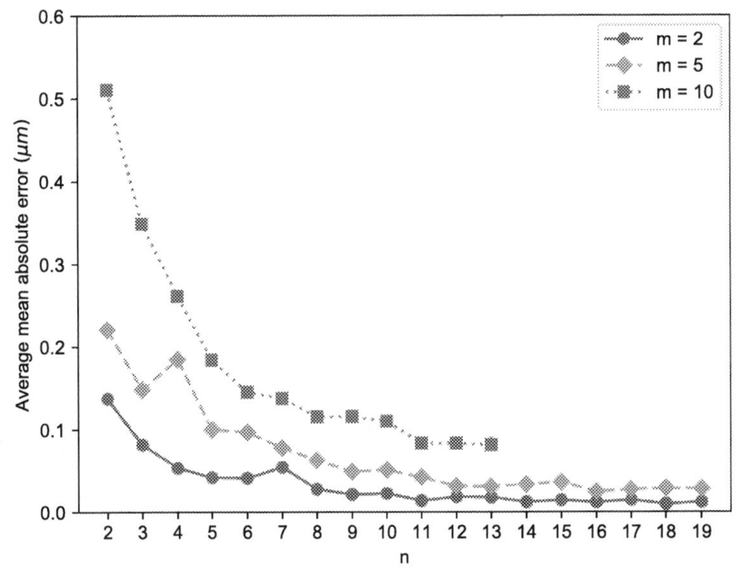

Figure 7.15 A performance comparison under various time ranges [37].

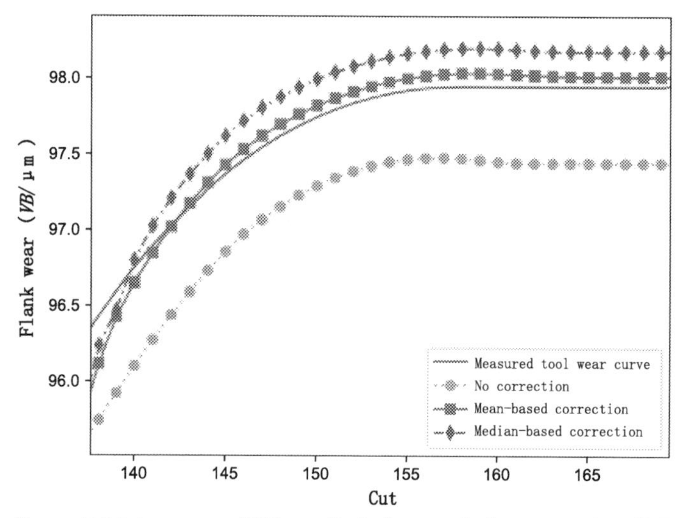

Figure 7.16 In-process TCF results before and after correction [37].

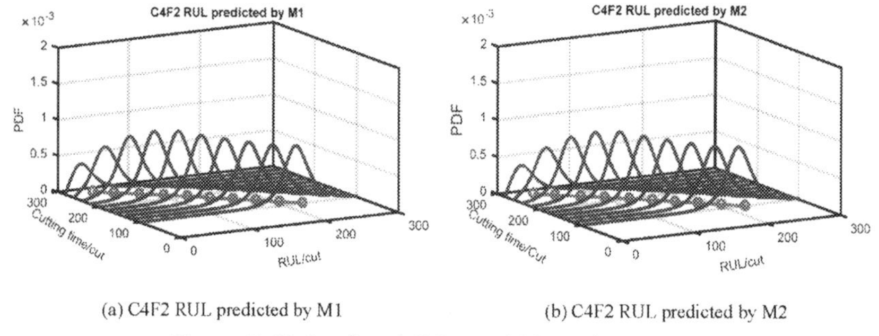

Figure 7.17 Predicted RULs and PDFs of C4F2 [40].

(3) DT-driven cutting tool RUL prediction service

Using IEEE PHM 2010 challenge dataset [42], DT-driven cutting tool RUL prediction service is implemented in MATLAB 2012. The failure threshold w is 0.20 mm. Parameters of the nonlinear model (abbr. M1) and variability model (abbr. M2) are estimated using MLE. Based on the Bayesian method, estimated parameters of M1 are updated to output the posterior estimation results. Using Kalman filter, prior estimations of M2's parameters are also updated.

As Fig. 7.17 and Fig. 7.18 show, predicted RULs fit measured ones well. The differences between predicted RULs and Real RULs decline

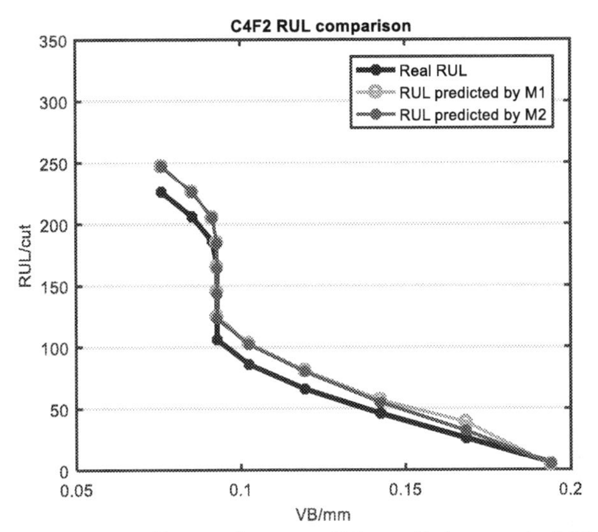

Figure 7.18 Difference between RUL prediction results [40].

when the cutting tool wear progresses. Because variability is considered, M2 is more accurate than M1.

The confidence intervals are decided by the PDFs. The upper confidence bounds gradually decrease when the confidence level increases. During the machining process, an appropriate confidence level should be selected according to specific machining requirements. A higher confidence level can be used to guarantee the machining quality at lower risk. A lower confidence level can be used to make use of cutting tool RULs further.

(4) DT-driven cutting tool selection service

DT-driven cutting tool selection service is implemented on the basis of RUL prediction results. In order to use cutting tools at controllable risk, 85% of average life is set as the wear criterion, according to engineering experience. If every cutting tool is used based on its specific RUL prediction, the situation is different. Under the confidence level $\alpha = 99\%$, the upper bound of every cutting tool life is obtained.

Better sustainability is achieved by using each cutting tool according to its specific RUL prediction. As to economic sustainability, the total MRV is improved by 15.81%. As to environmental sustainability, carbon emission for a flute production is decreased by 8.39%. As to social sustainability, cutting tools are used further at precisely controlled risk.

Detailed comparison between a unified bound and some specific bounds is given in Fig. 7.19.

In Fig. 7.20, a radar diagram is used to compare the two situations vividly. To present in "the higher the better" manner, the reciprocal of

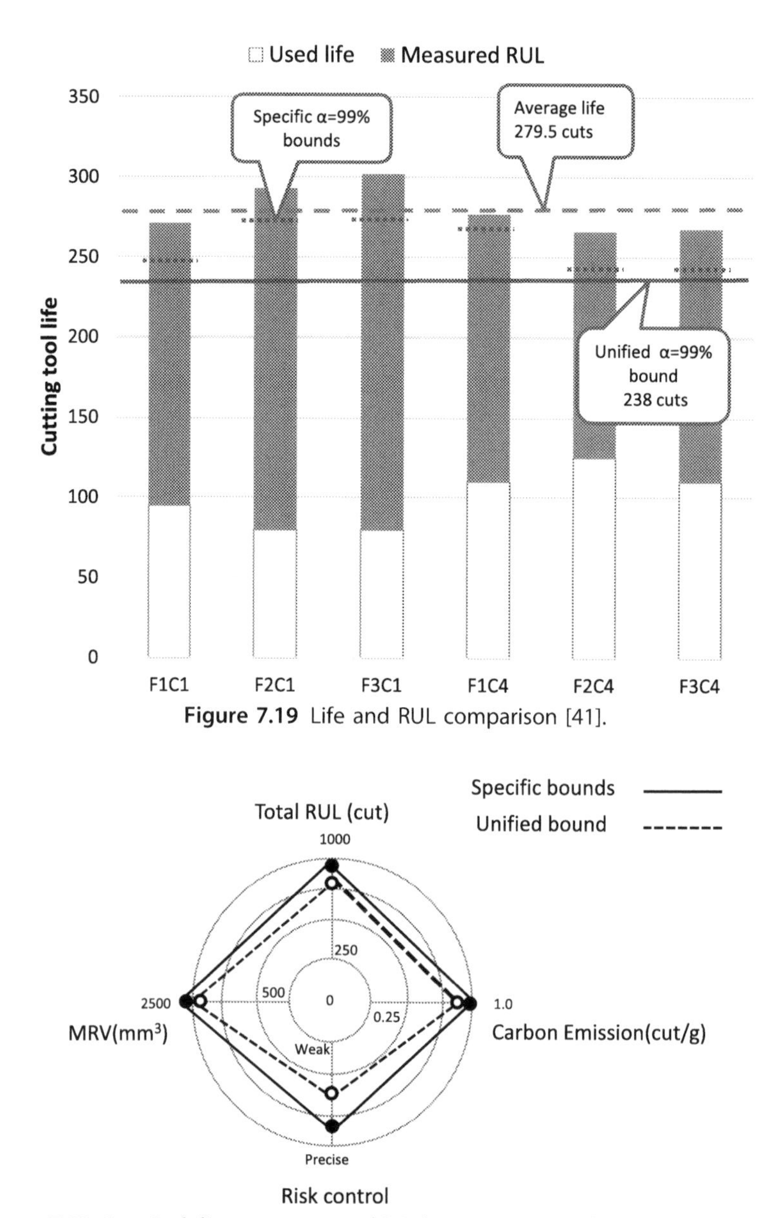

Figure 7.19 Life and RUL comparison [41].

Figure 7.20 Sustainability assessment of DT-driven cutting tool selection service [41].

carbon emission (cut/g) is used in the diagram. It can be seen that DT-driven cutting tool selection service based on the specific bounds is more sustainable. Under a constant MC, a cutting tool's life can be extended according to its specific RUL prediction. PDF, CDF, and confidence level α contribute to the balance between benefits and risks.

(5) Cutting tool service availability

Intangible cutting tool services are available ubiquitously via the mobile computing technology. Based on the Java Web architecture, Android Studio 1.3.2 is used to develop and simulate the mobile computing environment. Tomcat 7.0 is used to make the service available ubiquitously. As Fig. 7.21 shows, all services are available via Android applications (APPs). The Servlets on the server side accept the Hyper Text Transfer Protocol (HTTP) requests posted by Android APPs, and access the database via Java Database Connectivity interface. TCM, TCF, RUL prediction, and decision-making functions are implemented as jar packages. They are called by Servlets to enable the cutting tool services. The result is saved in the SQL Server 2008 database via the Internet. Finally, Servlets return the results to Android APPs by using HTTP. Both Android emulators and smart phones are also used to assess the services.

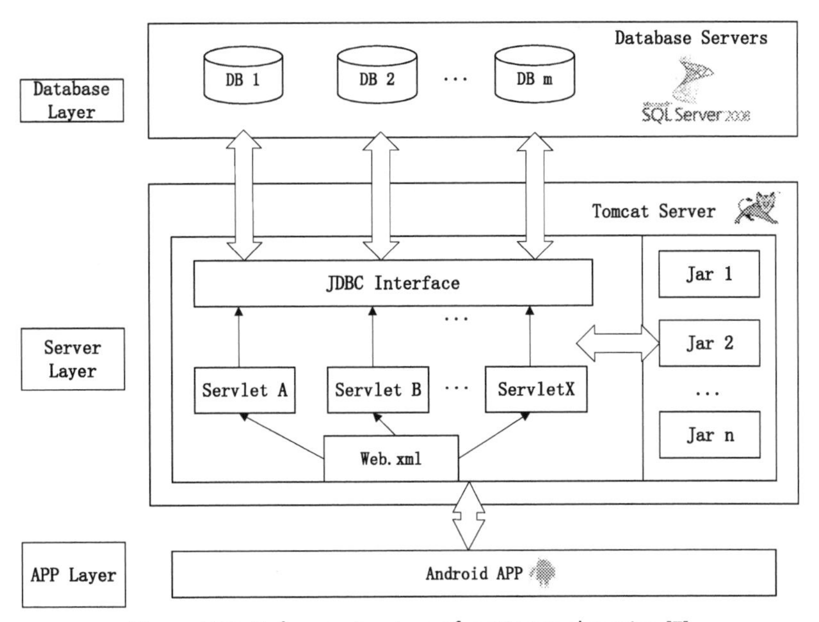

Figure 7.21 Software structure of cutting tool service [7].

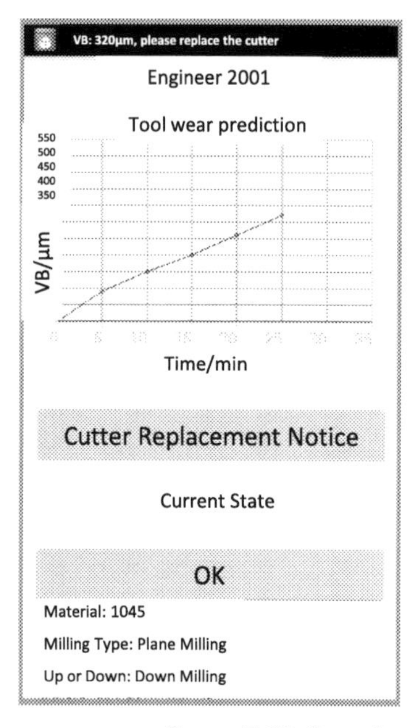

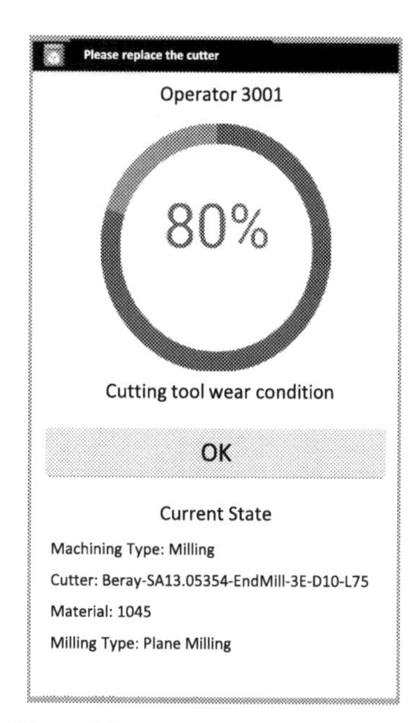

Figure 7.22 Snapshots of DT-driven TCM service [7].

As Fig. 7.22 shows, the DT-driven TCM service is available on smart phone emulator. The virtual model monitors the cutting tool conditions. It also sends cutting tool replacement suggestions. The operator does nothing other than following the suggestion. Without senior knowledge and skills, he is free from the professional TCM work and only needs to focus on the machining process.

7.6 Summary

As discussed above, DT-driven cutting tool service is investigated and demonstrated using the prototype. Intangible cutting services are integrated with physical cutting tools to satisfy consumer cutting service requirements. Consumers pay for the intangible cutting time without owning the physical cutting tools. Cutting tool service providers monitoring cutting tool conditions and ensuring the cutting time. A win-win situation could be achieved. The prototype could be referred by further implementation of more practical DT-driven cutting tool service.

Although great potential can be expected, DT-driven cutting tool service is far from large-scale application. The following limitations still exist.

(1) The roles designed in the prototype are used to demonstrate the service procedure. However, they are really new and may be infeasible. To advance the study and development of cutting tool service, roles and procedures should be redesigned according to practical requirements.

(2) The quality of service depends on the accuracy, reliability, and robustness of TCM, TCF, and RUL prediction. In fact, functions discussed above are too limited. More efforts should be considered to improve the practicality.

(3) Due to the complexity, cutting tool services are not available in real time. Some efforts should be made to improve the efficiency of signal processing and decision-making. The Internet of Things could also be integrated to improve the real-time services and enable cloud manufacturing service system.

(4) More services and functions could be developed and enabled. Typical examples include new cutting tool delivery service, worn cutting tool collection service, cutting tool delivery/collection route optimization, etc.

References

[1] Astakhov VP. In: Davim JP, editor. Improving sustainability of machining operation as a system endeavor chapter 1 in book: "measurement in machining and tribology". Springer; 2017. p. 1–29.

[2] Liu C, Li Y, Hua J, et al. Real-time cutting tool state recognition approach based on machining features in NC machining process of complex structural parts. Int J Adv Manuf Technol 2018;97(2):229–41.

[3] Li C, Tang Y, Cui L, et al. A quantitative approach to analyze carbon emissions of CNC-based machining systems. J Intell Manuf 2015;26(5):911–22.

[4] Zhou Y, Xue W. Review of tool condition monitoring methods in milling processes. Int J Adv Manuf Technol 2018;96(5–8):2509–23.

[5] Martinova LI, Grigoryev AS, Sokolov SV. Diagnostics and forecasting of cutting tool wear at CNC machines. Autom Rem Control 2012;73(4):742–9.

[6] Sun P, Jiang P, Cao W. Cutting tool Delivery method in the context of industrial product service systems. Concurr Eng Res Appl 2016;24(2):178–90.

[7] Zhang G, Sun H. Enabling a cutting tool iPSS based on tool condition monitoring. Int J Adv Manuf Technol 2018;94(9):3265–74.

[8] Zhang G, Sun H. Enabling cutting tool services based on in-process machining condition monitoring. Int J Internet Manuf Serv 2018;5(1):51–66.

[9] Lindahl M, Sundin E, Sakao T. Environmental and economic benefits of integrated product service offerings quantified with real business cases. J Clean Prod 2014;64(2):288–96.

[10] Qi Q, Tao F, Hu T, et al. Enabling technologies and tools for digital twin. J Manuf Syst 2021;58:3–21.

[11] Cheng J, Zhang H, Tao F, et al. DT-II: digital twin enhanced Industrial Internet reference framework towards smart manufacturing. Robot Comput Integrated Manuf 2020:62.

[12] Cheng Y, Zhang Y, Ji P, et al. Cyber-physical integration for moving digital factories forward towards smart manufacturing: a survey. Int J Adv Manuf Technol 2018;97(1–4):1209–21.

[13] Tao F, Cheng J, Qi Q, et al. Digital twin-driven product design, manufacturing and service with big data. Int J Adv Manuf Technol 2018;94(9):3563–76.
[14] Tuegel EJ, Ingraffea AR, Eason TG, et al. Reengineering aircraft structural life prediction using a digital twin. Int J Aerospace Eng 2011;2011:1–14.
[15] Glaessgen E, Stargel D. The digital twin paradigm for future NASA and US air force vehicles. In: 53rd AIAA/ASME/ASCE/AHS/ASC structures, structural dynamics and materials conference 20th AIAA/ASME/AHS adaptive structures conference 14th AIAA; 2012.
[16] Grieves M. Digital twin: manufacturing excellence through virtual factory replication. 2015.
[17] Qi Q, Tao F, Zuo Y, et al. Digital twin service towards smart manufacturing. Procedia CIRP 2018;72:237–42.
[18] Tao F, Zhang M, Liu Y, et al. Digital twin driven prognostics and health management for complex equipment. CIRP Ann Manuf Technol 2018.
[19] Zaccaria V, Stenfelt M, Aslanidou I, et al. Fleet monitoring and diagnostics framework based on digital twin of aero-engines. In: ASME Turbo Expo; 2018.
[20] Cai Y, Starly B, Cohen P, et al. Sensor data and information fusion to construct digital-twins virtual machine tools for cyber-physical manufacturing. Procedia Manuf 2017;10:1031–42.
[21] Reim W, Parida V, Ortqvist D. Product-Service Systems (PSS) business models and tactics - a systematic literature review. J Clean Prod 2015;97:61–75.
[22] Meier H, Roy R, Seliger G. Industrial product-service system. Int J Adv Manuf Technol 2010;52(2):1175–91.
[23] Mu H, Jiang P, Leng J. Costing-based coordination between mt-iPSS customer and providers for job shop production using game theory. Int J Prod Res 2016;55(2):430–46.
[24] Cao W, Jiang P. Modelling on service capability maturity and resource configuration for public warehouse product service systems. Int J Prod Res 2013;51(6):1898–921.
[25] Zhu Q, Jiang P. Machining capacity measurement of an industrial product service system for turning process. Proc IME B J Eng Manufact 2011;225(B3):336–47.
[26] Biffl S, Luder A, Gerhard D. Multi-disciplinary engineering for cyber-physical production systems. Switzerland: Springer International Publishing AG; 2017.
[27] DMG MORI Messenger. https://www.dmgmorimessenger.com/; 2016.
[28] SKF @ptitude Analyst. http://www.skfmaintenanceservices.de/uploads/media/aptitude_analyst_englisch.pdf; 2016.
[29] Sakao T, Shimomura Y. Service engineering: a novel engineering discipline for producers to increase value combining service and product. J Clean Prod 2006;15(6):590–604.
[30] COMET ToolScope. http://www.kometgroup.com/en/plus-avigation/plus/process-monitoring.html; 2016.
[31] Tool Consulting & Management Group. http://www.tcm-international.com/en/; 2016.
[32] Botkina D, Hedlind M, Olsson B, et al. Digital twin of a cutting tool. Procedia CIRP 2018;72:215–8.
[33] Qiao Q, Wang J, Ye L, et al. Digital twin for machining tool condition prediction. Procedia CIRP 2019;81:1388–93.
[34] Xie Y, Lian K, Liu Q, et al. Digital twin for cutting tool: modeling, application and service strategy. J Manuf Syst 2021;58:305–12.
[35] Sun H, Pan J, Zhang J, et al. Digital twin model for cutting tools in machining process. Comput Integr Manuf Syst 2019;25(6):1474–80.
[36] Sun H, Cao D, Zhao Z, et al. A hybrid approach to cutting tool remaining useful life prediction based on Wiener process. IEEE Trans Reliab 2018;67(3):1294～1303.

[37] Sun H, Zhang J, Mo R, et al. In-process tool condition forecasting based on a deep learning method. Robot Comput Integrated Manuf 2020:64.
[38] Zhang J, Wang P, Yan R, et al. Long short-term memory for machine remaining life prediction. J Manuf Syst 2018;48:78–86.
[39] Wang J, Yan J, Li C, et al. Deep heterogeneous GRU model for predictive analytics in smart manufacturing: application to tool wear prediction. Comput Ind 2019;111:1–14.
[40] Sun H, Pan J, Zhang J, et al. Non-linear Wiener process–based cutting tool remaining useful life prediction considering measurement variability. Int J Adv Manuf Technol 2020;107(2).
[41] Sun H, Liu Y, Pan J, et al. Enhancing cutting tool sustainability based on remaining useful life prediction. J Clean Prod 2020:244.
[42] 2010 PHM Society Conference Data Challenge, https://www.phmsociety.org/competition/phm/10, (accessed 17 December 2017).

CHAPTER 8

Digital twin-driven prognostics and health management

Jinsong Yu and Diyin Tang
School of Automation Science and Electrical Engineering, Beihang University, Beijing, China

8.1 Introduction

With the development of modern industry and the advancement of science and technology, the structure of modern equipment is becoming more complex, and the scale is getting increasingly larger, especially products and equipment in high-tech fields such as aerospace. Aerospace products and equipment are expensive, complex in structure, and usually operate in harsh environments. As these complex systems undertake critical missions in the military, national economy, and other fields, a huge logistics support system must be provided to ensure the safe operations of such equipment. Currently, a periodically planned maintenance support system based on statistical reliability indicators is widely used, regardless of the actual health of the equipment, which not only causes waste of early maintenance manpower and funds but may also cause safety hazards and accidents due to failure.

Prognostics and health management (PHM) is proposed based on above background. Under the premise of ensuring the safety and reliability of high-tech complex equipment and mission capabilities, it supports rapid and accurate fault location, maintenance decision, and mission decision based on actual health status. It significantly reduces the maintenance cost of end users by realizing condition-based maintenance (CBM) based on the current health status of the equipment. The core of health management technology is predictive technology, so it is also called prognostic and health management (PHM).

In recent years, the technology of digital twin has begun to appear in the field of PHM. According to news reports, the U.S. Department of Defense has introduced the concept of digital twins into the health maintenance of spacecraft, and defined it as a simulation process that integrates multiple physical quantities, multiple scales, and multiple probabilities [1]. The digital twin builds a virtual model of its complete mapping based on the

Digital Twin Driven Service
ISBN 978-0-323-91300-3
https://doi.org/10.1016/B978-0-323-91300-3.00005-X

physical model of the aircraft and uses historical data and real-time updated data from sensors to describe and reflect the full life cycle process of physical objects [2]. Some inspiring results are reported. In Ref. [3], a diagnosis and prediction method of system faults was proposed for complex systems based on the combination of physical system with its equivalent virtual system, which was verified in some important systems by NASA. In Ref. [1], aircraft virtual models with ultrahigh-precision are combined with structural deviation and temperature calculation models to support the life prediction of aircraft structures. The technical advantages and application difficulties of digital twins are also summarized in this paper. In 2015, General Electric Company plans to implement real-time engine monitoring, timely inspection, and predictive maintenance through its own cloud service platform Predix, using big data, Internet of Things, and other advanced technologies based on digital twins. All of these applications reflect the effectiveness of the digital twin idea in dealing with complex system health monitoring, diagnosis, and prognostic problems.

8.2 Digital twin-driven PHM framework

Traditionally, digital twins have relied on deterministic physics-based simulations to approximate the system being monitored. Unfortunately, for complex systems, it is generally not possible to simulate each individual asset within a fleet of homogeneous assets due to manufacturing and material uncertainties.

Digital twin is essentially a unique living model of the physical system with the support of enabling technologies including multiphysics simulation, machine learning, etc. It allows the continuous adaptation to the changes in the environment or operations and delivers the best business outcome. However, there are some open questions remain, such as (1) how to build the models to describe the complex equipment thoroughly and accurately, (2) how to establish the interaction mechanism between the equipment and its digital mirror model to make them evolve jointly and support PHM together, and (3) how to fuse the data from both physical and virtual spaces and generate valuable information for PHM.

The system structure standard of system-level health management adopts the open system architecture (OSA)/CBM [4,5], as shown in Fig. 8.1. The standard was initiated by Boeing, and later accepted by the industry, and was named the ISO13374 standard by the International Organization for Standardization. The standard defines the functional

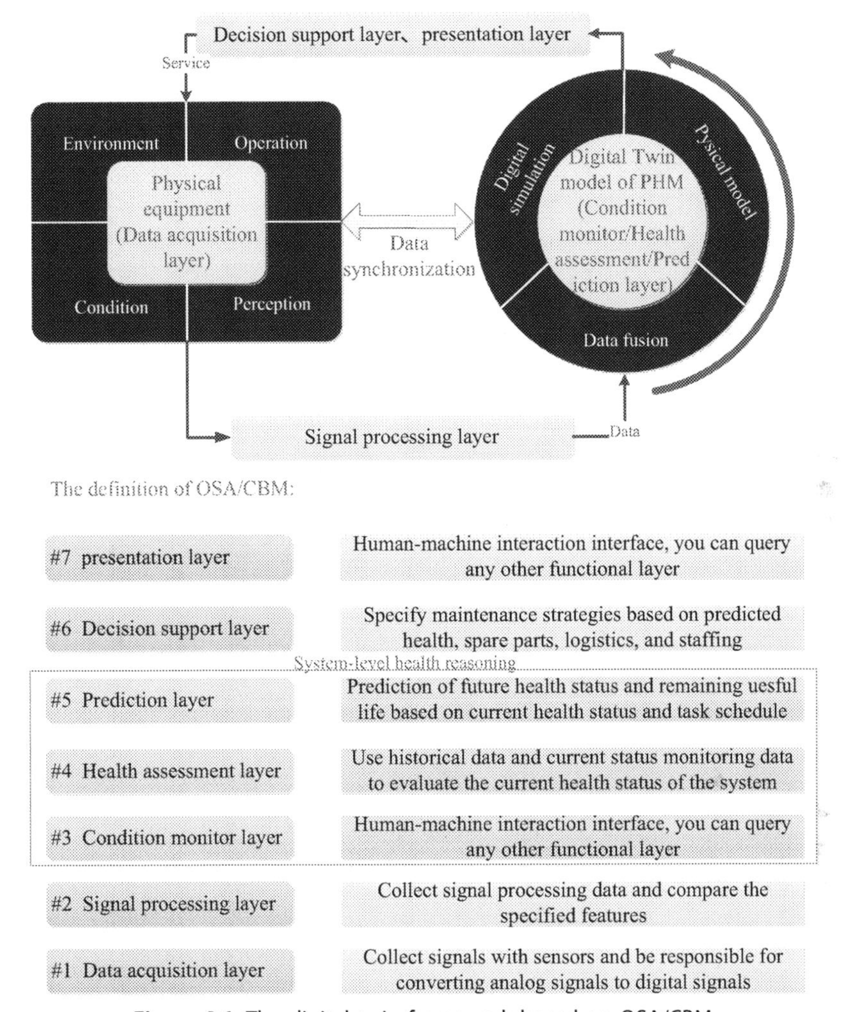

Figure 8.1 The digital twin framework based on OSA/CBM.

hierarchy of the logical structure of the health management system from the information hierarchy and integrates the existing sensor technology, information processing technology, diagnosis technology, prediction technology, and decision-making technology into a unified framework. The standard defines the information exchange interfaces between various intelligent technologies, which can meet the needs of information interaction, transplantation, sharing, and reuse in the entire health management engineering field, laying the foundation for the expansion, upgrade, and interoperability of various intelligent technologies in the health management field. At present, this standard has gradually permeated the PHM system of a new generation of military and civilian aircraft and spacecraft [1,3,6,7].

Digital twin framework mainly includes four parts: physical devices, virtual digital twin models, services provided by virtual models to physical devices, and data synchronization between physical equipment and virtual models, which can be explained by PHM's OSA/CBM model as the figure shown below. Among them, the physical device is equivalent to the data acquisition layer, which can be used to collect various information such as environment, working conditions, status, and measurement point perception; then the data are transferred to the virtual model after the physical device passes through the signal processing layer; afterward the virtual model realizes the condition monitoring layer, health assessment layer, and prediction layer of OSA/CBM using technologies such as digital simulation, physical modeling, and data fusion technologies; finally, the virtual model provides the various PHM services to the physical devices through the decision support layer and the presentation layer. Data synchronization between the physical device and the virtual model can be achieved through technologies such as artificial intelligence, machine learning, and data mining, and different data synchronization methods can be used for different physical devices.

8.2.1 Understanding of physical equipment

A standardized physical system is a prerequisite for a digital twin model. It can be any real-world system, such as the power system, the water supply system, or the spacecraft system. Compared with the traditional system, the physical system represented by the digital twin must have the ability to collect and transfer real-time data.

In order to build digital twin models for PHM, the physical system collects data in a bottom-up manner. The physical system first obtains various parameter data of the subsystem and then collects system-level observation data. Finally, these data are merged from subsystems to system as health indicators (HI) of the entire system. This bottom-up approach facilitates the subsequent establishment of the digital twin model based on physical mechanisms.

Furthermore, the physical world needs to transfer many types of data to the virtual world, such as operating environments, working conditions, sensor data, etc. The physical system requires standard data communication devices to achieve uniform data packaging strateties and communication interfaces (or protocols). Based on these communication devices, the multitype and multiscale data are standardized, cleaned, and packaged by the physical system and then uploaded to the digital twin model in the virtual world. This greatly improves the operability of data in the virtual world.

8.2.2 Construction of virtual equipment

Each component in the physical equipment degrades over time and the focus of virtual equipment modeling is to establish a virtual degradation model under the coupling influence of the degradation process of each component. At present, there are generally two categories of complex system modeling methods in the PHM field. The firsty category directly analyzes the overall performance degradation model of the system through the overall output of the system without taking the degradation of subsystems into consideration. This method employs single or several performance indicators to describe the degradation process and builds a data-driven degradation model based on the degradation data of the overall system performance. For example, [8] employs historical data obtained by different types of sensors to form performance indicators that are directly used to describe the system-level degradation state of wind farms and then uses pure data-driven algorithms to establish a health model for monitoring the state of wind farms. However, for most complex systems, the above method must be basically completed based on sufficient data because many factors affect the overal performance.

The second category of methods considers subsystems or components. Firstly, the performance degradation model of each subsystem and component is established; then the degradation modeling is carried out according to the correlation relationship between the subsystems and components, and finally the model is used to predict the overall system performance [9]. In this method, the core idea is to establish the mathematical relationship between the degradation indicators of subsystems or components and the overall system degradation indicators [9], while the core step is to analyze and establish the relationship between the various degradation processes. The approaches of describing the relationship between the degradation processes of each component can be roughly divided into two types in the existing literature.

The first type of approaches that uses a fixed data model has lower modeling requirements, and the main work is focused on the parameter estimation of the model. There are two main kinds of these models. The first kind of model directly uses the defined distribution function to describe the joint distribution between the two degradation modes, as in Ref. [10]. The function *copula* is a typical representative of these distribution functions. The function *copula* can adapt to many kinds of associated degradation situations through various transformations [11]. This kind of association relationship modeling method describes the quantitative relationship between the degradation modes. Nevertheless, it requires a large amount of data to complete the parameter estimation of the model and the assumed quantitative relationships may not be consistent with the actual situation because of ignoring the system structure and operation rules. The second kind of model divides the association relationship between subsystems or components into multiple categories, such as competition, accumulation, acceleration and competition, trigger, and competition, etc. [12] and then establishes an association model based on these known relationships. This association relationship model can correspond to the actual system conditions, and the parameter calculation is relatively simple, but the types of association relationships that can be described are very limited and can only be adapted to simple systems.

Another approach is to describe the association relationship based on the system operation law or failure mechanism, which is more suitable for the performance degradation modeling of more complex systems [13,14]. There are some simple model methods in the references, such as hierarchical description. For instance, Ref. [15] divides the satellite power system into four levels, namely, the top-level satellite power system health layer, the satellite power system component layer, the power component index layer, and the telemetry data layer. The state of health is obtained step by step through fuzzy judgment matrix and expert knowledge, and finally the state of health of the entire power system is calculated. This hierarchical description method is simple to model but the application is very limited, because for systems with complex structures, it cannot consider the coupling degradation effects between the same levels. For complex systems, existing papers mainly focus on traditional system modeling methods, such as bond graphs and system dynamic equations. Matthew et al. [16–18] from NASA Arms Research Center proposed a system-level modeling method based on a structural model decomposition framework. This method employs bond graphs as a system modeling tool and transforms the centralized system state reasoning issue into a distributed one to address by model structure decomposition. Some scholars, such as Biswas from Vanderbilt University and his team [9,19] focused on the use of system dynamic equation modeling. The performance of the entire system is usually described by one or more dynamic variables. One can obtain the evolution law of the overall system performance by establishing the mathematical relationship between the overall system performance variables and the multiple dynamic variables of the system. Both methods have achieved good results on system-level prediction, but they are highly dependent on the physical operation principle of the system, resulting in a high demand of expert knowledge of the system. In addition, they have some limitations in describing system uncertainty and degradation uncertainty.

8.2.3 Determination of PHM services (Ss)

With the practical applications of PHM, the standard research on PHM at home and abroad is also developing vigorously. Currently international standards related to fault diagnosis and integrated health management include: IEEE1232 (Artificial Intelligence Information Interactions and Services Related to All Test Environments) [20], IEEE1451 (Intelligent Transmitter Interface Standard) [21], IEEE 1522 (measurable test quality indicators) [22], IEEE 1856 (Electronic System Health Management Standard Framework) [23], ISO13374 Standard (Conditional Maintenance Architecture) [24], OSA-CBM Standard (Conditional Maintenance Open Architecture) [25], etc.

The IEEE1232 standard defines interface boundaries, develops switching formats, and specifies standard services through encapsulating issues. AIES-TATE provides a way to develop diagnostic systems that are interoperable and independent of specific vendor and product solutions, the hierarchical structure of which is shown in Fig. 8.2. The IEEE1451 standard mainly stipulates the information model of the application processor with networking capability, the transmission microprocessor communication protocol and the transmitter spreadsheet format, the hardware interface specification of the transmitter, etc. IEEE1522 is described as an information model that enables information interaction with the IEEE-std-1232 standard. In particular, it is worth pointing out that the 1522 standard puts forward the concept of diagnostic, more clearly defined test objectives, etc. These standards and manuals define the test parameters on a quantity level, such as fault detection rate, fault isolation rate, as well as service-related parameters such as average service time, percentage of isolation to a replaceable unit, percentage

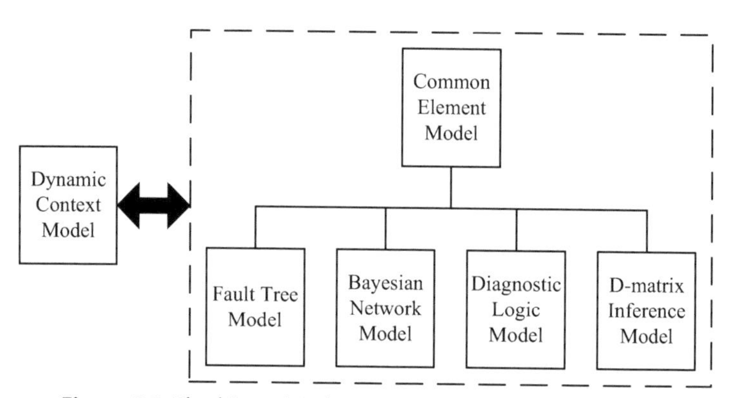

Figure 8.2 The hierarchical structure in the AI-ESTATE model.

of isolation to a set of replaceable units, maintenance ratio (i.e., the number of maintenances required for a working hour), etc.

ISO13374 standard deals with mechanical condition monitoring and diagnosis problems, which provides the basic requirements of open software specifications for data and information, establishes comprehensive guidance on data processing, communication and mechanical condition monitoring and diagnostic information expression software specifications, and defines the conceptual framework of processing models from data acquisition and analysis to health assessment, prediction assessment, and consulting recommendations using six different hierarchical processing modules and common inputs and outputs. The OSA-CBM standard is a CBM OSA (OSA for CBM, OSA-CBM) developed by an industry group of Boeing, Caterpillar, Rockwell Automation, Rockwell Scientific Company, and others. The management and release of OSA-CBM standards is currently the responsibility of MIMIMOSA. OSA-CBM recommended standard covers all functions of CBM system, including hardware and software units. It is a nonpatental standard, which is conducive to the selection of system modules in a wider range of optimized combinations, is conducive to the realization of system unit upgrades, and also conducive to the rapid development of CBM technology. Unlike MIMIMOSA's OSA.EAI, the main aim of OSA-CBM is to develop an open and standardized architecture for distributed CBM software modules, in order to make hardware and software unit components from different vendors interchangeable and enhance system integration capabilities.

Based on the above standards, the standardization and normalization development of the fault diagnosis knowledge model, as well as the new diagnostic reasoning machine architecture is advancing continuously. For example, in order to cope with the work and hierarchical distribution of different types of reasoning machine systems, NASA's ESR-T (Exploratory Systems and Technology) project group proposed a theoretical framework for hybrid reasoning machines [25], which is based on the OSA-CBM standard.

8.2.4 Strategy of DT data synchronization

The process of data synchronization can be described as the physical system uses the model updating strategy and real-time observation data to track and predict the DT model, and then the DT model returns the health monitoring data to the physical system to facilitate the subsequent CBM.

Digital twin technology first establishes a virtual model of the physical system based on the physical mechanism, and the physical system sends historical data and real-time monitoring data into the virtual model. The virtual model trains, monitors, and predicts the above data in real time and analyzes the current state of health or performance degradation of the physical system, then transmits the analysis results back to the physical system to provide services such as health monitoring, fault diagnosis, remaining useful life prediction, and maintenance assistance decision. The operation process of digital twin is the process of synchronous evolution of the physical system and the virtual model, of which the most important part is the data synchronization between the physical system and the virtual model. The current data synchronization algorithms mainly include Bayesian update algorithm, random filter algorithm, and machine learning algorithm, etc.

Bayesian updating algorithm is mainly used in statistical models. The degradation process of the physical system can be described by virtual models such as random-coefficient models [26], Wiener models [26], Gamma models [27], Inverse-Gaussian models [28], and then combined with the physical system through the Bayesian update algorithm. The biggest feature of this method is that it can form a convincing representation of the uncertainty of the degradation process based on probability theory, thereby enhancing the robustness and fault tolerance of virtual equipment.

Random filtering algorithms are mainly used for virtual models described by system dynamic equations (state-space), including Kalman filtering, unscented Kalman filtering, particle filtering, etc. The virtual model usually has one or more dynamic state variables. The change law of the key performance variables of the physical system can be obtained by tracking the system state variables, through establishing the mathematical relationship between the overall system performance variables and the multiple state variables of the system. For example, [29] proposed a real-time model update scheme based on parameter sensitivity analysis to complete the synchronous evolution between the physical system and the virtual model. The random filtering algorithm has stronger uncertainty management capabilities for the virtual equipment than the Bayesian update algorithm. It can integrate the uncertainty representations of multiple state variables in the system more conveniently and has the ability to manage multiple uncertain factors such as models, current status, and future operating conditions.

In recent years, the application of machine learning algorithms in digital twin technology has become more widespread. It is helpful to achieve more efficient and faster data synchronization and virtual model updating by combining the data synchronization of the digital twin with machine learning. At the same time, machine learning methods can fuse various historical data, real-time data, and simulation data to achieve high-fidelity digital twin. [30] used the deep stacked gated recurrent unit algorithm to analyze CNC data to obtain real-time fault diagnosis and prediction results, thereby providing the remaining service life prediction service of the physical system. [31] proposed a deep digital twin method based on a generative adversarial network, which learns the distribution of health data through training historical data, generating an adversarial network, then uses the discriminator of generative adversarial network as a HI to track and monitor the degradation of the physical system during its life cycle, and to identify various failure modes. [32] proposed domain adaptation digital twin (DANN) to realize the adaptability between target simulation in virtual model and actual data in physical system. The DANN model takes simulation data as the source domain and real data as the target domain, so RUL can be predicted without any prior knowledge of labeling information.

8.3 Case study

8.3.1 Example of aerospace control moment gyro

8.3.1.1 Problem description

Control Moment Gyro (CMG) is a key actuator for fast maneuvering and high-precision attitude control of large spacecraft [33], its severe degradation and accidental failure may greatly affect the spacecraft attitude control and even threaten service life of the entire vehicle. In CMG device, the rotor bearing is the core component of system, which directly determines the life of the CMG. In long-term service, high-speed rotor bearings inevitably degrade under the condition of abrasion, corrosion, fatigue spalling, etc., which affect the stable output of gyro torque. Therefore, it is of great significance to estimate and predict the degradation state of rotor bearings for fault early detection of CMG.

8.3.1.2 The framework of DT-driven solution

In this case study, a DT-driven framework of bearings RUL prediction is proposed. It is different from the general prediction method for the high-speed rotor bearings of CMGs, which work in a vacuum environment with poor heat dissipation conditions. Based on the real-time data of current, the established DT-driven framework can be used to achieve the RUL prediction of rotor bearings. In the first place, the physical model of CMG is built for degradation analysis and furthermore for feature extraction. Then, due to low-frequency collected data, compressed sensing component is introduced for data recovery. Finally, CRNN is trained for RUL prediction. Fig. 8.3 describes the whole framework of bearing RUL prediction.

8.3.1.3 Physical analysis of CMG

The CMG is mainly divided into single gimbal CMG (SGCMG) and double gimbal CMG with a constant rotor speed, and variable speed CMG with a variable speed. This case study focuses on SGCMG as the research object.

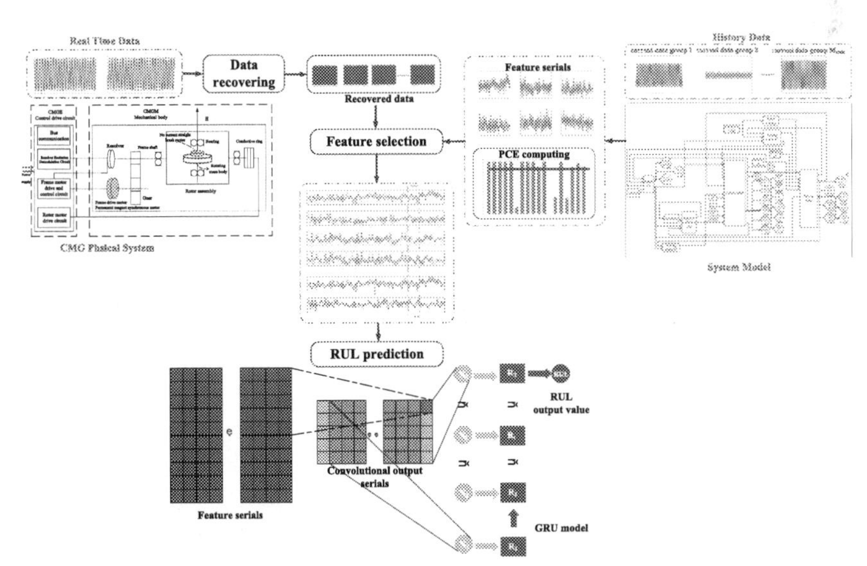

Figure 8.3 The digital twin framework of bearing RUL prediction.

As shown in Fig. 8.4, CMG is composed of Control Moment Gyro Mechanism (CMGM) and Control Moment Gyro Electric. CMGM includes the rotor gimbal part (high-speed component) and the gimbal-driven base part (low-speed component). When the gimbal of the CMG drives the rotor at a constant rotation speed under a specific angular velocity, the direction of the rotor angular momentum also changes. The cross product of the rotor angular momentum and the gimbal angular velocity is the resulting gyro torque that is much larger than the driving torque. This torque drives the base of the gyro gimbal frame. At a certain instant, the output of the SGCMG is single degree of freedom, while multiple SGCMGs are often used to form a redundant SGCMG cluster.

Generally, bearing degradation analysis often resorts to vibration data or time-frequency domain characteristics, which can be seen in DT solutions of bearings in other products. However, since the structure limitation of high-speed rotor, the acceleration sensor or force sensor cannot be installed in rotor subsystem, which makes such method unable to be applied in data synchronization of DT solution. Moreover, due to the limitation of the communication bandwidth of the omnidirectional measurement and control antenna, the low sampling frequency of various original signals during CMG operation is also an unavoidable problem.

As an effect of bearing aging, stator current signals can be regarded as noninvasive monitor of spacecraft over time which provide information about degradation before significant changes of bearing [34]. In diagnostic detection of bearings, research has proved that stator current signals could be a replacement of vibration signals due to the similar frequential characters

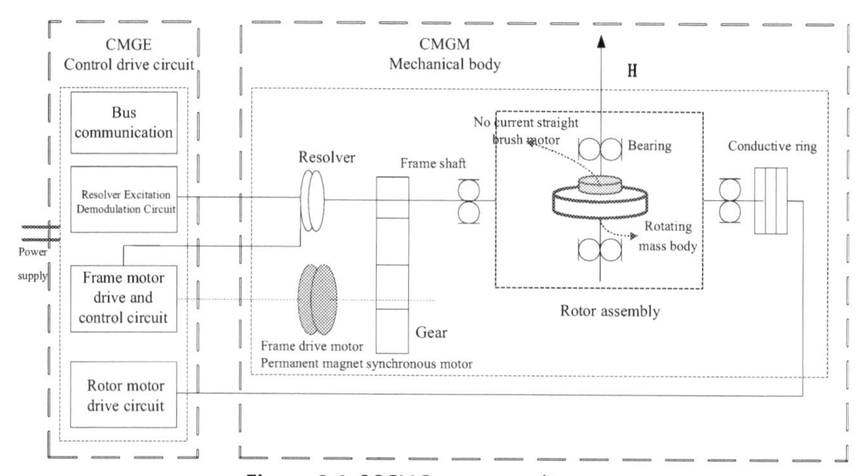

Figure 8.4 SGCMG structure diagram.

under different fault types [34,35]. In Ref. [34], Singleton et al. considered electric discharged current as a label of state transform from health states to accelerated aging states, which indicates the start for RUL prediction. Though in Ref. [36], vibration methods performed better than current methods, Tomasz et al. drew conclusion that stator current is worth of further research and development for bearing health monitoring. Inspired by those works, in our case, current signals are firstly considered as foundational signals for RUL prediction of bearing.

8.3.1.4 Construction of virtual control moment gyro

In this section, the digital twin model is constructed using of the CMG system to realize bearing RUL prediction. Fig. 8.5 describes the data flow of bearing RUL prediction in DT model. As proved in previous works, only one feature used for prediction brings low accuracy. Multiple-feature vectors as input vector increase more information and more robustness. Features are extracted from recovered historical data. Performance evaluation criterion (PEC) corresponding to each feature based on CMG digital model is calculated and ranked from high to low. Then features are selected given specific PEC threshold and number threshold of selected features. Finally, selected features are composed into one vector and a two-dimensional matrix is composed with multitime serials. The matrix is feed into CRNN model as input and RUL label is estimated as output.

Feature extraction and selection

In general, raw current data cannot be used directly for RUL prediction. Hidden information comes from extracted features. Time domain,

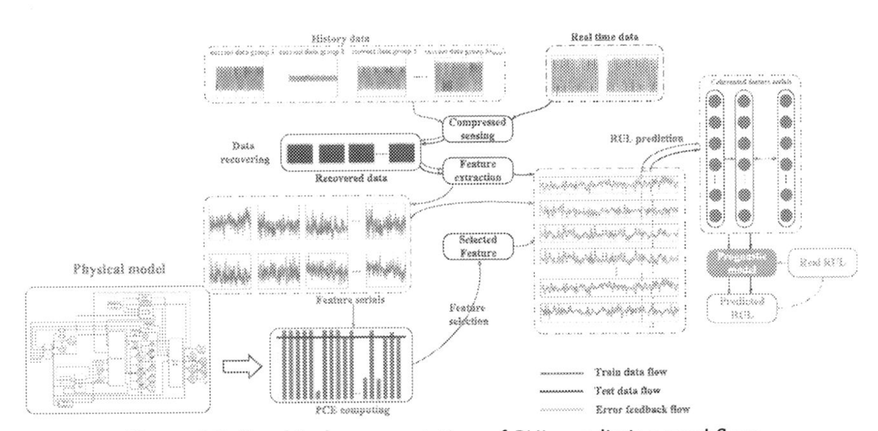

Figure 8.5 Graphical representation of RUL prediction workflow.

frequency domain, and time-frequency domain are three types of feature extraction. To describe current signals from temporal sight, typical features are extracted from time domain. Because of spectral changing as bearings aging, frequential features should be considered. Moreover, energy domain can reflect friction coefficient increasing. Thus, energy from fast Fourier transform (FFT) and wavelet packet decomposition is considered.

Any features could be selected as HI, as long as the demand of high prediction accuracy is reached. However, fixed feature combination-based prior knowledge decreases prediction accuracy. This combination could be changed as different bearing targets and various working conditions, which are reflected by collected data. The selection should be adaptive with various bearings and datasets. Based on the principle that remained useful life has solid but end-to-end relationship with HI, an autoregressive model (AR) and backpropagation neutron network is used for feature selection.

A basic workflow is illustrated in Fig. 8.6. Firstly, stator current data are collected after data recovery, and after denoising using discrete wavelet transform, candidate features are extracted from time serials. Various feature trajectories are obtained. In this paper, the model parameters are determined using the Yule-Walker methods [37]. Before that, to optimize the size of the order of AR model, Akaike information criteria is selected as a metric to represent performance of parameter evaluation.

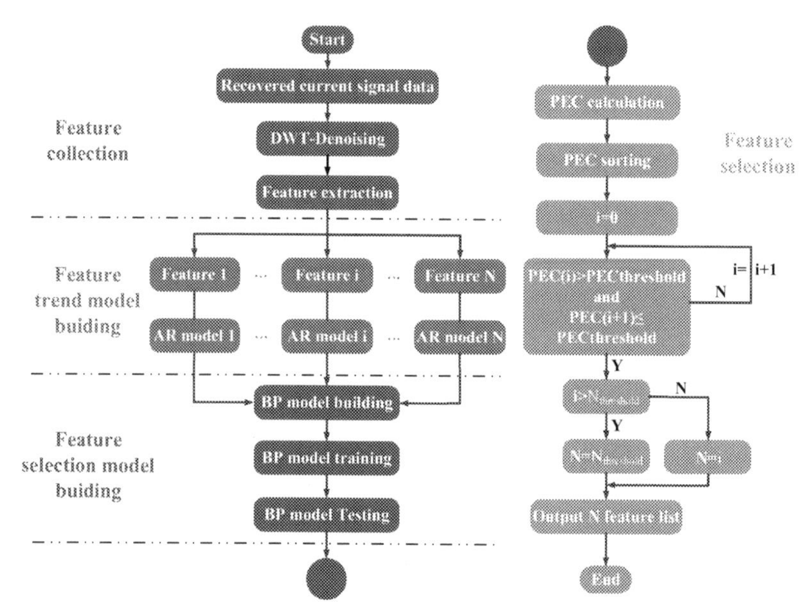

Figure 8.6 Feature selection workflow.

Features beyond threshold are selected. Then, to avoid large number of features slowing down the speed of computing, if number of selected features are beyond the specific value, top features are selected. Before experiments, threshold of PEC and upper limit number of features is decided according to grid research.

RUL calculation via CRNN

Considering temporal relationship in stator current data, recurrent neutral network (RNN) is used as the RUL prediction model, whose basic schedule is depicted in Fig. 8.7A.

Under the foundation of RNN unit, extended model has been researched in recent decades. Long—short-term model is widely used in natural language processing, speech recognition, and fault detection. As simplification, Gate Recurrent Unit (GRU) is easier to use with less limitation of initial hidden state, and more suitable for RUL prediction because time point is random to state prediction. A GRU cell is shown in Fig. 8.7B. RNN model is composed with same cells for historical information storing and hidden state transformation. Computing schedule of GRU could be described as

$$y_t = g(V \cdot h_t) \tag{8.1}$$

$$h_t = f(U \cdot X_t + W \cdot h_{t-1}) \tag{8.2}$$

where V, U, W represent weight matrix of GRU, h_t means the hidden state at time slice t, and y_t means the output vector at the time slice t. X_t represents the input vector in time slice t, whose computing flow could be described as

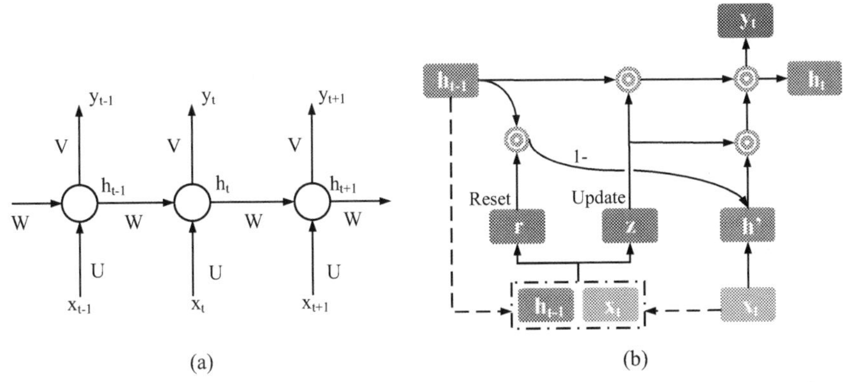

Figure 8.7 Basic structure of RNN model (A) Typical RNN schedule (B) GRU cell.

$$r_t = \sigma(U_r x_t + W_r h_{t-1} + b_r) \tag{8.3}$$

$$z_t = \sigma(U_z x_t + W_z h_{t-1} + b_z) \tag{8.4}$$

$$h_t = z_t \odot h_{t-1} + (1 - z_t) \odot \tanh(U_c x_t + r_{t\odot(W_c h_{t-1})} + b_c) \tag{8.5}$$

where r_t is reset gate and z_t is forgiven gate, and σ is sigmoid function as activation function.

Considering RUL information hidden in multiple features, a CNN model is connected before GRU model, which is shown in Fig. 8.8. When feature vectors are fed into RUL prediction model, a conventional operator extracts hidden information into a low dimension vector, which is regarded as the input of GRU at a single time slice. And two linear layers are used before RUL label for dimension compression.

8.3.1.5 Data recovery via compressed sensing

Considering Shannon sampling principle, current data collected from spacecraft are difficult for traditional frequency analysis under low-frequency collection. The compressed sensing technology was adopted to transform actual data from dense domain to sparse domain. The length of raw data is enlarged and missing data are estimated.

Generally, typical relationship between dense domain and sparse domain data could be described as

$$y = \Phi x \tag{8.6}$$

where y represents obtained current data in orbit, x means recovered data before RUL prediction and Φ is an operator that transforms x to y.

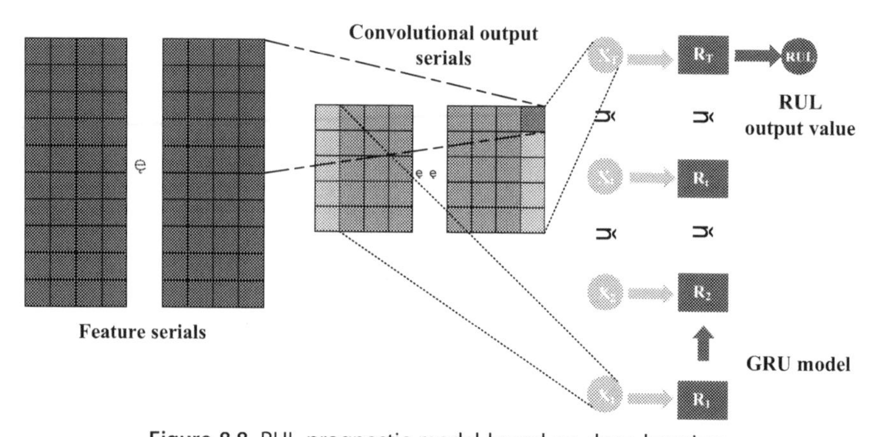

Figure 8.8 RUL prognostic model based on deep learning.

Mathematically, x could be decomposed into ψ, which is called sparse framework that contains several basic vectors, and s, which is sparse elements that connects sparse frameworks.

$$y = \Phi\psi s = Au \tag{8.7}$$

where A is called sensing matrix and u is called sparse coefficient in compressed sensing theory. To recover current data from low-frequency downlink, the problem can be represented as

$$u_{opt} = \min_u \mu_1 \|u\|_1 + \frac{1}{2}\|Au - f\|_2^2 \tag{8.8}$$

where f represents obtained signal and u_{opt} is regarded as expected current data serial. $\|\cdot\|_1$ means the function of one-norm. Specially, for sparer matrix, u_{opt} also can be described as

$$u_{opt} = \min_u \mu_0 \|u\|_0 + \mu_1 \|u\|_1 + \frac{1}{2}\|Au - f\|_2^2 \tag{8.9}$$

where $\|\cdot\|_0$ means L_0 normalization while $\|\cdot\|_1$ represents L_1 normalization. μ_0 and μ_1 are specific hyperparameters.

Considering L_0, L_1 and L_2 normalization, when u is sparse, L_0 is equivalent to L_1. Adding L_2, linear regression is used to estimate dictionary as

$$u = (A^T A + I)'(A^T y - \lambda) \tag{8.10}$$

$$u^{k+1} \leftarrow shrink\left(u^k, \max\left(u^k - {}^1/_\lambda\right)\right) \tag{8.11}$$

where $shrink\left(u^k, \max\left(u^k - {}^1/_\lambda, 0\right)\right)$

$$= sign(u^k)\max\left\{\left|u^k\right| - \max\left(u^k - {}^1/_\lambda, 0\right)\right\}$$

$$= \begin{cases} u^k - \max\left(u^k - {}^1/_\lambda, 0\right), u^k \in \left(\max\left(u^k - {}^1/_\lambda, 0\right), +\infty\right) \\ 0, u^k \in \left(-\max\left(u^k - {}^1/_\lambda, 0\right), +\max\left(u^k - {}^1/_\lambda, 0\right)\right) \\ u^k + \max\left(u^k - {}^1/_\lambda, 0\right), u^k \in \left(-\infty, -\max\left(u^k - {}^1/_\lambda, 0\right)\right) \end{cases} \tag{8.12}$$

u^k and u^{k+1} represent estimated current data at the kth and $(k+1)$th iteration. λ is specific as the hyperparameter.

In order to get recovered current data serials, analysis operator could be set priorly. Under the situation of low-frequency downlink in spacecraft, analysis operator is set as a matrix $\mathbf{\Phi}$ in which nonzero elements are assigned at regular intervals [38], which could be described as

$$\mathbf{\Phi} = \begin{pmatrix} 1 & \dots & 0 & 0 & \dots & 0 \\ 0 & \dots & 1 & 0 & \dots & 0 \\ 0 & 0 & \dots & 1 & \dots & 0 \end{pmatrix} \tag{8.13}$$

Then original data serials are collected reversely. Inspired by Ref. [39], Parseval's law is considered and discrete cosine transform, shown in Eq. (8.9), is utilized as sparse framework. SBIL1 [40] algorithm is used to recover signals.

$$F = \begin{pmatrix} \frac{1}{\sqrt{N}} & \frac{1}{\sqrt{N}} & \frac{1}{\sqrt{N}} & \dots & \frac{1}{\sqrt{N}} \\ \sqrt{\frac{2}{N}} & \sqrt{\frac{2}{N}} & \sqrt{\frac{2}{N}} & \dots & \sqrt{\frac{2}{N}} \\ \sqrt{\frac{2}{N}}\cos\left(\frac{2\pi}{2N}\right) & \sqrt{\frac{2}{N}}\cos\left(\frac{3\times 2\pi}{2N}\right) & \sqrt{\frac{2}{N}}\cos\left(\frac{5\times 2\pi}{2N}\right) & \dots & \sqrt{\frac{2}{N}}\cos\left(\frac{(2N-1)\times 2\pi}{2N}\right) \\ \vdots & \vdots & \vdots & \ddots & \vdots \\ \sqrt{\frac{2}{N}}\cos\left(\frac{(N-1)\pi}{2N}\right) & \sqrt{\frac{2}{N}}\cos\left(\frac{3(N-1)\pi}{2N}\right) & \sqrt{\frac{2}{N}}\cos\left(\frac{5(N-1)\pi}{2N}\right) & \dots & \sqrt{\frac{2}{N}}\cos\left(\frac{(2N-1)(N-1)\pi}{2N}\right) \end{pmatrix} \tag{8.14}$$

where N represents the length of sparse coefficient vector.

It is noteworthy that the sparse frame is designed as a tall matrix [39]. It brings a problem that the frame matrix is not an overcomplete matrix, and L_1 normalization is considered due to underfitting. A sparse dictionary would be obtained to reflect frequency distribution.

8.3.1.6 Real bearing for RUL prediction services

(1) Dataset description: To validate the effectiveness in aircrafts, data collected from CMG were tested. A typical structure of CMG is shown in Fig. 8.4. As the key component that supports gimbal-driving motors and rotor, bearings have a significant effect on stability and maintenance of CMG.

In this case, a life test of CMG was performed. Due to bandwidth constraints of the antennas, the spacecraft, especially satellites, downlink sensor data, or telemetry through an S-Band Antenna at a rate of 8–10 Hz to stations. In practical application, there are more than 10,000 parameters in a subsystem being constantly downlinked at this rate. In order to simulate the situation of low frequency, the collected data of case were limited at 8 Hz and were measured crossed about 2 years in a row. At the end of aging test, the bearing reached its life and the experiment stopped. One day was selected as duration of a cycle triggering the RUL prediction.

Previous work about RUL prediction of satellites has been done and published in Ref. [41], and due to absence of vibration sensors in predesign, stator current signals were selected as raw sensorial data. The test bench in case is shown in 13. A CMG was used for life test and the test equipment was used to receive sensor data and send control instructions. In CMG of satellites, stator current signals come from Brushless DC motor and Hall current sensors are installed to collect current signals. In this experiment, high-speed motor current is used for RUL prediction. In this case, data collected in the last 100 days were divided into 10 groups. Based on the principle of hold-out, these 10 data groups were randomly arranged into seven trained datasets and three test datasets. For convenient of description, trained datasets were labeled with data group 1 ~ data group 7, while tested dataset are labeled with data group 8, data group 9, and data group 10. The RUL prediction was performed at the beginning of trained datasets and test datasets.

(2) Experimental result: In design of CMG, due to consideration of cost and other specific industrial demand, vibration signals are difficult to access. It is critical to replace vibration with suitable signals. As mentioned above, current signal has similar feature with vibration signals. Thus, current signals are chosen as observed data. In this case, the threshold of PEC is specific as 0.95 and the upper limit number of selected features is specific as 10. All features extraction from stator current signals are showed from 14 to 16. According to PCE result depicted in Fig. 8.9, 10 features are selected as HI, which are RMS, Peak value, Shape factor, Variance, Crest factor, Impulse factor, Kurtosis, Line integral, Shannon entropy, Energy of FFT. The specific value and structure of selected features extraction are listed in Table 8.1.

The prediction result of proposed model is depicted in Fig. 8.10, and the error of prediction is shown in Fig. 8.11. MSE and NRMSE reflect that the RUL prediction accuracy of proposed approach is better than other methods, such as DCNN [42], RNNAE [43], and RCNN [44]. MAE reflects that the stability of RUL prediction accuracy is better than other methods. From the case, it could be learned that RNN-based model performs better than CNN-based model. It is reasonable because temporal

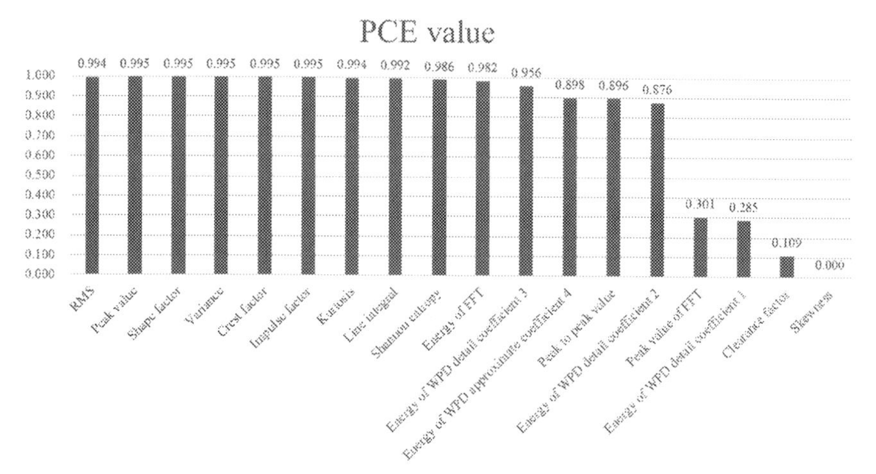

Figure 8.9 PEC in this case.

relationships are considered into deep learning models. In DT prediction framework, CS-CRNN model mines the information from down-sampled data and push them into deep learning model, which increases the accuracy of RUL prediction.

8.3.2 Example of lithium-ion battery

8.3.2.1 Problem description

Lithium-ion battery perhaps is the most popular electrical power source used in products which has strict weight requirements, such as cell phones, laptops, and small unmanned aerial vehicles (UAVs). It owns many merits such as high energy/power density, long lifetime, and environmental friendliness [45]. However, when it runs out of power without immediate recharge, the consequence might be troublesome (e.g., cell phones and laptops) or disastrous (e.g., UAVs).

Since the advancement of lithium-ion batteries is still in an early stage, their capacity is very limited in a small volume. For example, a 1.5 kg quadrotor can only fly up to 30 min (depending on vehicle, payload, and wind conditions) with a 5200 mAh lithium-ion battery [46]. Thus, running out of power caused by improper energy management is very likely to happen for UAVs in long missions (such as reconnaissance, photography, plant protection, search, and rescue, etc.), and the resulting mission failure or equipment damage is the operators really want to avoid. To ensure energy safety of UAV flight missions, it is necessary to consistently estimate the battery state and predict its end of discharge (EOD) time during each mission.

Table 8.1 Features extracted from denoising current signals.

Feature name	RMS	Peak value	Shape factor	Variance	Crest factor
Feature type	Time domain features				
Mathematical represent	$\sqrt{\frac{1}{n}\sum_{i=1}^{n} x_i^2}$	$\max(\lvert x_i \rvert)$	$\frac{RMS}{\frac{1}{n}\sum_{i=1}^{n}\sqrt{\lvert x_i \rvert}}$	$\frac{1}{n}\sum_{i=1}^{n}\left(x_i - \widetilde{x}\right)^2$	$\frac{Pv}{RMS}$
Feature trajectory		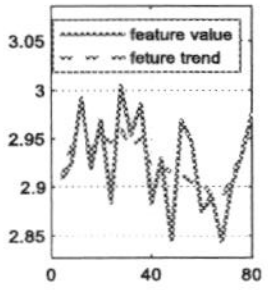	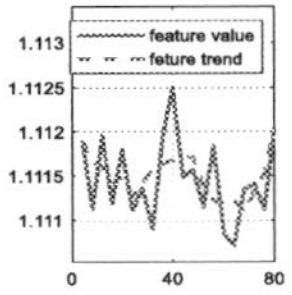	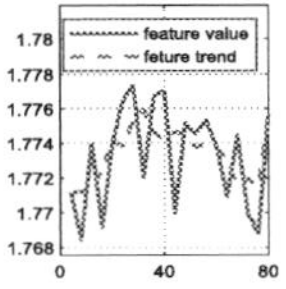	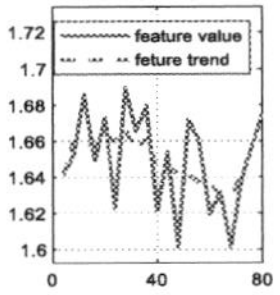
Feature name	Impulse factor	Kurtosis	Line integral	Shannon entropy	Energy of FFT
Feature type	Time domain features				Frequency domain
Mathematical represent	$\frac{Pv}{\frac{1}{n}\sum_{i=1}^{n}\sqrt{\lvert x_i \rvert}}$	$\frac{\sum_{i=1}^{n}\left(x_i - \widetilde{x}\right)^4}{n \times var} - 3$	$\sum_{i=0}^{n}\lvert x_{i+1} - x_i \rvert$	$-\sum_{i=1}^{n} x_i^2 \log\left(x_i^2\right)$	$En = \sum_{k=1}^{N} r_k$
Feature trajectory	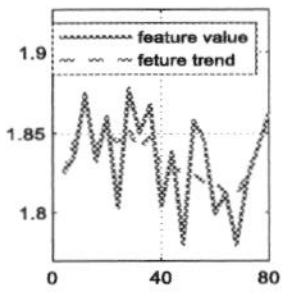	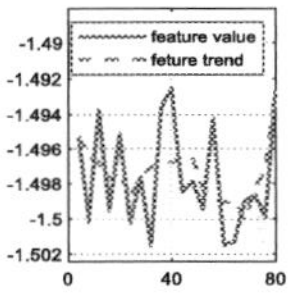	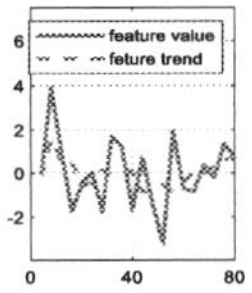	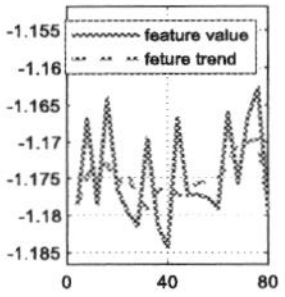	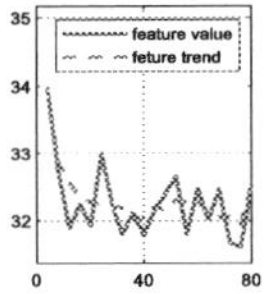

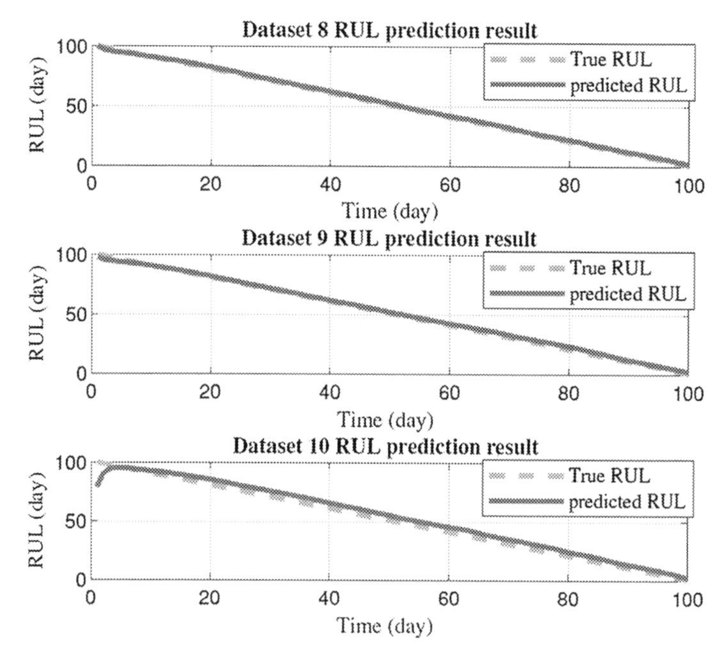

Figure 8.10 RUL prediction result for this case.

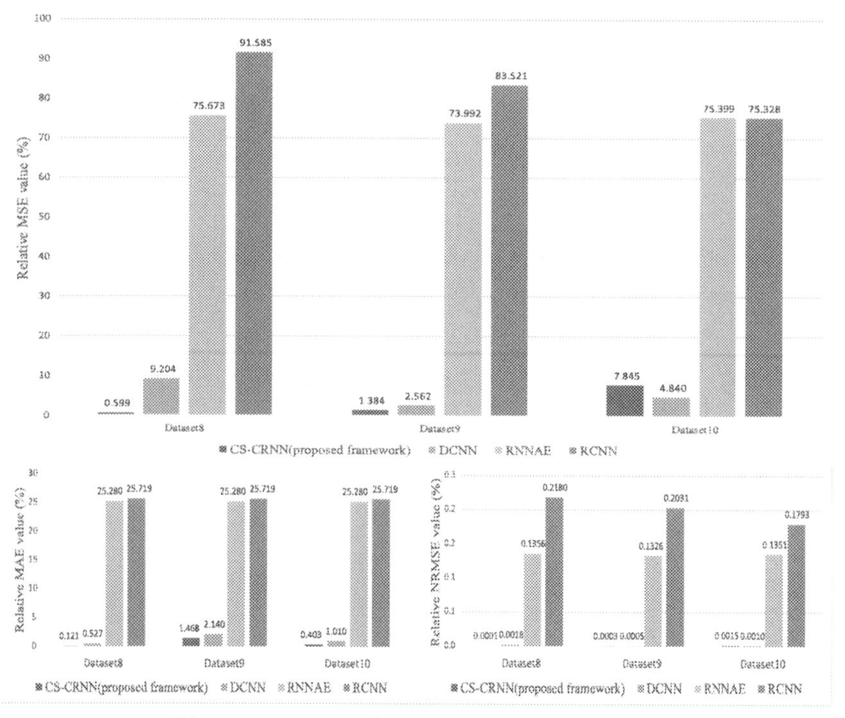

Figure 8.11 Error of RUL prediction in this case.

8.3.2.2 The framework of DT-driven solution

In this case study, a DT-driven framework of small-size rotatory-wing UAVs is proposed to illustrate the application of digital twin in battery prognostics. Its structure is shown in Fig. 8.13. In this framework, the lithium-ion battery and rotatory-wing UAVs are regarded as the physical equipment, correspondingly, the battery model, and UAVs power consumption model are the virtual equipment. Based on the measurements of physical equipment, the established models can be used to achieve the battery state estimation and EOD time prediction. The obtained results can be further used for future task planning and risk assessment of UAVs. This indicates the interaction between physical equipment and virtual equipment (Fig. 8.12).

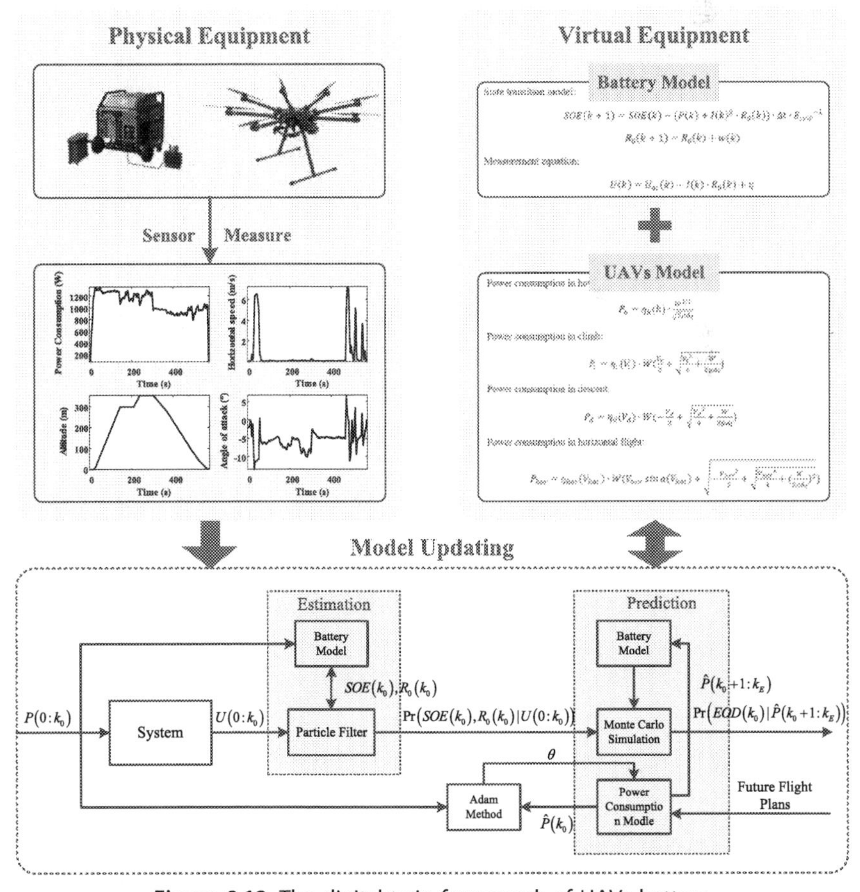

Figure 8.12 The digital twin framework of UAVs battery.

8.3.2.3 Physical analysis of battery health status

State of Energy (SOE), which is defined as the normalized remaining available energy of lithium-ion batteries, is a critical evaluation index for energy optimization and management of the battery system. Generally, EOD time can be considered as the time point that SOE reaches a certain threshold [47]. Therefore, the estimation of SOE is of great significant for the EOD time prediction of UAVs battery. Among the solutions of SOE estimation, Bayesian-based method has been widely studied because it is able to estimate SOE in real time for dynamic load and assess the estimation uncertainty using a Probability Density Function, which is expected in real applications. In addition, the EOD prediction is also affected by the future operational conditions of UAVs, if the future operational conditions can be predicted, combining them in the EOD prediction model ought to be of great help for prediction accuracy. In order to simulate the consumption process of battery and predict the dynamic load determined by the future operational conditions of UAVs, the digital twin models need to be established.

8.3.2.4 Construction of virtual battery

In the proposed DT framework, a Rint-based equivalent circuit model of battery [48] is first established to link unobservable SOE and measurable voltage and current, which is defined as Eqs. (8.1)−(8.5).

State transition model:

$$\mathrm{SOE}(k+1) = \mathrm{SOE}(k) - \left(P(k) + I(k)^2 \cdot R_0(k)\right) \cdot \Delta t \cdot E_{crit}^{-1} \tag{8.15}$$

$$R_0(k+1) = R_0(k) + w(k) \tag{8.16}$$

Measurement equation:

$$U(k) = U_{oc}(k) - I(k) \cdot R_0(k) + \eta \tag{8.17}$$

where:

$$U_{oc}(k) = v_L + \lambda \cdot e^{-\gamma \cdot \mathrm{SOE}(k)} \tag{8.18}$$

$$I(k) = \frac{U_{oc}(k) - \sqrt{U_{oc}(k)^2 - 4 \cdot R_0(k) \cdot P(k)}}{2 \cdot R_0(k)} \tag{8.19}$$

where the sampling interval Δt is a constant value; the power consumption $P(k)$ in Eq. (8.15) is the only input in current time epoch, which will be calculated using a power consumption model in the prediction process;

and the voltage $U(k)$ in Eq. (8.17) is the only observation in the battery model; the open circuit voltage (OCV) $U_{oc}(k)$, and the current $I(k)$ are the intermediate variables; The normalized remaining battery energy SOE(k) in Eq. (8.15) and battery internal resistance $R_0(k)$ in Eq. (8.16) are the states; $w(k)$ and η are defined as process noise and measurement noise, respectively, and both obey Gaussian distribution.

The future load of battery is the key to predict EOD time. By establishing the power consumption model of UAV under different flight states, the future load can be estimated. Based on helicopter aerodynamic theory [49], an approximate power consumption model of rotatory-wing UAVs is established as Eqs. (8.20)–(8.26).

Power consumption in hovering:

$$P_h = \eta_h(h) \cdot \frac{W^{3/2}}{\sqrt{2\rho A_t}} \tag{8.20}$$

Power consumption in climb:

$$P_c = \eta_c(V_c) \cdot W\left(\frac{V_c}{2} + \sqrt{\frac{V_c^2}{4} + \frac{W}{2\rho A_t}}\right) \tag{8.21}$$

Power consumption in descent:

$$P_d = \eta_d(V_d) \cdot W\left(-\frac{V_d}{2} + \sqrt{\frac{V_d^2}{4} + \frac{W}{2\rho A_t}}\right) \tag{8.22}$$

Power consumption in horizontal flight:

$$P_{hor} = \eta_{hor}(V_{hor}) \cdot W\left(V_{hor}\sin\alpha(V_{hor}) + \sqrt{-\frac{V_{hor}^2}{2} + \sqrt{\frac{V_{hor}^4}{4} + \left(\frac{W}{2\rho A_t}\right)^2}}\right) \tag{8.23}$$

Power consumption in oblique upward flight:

$$P_{horc} = \eta_{horc}(V'_{hor}) \cdot \left[W(\frac{V'_c}{2} + \sqrt{\frac{V'^2_c}{4} + \frac{W}{2\rho A_t}} + W(V'_{hor}\sin\alpha_{horc}(V'_{hor}) + v)\right] \tag{8.24}$$

Power consumption in oblique downward flight:

$$P_{hord} = \eta_{hord}(V'_{hor}) \cdot \left[W(-\frac{V'_d}{2} + \sqrt{\frac{V'^2_d}{4} + \frac{W}{2\rho A_t}} + W(V'_{hor}\sin\alpha_{hord}(V'_{hor}) + v)\right] \tag{8.25}$$

where the induced velocity in horizontal flight:

$$v = \sqrt{-\frac{V'^2_{hor}}{2} + \sqrt{\frac{V'^4_{hor}}{4} + \left(\frac{W}{2\rho A_t}\right)^2}} \tag{8.26}$$

where ρ is the air density, W is the total weight of the UAV, A_t is the rotor area, h is the flying altitude, V_c is the climbing speed, V_d is the descent speed, V_{hor} is the horizontal flight speed, $\alpha(V_{hor})$ is the angle of attack. $\eta_h(h)$, $\eta_c(V_c)$, $\eta_d(V_d)$ and $\eta_{hor}(V_{hor})$ are empirical coefficients. When calculating the power consumption using Eqs. (8.24)–(8.26), the velocity of the UAV needs to be decomposed first. V'_{hor} is the horizontal flight component, V'_c is the climb component and V'_d is the descent component. $\alpha(V_{hor})$, $\eta_{horc}(V'_{hor})$ and $\eta_{horc}(V'_{hor})$ are empirical functions of V'_{hor}.

Finally, by analyzing the real flight data, the empirical models of α and η are established (Eq. 8.25). In the equation, $a_0, a_2, a_3, b_0, b_1, b_2, c_0, c_1, c_2, d_0, d_1, d_2, e_0, e_1, e_2, f_0, f_1, f_2, g_0, g_1, g_2$ are empirical coefficient, which can be obtained using least square fitting.

$$\begin{gathered}
\eta_h(h) = a_0 + a_1\cos(a_2 \cdot h) + a_3\sin(a_2 \cdot h),\\
\eta_c(V_c) = b_0e^{-b_1 \cdot V_c} + b_2,\ \eta_d(V_d) = c_0e^{-c_1 \cdot V_d} + c_2,\\
\alpha_{horc}\left(V'_{hor}\right) = d_0e^{-d_1 \cdot V'_{hor}} + d_2,\\
\alpha_{hord}\left(V'_{hor}\right) = e_0e^{-e_1 \cdot V'_{hor}} + e_2,\\
\eta_{horc}\left(V'_{hor}\right) = f_0e^{-f_1 \cdot V'_{hor}} + f_2,\\
\eta_{hord}\left(V'_{hor}\right) = g_0e^{-\left(\frac{V'_{hor} - g_1}{g_2}\right)^2}.
\end{gathered} \tag{8.27}$$

8.3.2.5 Realization of battery health estimation and prediction services

Based on the models established in Section 8.3.2.4, a prognosis method for EOD time of UAV in flight was developed. At current time $k_0\Delta t$, the following two processes are conducted in sequence to perform prediction: (i) The estimation process, which uses the measured voltage from physical equipment $U(0:k_0)$ and state-space model (virtual equipment) developed in Section 8.3.2.4 to estimate the posterior probability distribution $Pr(\mathrm{SOE}(k_0), R_0(k_0)|U(0:k_0))$ at current moment $k_0\Delta t$; (ii) The prediction process, which predicts EOD distribution $Pr\left(EOD(k_0)\middle|\widehat{P}(k_0+1:k_E)\right)$

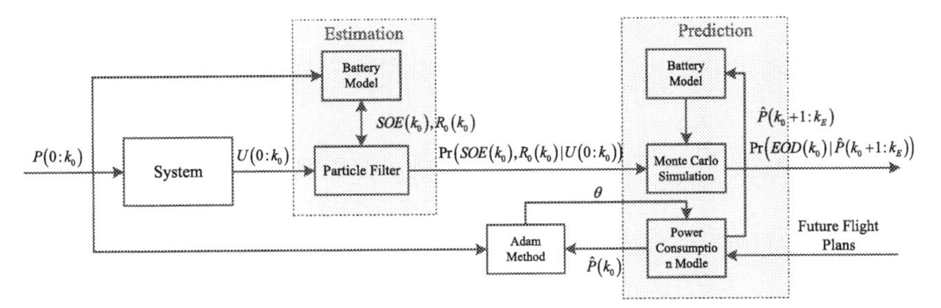

Figure 8.13 Framework of SOE estimation and EOD time prediction.

using $Pr(\mathrm{SOE}(k_0), R_0(k_0)|U(0:k_0))$ and the prediction results of future power consumption of UAV $\widehat{P}(k_0+1:k_E)$. Note that $\widehat{P}(k_0+1:k_E)$ is obtained by the UAV future flight plan and the power consumption model (virtual equipment) developed in Section 8.3.2.4, and the EOD time is determined by $\mathrm{SOE}(k)$ reaching a predetermined threshold.

The framework of proposed method is shown in Fig. 8.13. Particle Filter (PF) was used for the estimation process and those particles obtained by PF are used in a Monte Carlo simulation for prediction purpose. The posterior distribution of battery states can be updated using the measured voltage, and the Adam optimization algorithm was introduced to modify the power consumption model by the measured power consumption, which reflect the information exchange between physical equipment and virtual equipment. Considering the differences in working conditions and external environment when UAV performs different tasks, the offline training results of hyperparameters in the power consumption model (virtual equipment) may not be accurate enough. Therefore, an online update method is proposed to modify the model hyperparameters based on the Adam optimization algorithm. The measured power consumption P is the input of the Adam algorithm. The updated model hyperparameters, which used for the prediction of EOD time, are the output. The process of this method is shown in Table 8.2.

8.3.2.6 Experimental results using real flight data

In this section, the feasibility and effectiveness of the prediction method is validated using the measured data provided by Zhongshan Hankun Intelligent Technology. Six flight experiments are conducted by DJI S1000, and its parameters are summarized in Table 8.3. Among the collected data sets, five sets are selected for model training and one for method verification. The flight data recorded in each flight plan include battery output

Table 8.2 Parameters update by Adam algorithm.

Input: α: Stepsize, $\beta_1, \beta_2 \in [0, 1)$: Exponential decay rates for the moment estimates, $L(\theta)$: Loss function with parameters θ, θ_0: Initial parameter vector, P: Measured power consumption
Output: θ_t: Resulting parameters

Initialize: $m_0 \leftarrow 0,\ v_0 \leftarrow 0,\ t \leftarrow 0$
while new P comes **do**
$t \leftarrow t + 1, g_t \leftarrow \nabla_\theta L(\theta_{t-1})$
$m_t \leftarrow \beta_1 \cdot m_{t-1} + (1 - \beta_1) \cdot g_t,\ \widehat{m}_t \leftarrow m_t / (1 - \beta_1^t)$
$v_t \leftarrow \beta_2 \cdot v_{t-1} + (1 - \beta_2) \cdot g_t^2,\ \widehat{v}_t \leftarrow v_t / (1 - \beta_2^t)$
$\theta_t \leftarrow \theta_{t-1} - \alpha \cdot \widehat{m}_t / (\sqrt{\widehat{v}_t} + \varepsilon)$
end while
In the algorithm, the settings are $\alpha = 2 \times 10^{-4}$, $\beta_1 = 0.9$, $\beta_2 = 0.999$, $\varepsilon = 10^{-8}$ and $L(\theta) = \left(P - \widehat{P}(\eta(\theta))\right)^2$.

Table 8.3 Parameters of DJI S1000.

Parameter	Symbol	Value
Total disc actuator area (m^2)	A_t	0.912
Total weight (N)	W	96
Air density (g/L)	ρ	1.3

voltage U, current I, flying altitude h, horizontal flight speed V_{hor}, and angle of attack α. The power consumption P can be calculated by $U \cdot I$ and the vertical speed V_c and V_d can be obtained by the derivative of h. An example of the raw flight data of the physical equipment is shown in Fig. 8.14.

The experiments are conducted in the following steps: (1) estimate the parameters of the battery state-space model (Eqs. 8.15–8.19); (2) estimate the parameters of the power consumption model (Eqs. 8.20–8.27) and validate the effectiveness of the Adam-based online update method; (3) predict the EOD time based on the framework shown in Fig. 8.13 and analyze prediction performance.

Firstly, the measured voltage U and current I were used to calculate the OCV of each training set, thus the relationship between U_{oc} and SOE can be obtained.

In the second step, the parameters of the power consumption model of UAV (Eqs. 8.20–8.27) are estimated. The verification results of the offline model are shown in Fig. 8.15A. According to the results, the power consumption calculated using the offline model has the same trend with the

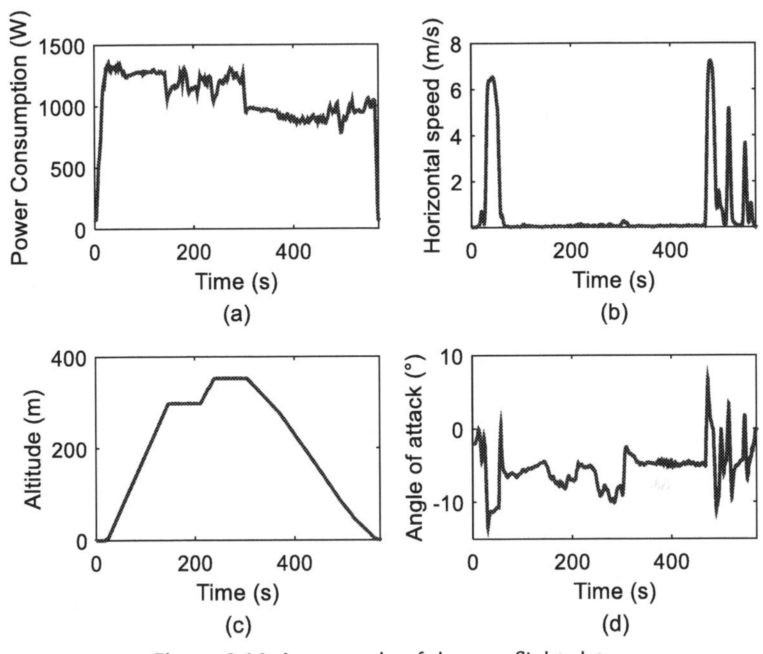

Figure 8.14 An example of the raw flight data.

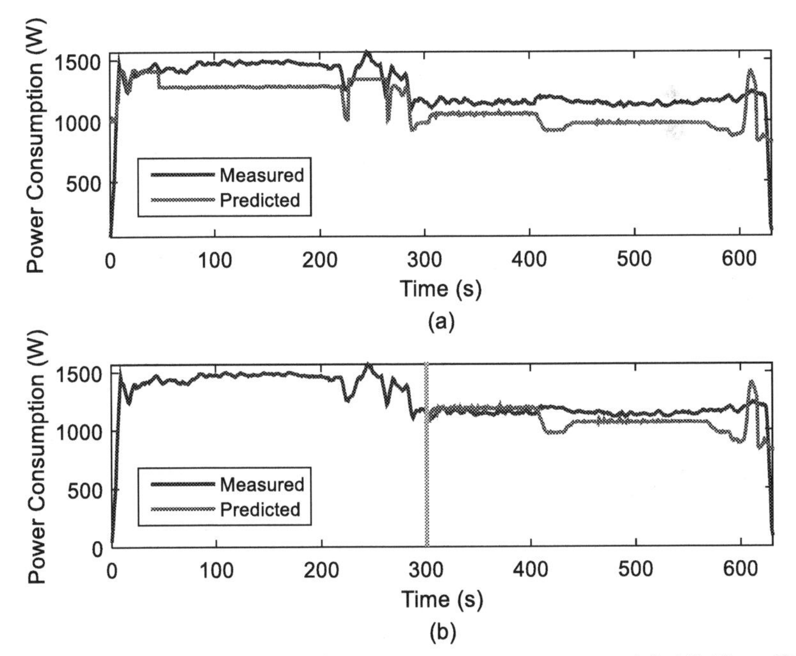

Figure 8.15 Verification results of the power consumption model. (A) The offline model (B) The online update results.

actual value. However, it is not accurate enough to ensure prediction accuracy. Fig. 8.15B shows the prediction results modified by the Adam-based online update method using consumed power observations before 300s. Comparing with Fig. 8.15A, it can be found that the online update method can reduce the prediction errors effectively.

Finally, to prove the validity of the EOD prediction method, the SOE threshold is set to 0.65, and the true value of EOD time is 465 s. Fig. 8.16 shows the prediction results of EOD time from 30 to 450 s. With the increasing amount of measured data, the prediction error decreases from 17.90% at 30 s to 1.29% at 450 s, and the prediction of EOD time obtained by PF combined with Adam optimization algorithm is more accurate than using the conventional PF with hyperparameters computed by offline training, which indicates the necessity of real-time update of the power consumption model. In conclusion, the results of experiments show that the established virtual equipment and the information of physical equipment can provide the necessary knowledge for the future task planning and risk assessment of the physical equipment, which is very helpful for task implementation.

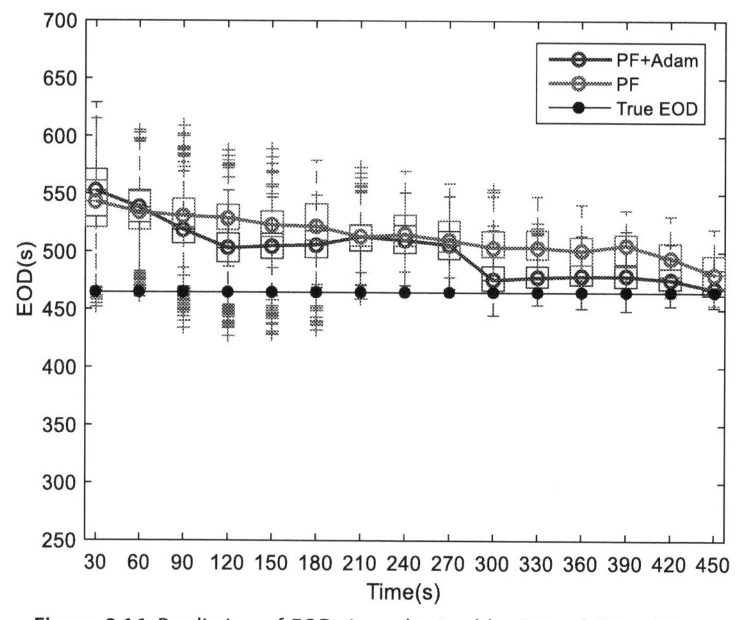

Figure 8.16 Prediction of EOD time obtained by PF and PF + Adam.

8.3.3 Example of infrared imaging system

8.3.3.1 Problem description

System designed to transform the incoming infrared (IR) radiation of targets and backgrounds into electrical signals for further applications such as temperature measurement, object detection, distance measurement, process monitoring, security alarm, and communication are called the IR imaging system [50]. Many IR systems generate IR images, usually referred to as IR imaging systems, are extensively used in real practice. For example, infrared search-and-track (IRST) system is designed to detect and/or track targets in the midst of background clutter and atmospheric effects, while forward looking infrared camera can be used to monitor environment, detect dangerous goods, and diagnose the health status of industrial process and important equipment [51].

As a typical electro-mechanical system, IR system contains many electronic and mechanical components which degrade gradually due to temperature, operation time, vibration, operational conditions, etc., leading to unavoidable system performance degradation. The degraded performance weakens the IR system ability of fulfilling its tasks. For instance, the target which could be discriminated from IR images when IRST system is in good state cannot be discriminated anymore when it has severe degradation [52]. When the IR system is in operation, especially for those implemented on other operation platforms (e.g., UAV, ship, and military vehicles), it is quite difficult for severely degraded system to be maintained or replaced in time. Therefore, performance monitoring of IR system has great significance to advance maintenance decisions before severe system degradation.

8.3.3.2 The framework of DT-driven solution

The digital twin model is constructed using the IR system to realize health monitoring in PHM services. The illustration of the digital twin model of the IR system is shown in Fig. 8.17.

The typical IR system contains optical subsystem, detector subsystem, signal processing subsystem, servo subsystem, and display subsystem. The optical subsystem is to filter and focus the optical signals in the field of view and then transmit the IR radiation of the target to the detector subsystem. The detector subsystem, which is the core subsystem in the IR system, is used to obtain IR images. It is used to sample the output signals of the optical subsystem and transforms the IR radiation into electrical signals using the electro-optic effect (or thermal effect). Then, these electrical

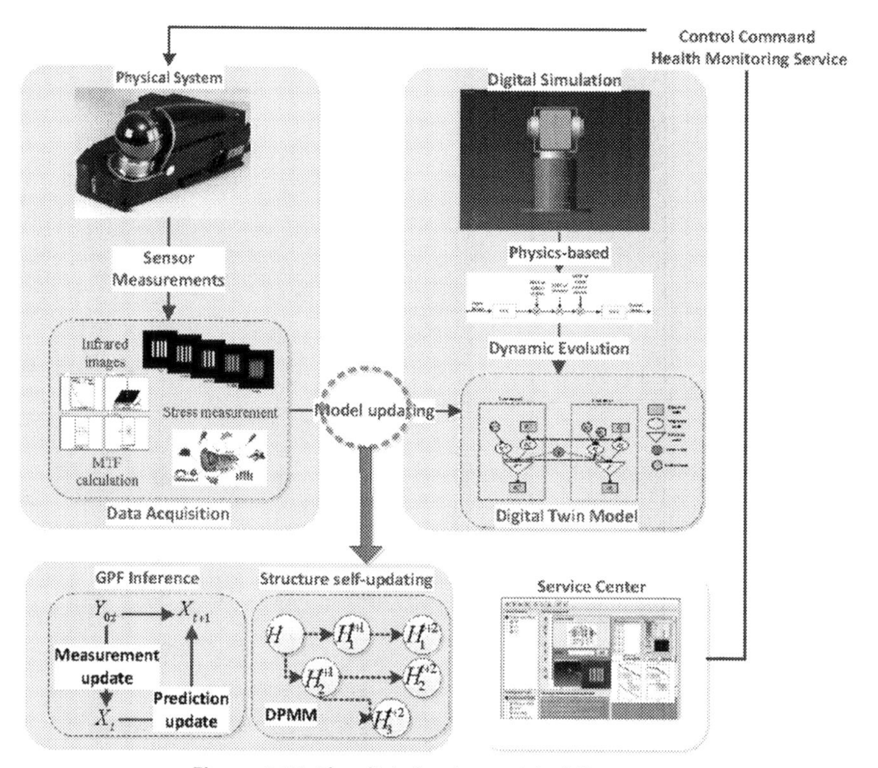

Figure 8.17 The digital twin model of IR system.

signals are processed by the signal processing subsystem in order for further use. The above three subsystems form the basic configuration of an IR system. The servo subsystem and the display subsystem are also commonly used, depending on the demand [53].

The virtual system uses Modulation Transfer Function (MTF) to indicate the health status of IR system, and then the nonparametric Bayesian network is proposed to describe the degradation process. In the process of data synchronization, the physical system uses the model updating strategy and real-time observation data to track and predict the digital twin model, and then the digital twin model returns the health monitoring data to the physical system to facilitate the subsequent CBM. The model updating strategy from the physical system to the digital twin model includes two parts: model inference based on the improved Gaussian Particle Filter (GPF) algorithm and model structure updating based on Dirichlet Process Mixture Model (DPMM).

8.3.3.3 Physical analysis of IR imaging system

The relationships between the components in IR system are shown in Fig. 8.18. The degradation of electronic and mechanical components in the IR system results in the degradation of subsystem performance. For instance, the degradation of sensitive material in the detector subsystem influences the signal transmission efficiency, and the drift of equivalent resistance and capacitance changes the time constant of the signal processing subsystem. In the servo subsystem, degradation in motors causes excessive vibration to introduce abnormal image motion, leading to additional blur in IR images. However, it is impossible and impractical to model all the physical degradations in a complex system. Therefore, MTF is considered to quantify the joint effect of component degradations in each subsystem.

8.3.3.4 Construction of virtual IR imaging system

The IR system can be treated as a spatial filter in the spatial filtering theory, and its transmission characteristic is represented by the optical transfer function (OTF) [54]. The OTF presents that the IR system has ability to deliver spatial energy information, which is given by:

$$H(\xi,\phi) = M(\xi,\phi)exp[jP(\xi,\phi)] \tag{8.28}$$

where $M(\xi,\phi)$ refers to the MTF and $P(\xi,\phi)$ refers to the Phase Transfer Function.

MTF is an important PEC used in the design stage of IR system. From the perspective of the IR system, the input is the original image of the target and its background environment, and the output is the image processed by the entire system. MTF can be obtained by calculating the ratio of the output image spectral amplitude to the input image spectral amplitude. As shown in Fig. 8.19, a series of changed MTF curves can quantitatively describe the degradation of the system imaging quality over time.

To clarify the physical mechanism of the IR system, the internal composition must be considered. The IR system is a complex system consisting of an optical subsystem, a detector subsystem, and a signal process

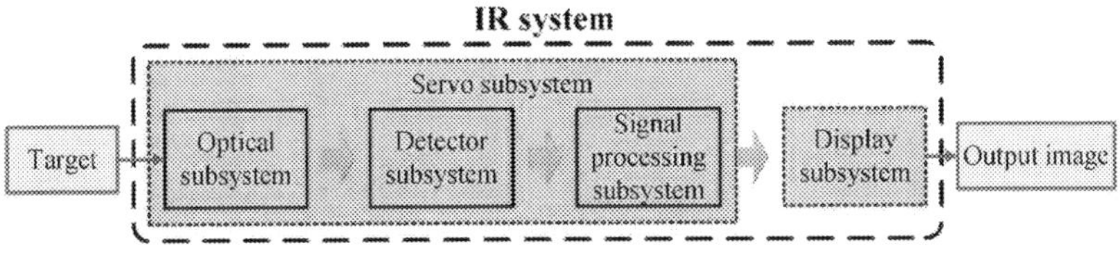

Figure 8.18 The structure and the signal transmission in a typical IR system.

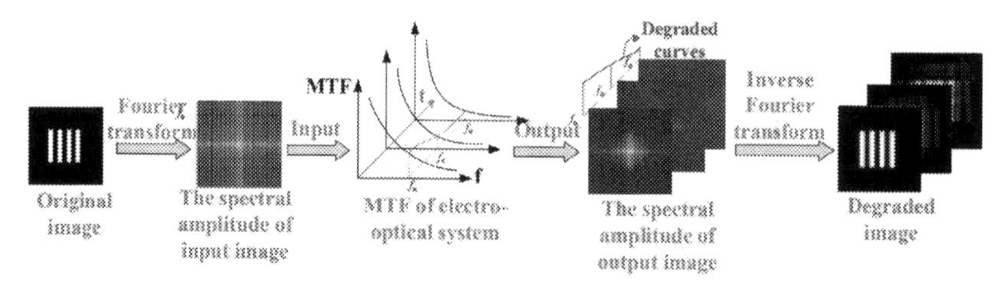

Figure 8.19 The impact of MTF on system imaging quality.

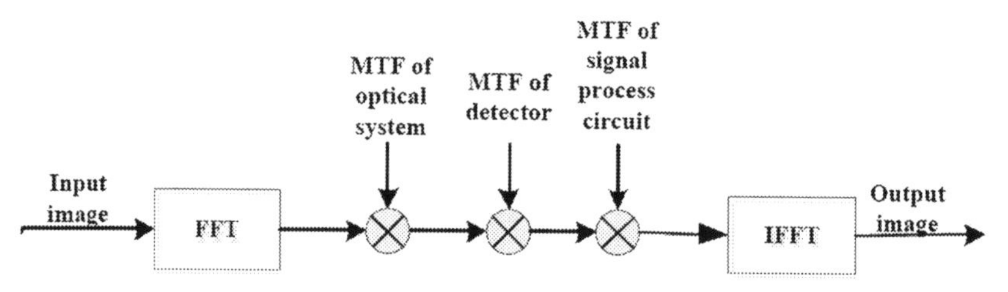

Figure 8.20 The spatial energy transfer of MTF.

subsystem, which interact with each other. Since the IR system can be regarded as linear time invariant system, MTF of the entire system is equal to the arithmetic product of MTFs of each subsystem in the spatial frequency domain, which significantly simplifies the complex coupling between subsystems. The spatial energy transfer process of MTF is shown in Fig. 8.20.

The spatial transfer characteristics of the IR system are mainly affected by diffraction of the optical subsystem MTF_{diff}, and aberration of the optical subsystem MTF_{aber}, the spatial filtering effects of the detector subsystem MTF_{CCD}, and the low-pass filtering effect of the signal process subsystem MTF_{elect}. The MTF of an IR system can be written as the product of all subsystem MTFs. Therefore, system-level MTF can be expressed as:

$$MTF_{system} = MTF_{diff} \cdot MTF_{aber} \cdot MTF_{CCD} \cdot MTF_{elect} \tag{8.29}$$

Then, the nonparametric Bayesian network modeling approach proposed in this Chapter not only can express the propagation of uncertainty between parameters and system states, but also adaptively estimate hidden variables to complete self-learning of the model structure by using real-time data.

To describe system changes in discrete time, the Bayesian network is extended to the dynamic Bayesian network [55], which contains two types

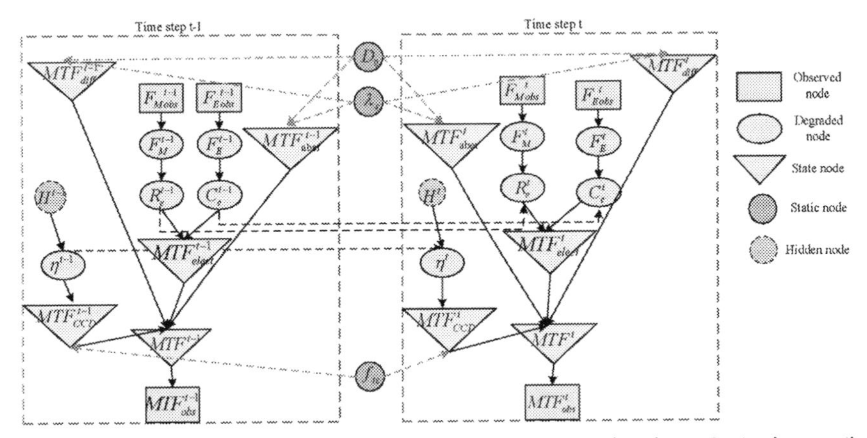

Figure 8.21 The illustrated nonparametric Bayesian network, where D_0 is the pupil diameter, λ_s is the central wavelength of incoherent light, η is the transmission efficiency, f_{tc} is the clock frequency, R_e is the equivalent resistance, C_e is the equivalent capacitor, F_M is mechanical stress, and F_E is electrical overstress.

of edges: (1) those connected nodes within one-time step; (2) those linked nodes across adjacent time steps. According to the physical mechanism, the established nonparametric Bayesian network of the IR system is shown in the figure. The entire dynamic nonparametric Bayesian network is shown as Fig. 8.21.

As shown in Fig. 8.21, the subscript t-1 or t denotes the time step. The solid arrows are used within a time step when the dashed arrows connect the nodes across two different time steps. The dotted node represents a hidden variable which probability distribution function is unknown. Moreover, the number of hidden variables is not fixed at different time steps and will change adaptively with the real-time data. The elliptical node is a degraded node, denoting the variable parameters which are stochastic for given values of parent nodes; the arrows toward it represent a conditional probability distribution. The rounded node is a static node, denoting the deterministic but probably unknown parameters which bring epistemic uncertainty. The triangular node is a system state node, indicating the health status of the system which is evaluated by the value of the parent nodes; the arrows toward it represent a deterministic function. The rectangular node is an observed node, representing the variables observed by the sensors.

After establishing a nonparametric Bayesian network model, two types of nodes in the network need to be updated in real time. One is the node which prior model structure is known, which is updated in real time by GPF inference algorithm [56]; the other is the node which prior model structure is unknown, and its hidden variables are learned by the DPMM algorithm [57], making the model have the ability to selfupdating the structure.

8.3.3.5 Strategy of DT data synchronization

The data synchronization contains two aspects: model inference based on the improved GPF and model selflearning based on DPMM. The details are discussed as follows.

Model inference based on the improved GPF

The nodes with known prior models in the Bayesian network need to be inferred after receiving the data for each time step. However, it is difficult to achieve accurate inference of multiple parameters for complex Bayesian networks. Therefore, the improved GPF algorithm is used for approximate inference. This Chapter estimates system states and parameters jointly based on GPF. The details are described as follows.

Inference goals of each time step are to update the model parameters and system states simultaneously with the help of newly available data. The state space equations are described as:

$$\begin{aligned} A_t &= A_{t-1} + N(0, \gamma_{t-1}) \\ X_t &= f(X_{t-1}, A_{t-1}, \nu_{t-1}) \end{aligned} \tag{8.30}$$

where A_t is the set of unknown model parameters at time step t; X_t is the system states at time step t, γ and ν are the noise components.

According to the Bayesian theory, the joint posterior density distribution to be estimated can be written as:

$$p(X_t, A_t | Y_{1:t}) \propto p(Y_t | X_t, A_t) p(X_t | A_t, Y_{1:t-1}) p(A_t | Y_{1:t-1}) \tag{8.31}$$

where Y is the observation vector.

For the first time step, the particles of the parameters and the system states are initialized, $x_0^i \sim p(\mathbf{X}_0)$, $a_0^i \sim p(A_0)$. For the time step t, the parameters update process is as follows:

Step 1: Measurement update

- Sample particle set $\{a_{t1}^i\}$ from $q(A_t|Y_{0:t}) = p(A_t|Y_{0:t-1}) = N(A_t; \mu_A^t, \sigma_A^t)$, where $q(A_t|Y_{0:t})$ is an important density function;
- Calculate important weight of $\{a_{t1}^i\}$ by $\omega_{at-1}^i \frac{p(Y_t|a_{t1}^i) p(a_{t1}^i|a_{t1-1}^i)}{q(a_{t1}^i|a_{0:t1-1}^i, Y_{0:t})}$, obtain $\{a_{t1}^i, \omega_{at}^i\}$;

- Calculate filtering density distribution $p(A_t|Y_{0:t}) = N\left(A_t; \widehat{\mu}^t_A, \sigma^t_A\right)$, where $\widehat{A}_t = \sum_{i=1}^{N} \omega^i_{at} a^i_{t1}, \sum_t = \sum_{i=1}^{N} \omega^i_{at}\left(\widehat{A}_t - a^i_{t1}\right)\left(\widehat{A}_t - a^i_{t1}\right)^T$;

Step 2: Prediction update

- Sample transition set $\left\{a^i_{t2}\right\}$ from $p(A_t|Y_{0:t}) = N\left(A_t; \widehat{\mu}^t_A, \sigma^t_A\right)$;
- Generate particle set $\left\{a^i_{t+1}\right\}$ based on $\left\{a^i_{t2}\right\}$ and $p(A_{t+1}|A_t)$;
- Calculate prediction density distribution $p(A_{t+1}|Y_{0:t}) = N\left(A_{t+1}; \widehat{\mu}^t_A, \sigma^t_A\right)$, where $\mu^{t+1}_A = \frac{1}{N}\sum_{i=1}^{N} a^i_{t+1}, \sigma^t_A = \frac{1}{N}\sum_{i=1}^{N}\left(\mu^{t+1}_A - a^i_{t+1}\right)\left(\mu^{t+1}_A - a^i_{t+1}\right)^T$;
- For the system states, just change all the $p(X|Y)$ in the above steps to $p(X|A, Y)$.

Model selflearning based on DPMM

DPMM is a nonparametric Bayesian model that can find a mixed model with an unknown number of hidden variables through the Dirichlet process (DP). The process of DPMM is shown in Fig. 8.22. x is the observation value of each time step, H is the hidden variable, μ and σ is the mean value and standard deviation of hidden variable, h is the hyperparameter.

For a parameter lacking a prior physical model (such as the transfer efficiency η), the number of hidden variables and the distribution function are unknown. Assume that the parameter observation data is x, x comes from a distribution with M hidden variables, and each hidden variable j follows a normal distribution $\theta_j = \left\{\mu_j, \sigma^2_j\right\}$. The probability of x can be calculated as:

$$p(x|\theta_j) = \sum_{j=1}^{M} N(x|\theta_j)\omega_j \tag{8.32}$$

where ω_j is the mixed weight of the jth hidden variable, and $\sum_{j=1}^{M} p_j = 1$.

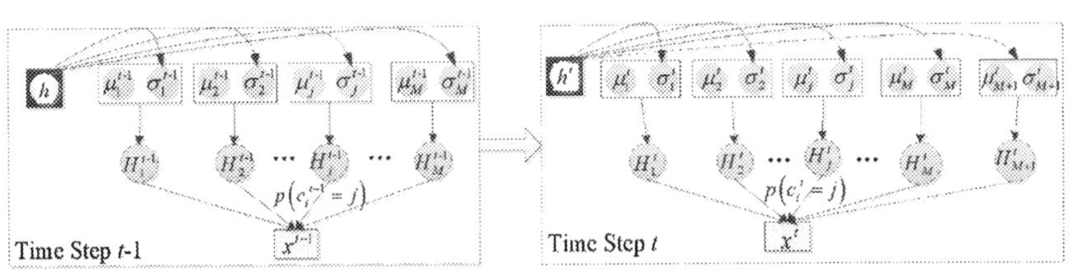

Figure 8.22 The process of DPMM.

The goal of the DPMM is to determine the appropriate number of hidden variables and the parameters θ_j of each hidden variable j. Specifically, the DPMM consists of two main steps: setting the DP prior distribution and using the Gibbs sampling method [58] to approximate the posterior distribution of the DP.

Step 1: Setting the DP prior

Given the previous $N-1$ time step data, the probability that a newly observation x_t comes from hidden variable j can be written as:

$$\begin{cases} p(c_t = j, j \in [1, M]) = \dfrac{n_j}{N-1+\alpha} \\ p(c_t = j, j \notin [1, M]) = \dfrac{\alpha}{N-1+\alpha} \end{cases} \tag{8.33}$$

where α is the concentration parameter that indicates the probability one expects to see a new type of hidden variable. n_j is the number of observations derived from the hidden variable j in the observations of previous $N-1$ time step, c_t represents x_t belongs to the hidden variable j.

In order to obtain the analytical solution of the posterior distribution and simplify the calculation, the conjugate distributions of μ_j and σ_j^2 are given by:

$$\begin{aligned} \mu_j \Big| \sigma_j^2 &\sim N\left(m, h\sigma_j^2\right) \\ \sigma_j^{-2} &\sim \Gamma(a, b) \end{aligned} \tag{8.34}$$

where m is the prior mean value of the observations and h is the hyperparameter. $\Gamma(a, b)$ is a gamma distribution with shape parameter a and scale parameter b.

Step 2: Calculating the DP posterior

Accurate inference over the posterior distribution $p\left(c_t, \mu_j, \sigma_j^2 \Big| x_t\right)$ is intractable. Instead, an iterative process, known as Gibbs sampling, is used to sample the relevant parameters following. To apply Gibbs Sampling for inference, one needs to repeatedly sample each conditional posterior distribution in turn. The procedure may be summarized as follows:

(1) Resampling c_t: Resampling c_t given the parameters $\theta_j = \left\{\mu_j, \sigma_j^2\right\}$. The probability that x_t derived from an instantiated hidden variable j is given by:

$$p\left(c_t = j \middle| \theta_j, x_t\right) \propto p\left(x_t \middle| \theta_j\right) \cdot \frac{n_j}{N-1+\alpha} \tag{8.35}$$

The probability that x_t comes from a new hidden variable θ is calculated by:

$$p(c_t = M + 1|\theta, x_t) \propto \left[\int p(x_t|\theta)d\theta\right] \cdot \frac{n_j}{N - 1 + \alpha} \tag{8.36}$$

(2) Resampling μ_j: Given c_t and variance σ_j^2, the mean value μ_j are resampled from conditional posterior distribution:

$$p\left(\mu_j\middle|\{c_t\}, \sigma_j^2, x_t\right) \sim N\left(m', h'\sigma_j^2\right) \tag{8.37}$$

where $m' = \frac{h\sum_{c_i=j} x_{c_i} + m}{hn_j + 1}$, $h' = \frac{h}{hn_j + 1}$.

(3) Resampling σ_j^{-2}: σ_j^{-2} is resampled according to the posterior distribution given the remaining parameters:

$$p\left(\sigma_j^{-2}\middle|\{c_t\}, \mu_j, x_t\right) \propto \prod_{c_t=j} p\left(x_t|\mu_j, \sigma_j^{-2}\right)p\left(\sigma_j^{-2}\right) \tag{8.38}$$

In summary, the model updating strategy based on GPF and DPMM not only can infer the system health state and model parameters, but also complete the self-learning of the model structure.

8.3.3.6 Realization of IR system estimation and prediction services

The experimental equipment to be tested is a 640 × 512 medium wave IR focal plane array detector system, and the test hardware devices include an IR target projector, an optical collimation system, a stress meter, and data acquisition boards. Proton irradiation experiments are conducted to accelerate the degradation of the electro-optical system, and 500 tests are performed for a long time. Parameters for each test include the MTF_{system} value at the Nyquist frequency, 10 captured images, mechanical stress, and electronic overstress. The results of experiments are shown as follows.

Model inference based on the improved GPF algorithm

In the first 400 time steps, the observation data of the physical system correct and update the digital twin model in real time (the estimation stage). In the last 100 time steps, the digital twin model is updated with the previous estimation values (the prediction stage) to verify that the proposed model updating strategy has the ability to ensure reliable operations without actual data.

When the shrinkage factor of kernel smoothing in the improved GPF algorithm is set to 0.3, the estimation and prediction results of nodes in digital twin model are shown in Figs. 8.23–8.25.

Figs. 8.23–8.25 show the estimation and prediction results of state nodes, degraded nodes, and fixed nodes. As seen in Fig. 8.23A, the estimated value (shown in red [gray in print version] dot line) and predicted value (shown in green [light gray in print version] dot line) for system state node MTF_{system} are nearly in accordance with the original value (shown in blue [dark gray in print version] line), and the estimation errors are mostly within 0.01. These results demonstrate that the improved GPF inference algorithm not only can perform real-time health monitoring well, but also accurately predict the health state without the original data, which indicates that the digital twin approach is efficient and robust.

Figs. 8.24 and 8.25 show that the estimated and predicted value of the degraded nodes and fixed nodes agree well with the original value, which reflects the strong ability of the improved GPF inference algorithm to

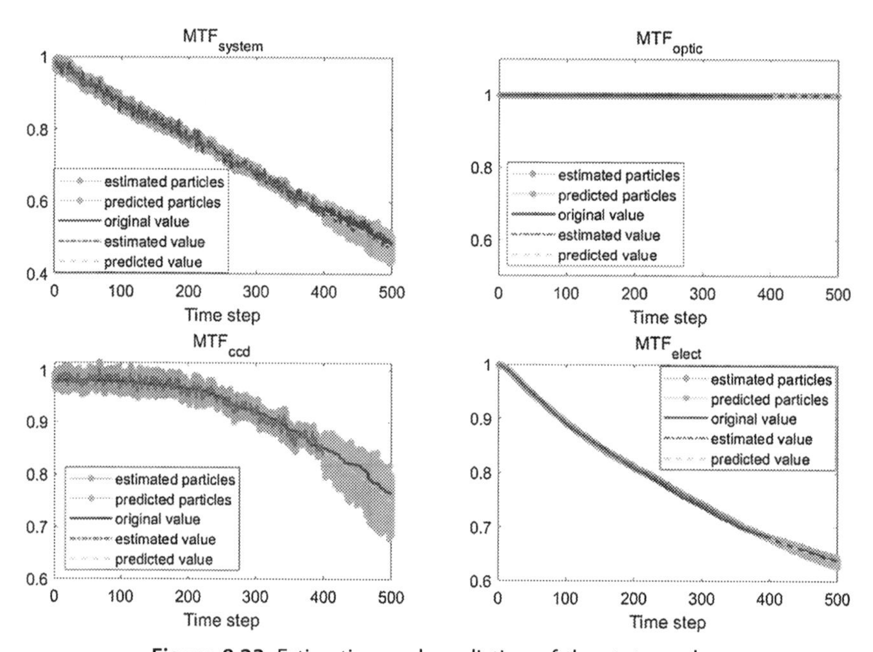

Figure 8.23 Estimation and prediction of the state nodes.

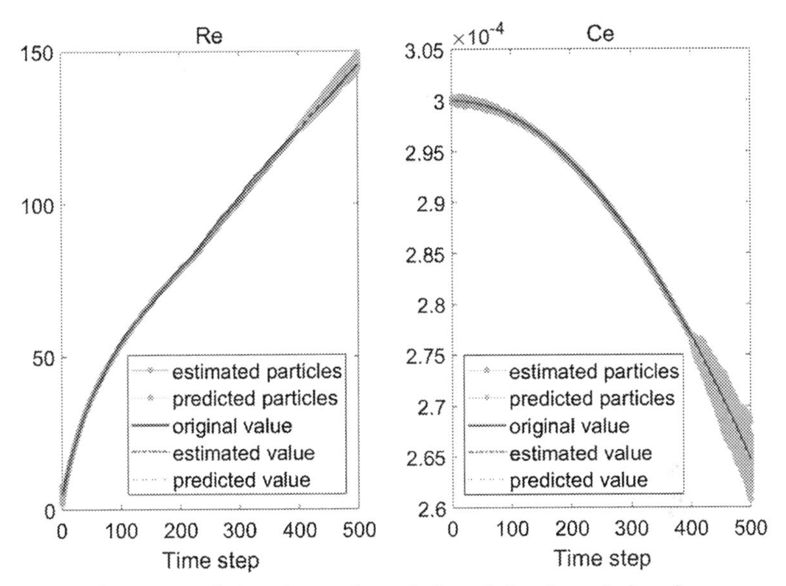

Figure 8.24 Estimation and prediction of the degraded nodes.

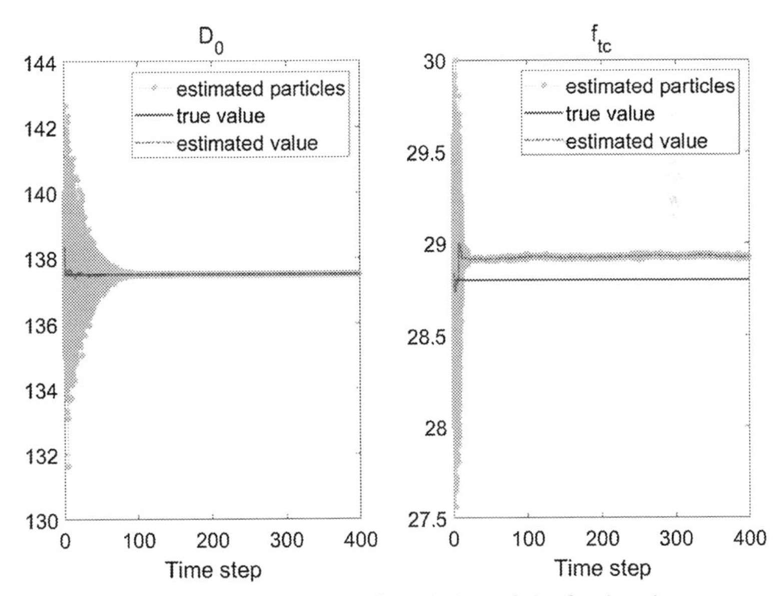

Figure 8.25 Estimation and prediction of the fixed nodes.

estimate model parameters. The estimated parameters can be used to indicate the health status of subsystems, such as equivalent resistance R_e and equivalent capacitance C_e can reflect the health status of the signal process subsystem.

Model structure updating based on DPMM

The priori model of parameter η is unknown, which makes the model updating process with great uncertainty. The nonparametric method based on DPMM obtains hidden nodes and new connection relationships by learning the data of η. The final estimated result of η is shown in Fig. 8.26.

Fig. 8.26 shows that after the hidden variables are learned by DPMM, the estimation accuracy of η is very high, and the error of prediction is relatively small, indicating that the DPMM method successfully learns the degradation process of η. In addition, the learned hidden variables of η are shown in Fig. 8.27.

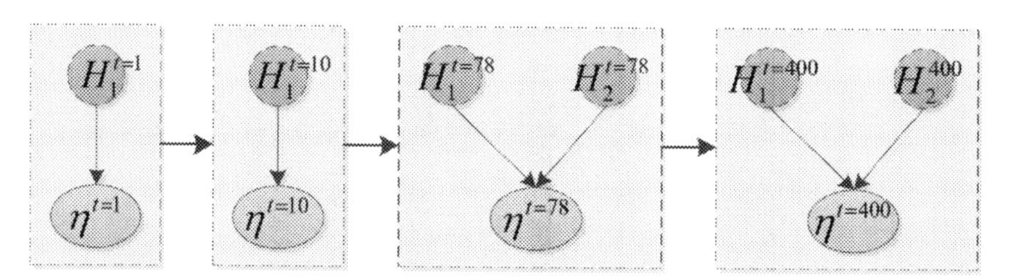

Figure 8.26 The estimated result of η.

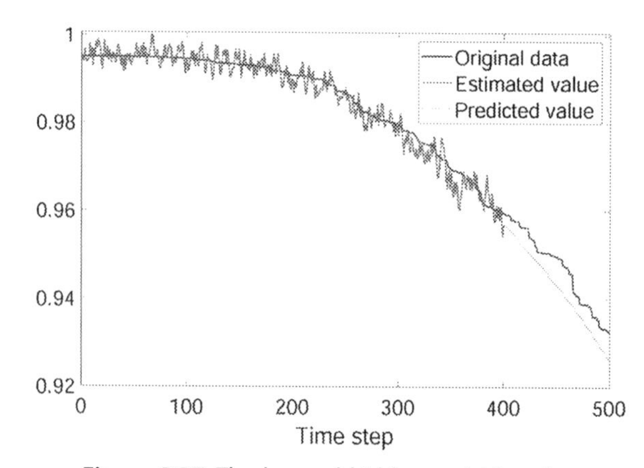

Figure 8.27 The learned hidden variables of η.

As can be seen from Fig. 8.27, in the first 78 time steps, η is represented by a set of hidden variables. Then, η is denoted by two sets of hidden variables and their values do not change much with time. At first, the selflearning ability of the model is not enough due to the lack of data. After that, the model structure tends to be stable, which indicates that the degradation mode of η is constant under a single working condition. The DPMM method increases the autonomous expression of uncertain degradation factors and realize the selflearning the model through iterative updating of hyperparameters.

8.4 Summary

This chapter introduces the PHM technology based on digital twins. First, the framework of the technology is illustrated in combination with the existing PHM standards; then the application of the technology in practical engineering problems is demonstrated through three typical cases. The first case takes the aerospace on-orbit CMG as an example, showing how to enhance system status monitoring and component RUL prediction services through digital twin technology; the second case aims at the prediction of the discharge time of the lithium-ion battery, and introduces how to use digital twin technology to realize the state estimation and trend prediction of the key components/ subsystems of the aircraft, so as to ensure the safety of the aircraft operation; the third case takes the IR imaging system as an example to show how to solve the problem of aircraft system-level PHM through digital twin technology.

References

[1] Tuegel E, Ingraffea A, Eason T, et al. Reengineering aircraft structural life prediction using a digital twin. Int J Aerospace Eng 2011:1−14.

[2] Glaessgen E, Stargel D. The digital twin paradigm for future NASA and U.S. air force vehicles. In: 53rd structures, structural dynamics and materials conference; 2012. p. 1−14.

[3] Grieves M, Vickers J, Grieves M, et al. Digital twin: mitigating unpredictable, undesirable emergent behavior in complex systems. In: Transdisciplinary perspectives on complex systems. Cham: Springer; 2017. p. 85−113.

[4] Srivastava A, Meyer C, Mah R. Integrated vehicle health management technical plan. Technical report. NASA; 2009.

[5] Sreenuch T, Tsourdos A, Jennions IK. Distributed embedded condition monitoring systems based on OSA-CBM standard. Comput Stand Interfac 2013;35(2):238−46.

[6] White J. Aviation safety program. NASA; 2009.

[7] Schwabacher M, Samuels J, Brownston L. NASA Integrated vehicle health management technology experiment for X-37. In: AeroSense 2002. International Society for Optics and Photonics; 2002. p. 49−60.

[8] Eickmeyer J, Li P, Omid G, et al. Data driven modeling for system-level condition monitoring on wind power plants. In: International workshop on the principles of diagnosis; 2015. p. 43–9.
[9] Khorasgani H, Biswas G, Sankararaman S. Methodologies for system-level remaining useful life prediction. Reliab Eng Syst Saf 2016;154:8–18.
[10] Liu S, Lu N, Cheng Y, et al. Remaining lifetime prediction for momentum wheel based on multiple degradation parameters. J Nanjing Univ Aeronaut Astronaut 2015;(03):360–6.
[11] Zhang X, Wilson A. System reliability and component importance under dependence: a copula approach. Technometrics 2017;59(2):215–24.
[12] Feng Y. A method for computing structural system reliability with high accuracy. Comput Struct 1989;33(1):1–5.
[13] Aizpurua JI, Catterson VM, Papadopoulos Y, et al. Supporting group maintenance through prognostics-enhanced dynamic dependability prediction. Reliab Eng Syst Saf 2017;168:171–88.
[14] Aizpurua JI, Catterson VM, Papadopoulos Y, et al. Improved dynamic dependability assessment through integration with prognostics. IEEE Trans Reliab 2017;66(3):893–913.
[15] Fang H, Shi H, Xiong Y, et al. The component-level and system-level satellite power system health state evaluation method. In: Prognostics and system health management conference. IEEE; 2014. p. 683–8.
[16] Daigle MJ, Bregon A, Roychoudhury I. Distributed prognostics based on structural model decomposition. IEEE Trans Reliab 2014;63(2):495–510.
[17] Roychoudhury I, Daigle M, Bregon A, et al. A structural model decomposition framework for systems health management. In: Aerospace Conference. IEEE; 2013. p. 1–12.
[18] Daigle M, Bregon A, Roychoudhury I. A distributed approach to system-level prognostics. Annu Conf Progn Health Manag Soc 2012;2012:71–82.
[19] Rodrigues L. Remaining useful life prediction for multiple-component systems based on a system-level performance indicator. IEEE/ASME Trans Mechatron 2017:141–50.
[20] Sheppard J, Kaufman M. IEEE 1232 and p1522 standards. In: 2000 IEEE autotestcon proceedings. IEEE systems readiness technology conference. Future sustainment for military aerospace (Cat. No. 00CH37057). IEEE; 2000. p. 388–97.
[21] Lee K. IEEE 1451: a standard in support of smart transducer networking. In: Proceedings of the 17th IEEE instrumentation and measurement technology conference [Cat. No. 00CH37066], vol. 2. IEEE; 2000. p. 525–8.
[22] Wattenberg M. Arc diagrams: visualizing structure in strings. In: IEEE symposium on information visualization, 2002. INFOVIS 2002. IEEE; 2002. p. 110–6.
[23] Bahler DR. A net-based approach to the synthesis of nondeterministic robot plans. In: Proceedings of the 1988 IEEE international conference on robotics and automation. IEEE; 1988. p. 1856–7.
[24] Swearingen K, Majkowski W, Bruggeman B, et al. An open system architecture for condition based maintenance overview. In: 2007 IEEE aerospace conference. IEEE; 2007. p. 1–8.
[25] Lee K, Gao RX, Schneeman R. Sensor network and information interoperability integrating IEEE 1451 with MIMOSA and OSA-CBM. In: IMTC/2002. Proceedings of the 19th IEEE instrumentation and measurement technology conference (IEEE cat. No. 00CH37276), vol. 2. IEEE; 2002. p. 1301–5.
[26] Zhang Z, Si X, Hu C, et al. Degradation data analysis and remaining useful life estimation: a Review on wiener-process-based methods. Eur J Oper Res 2018;271(3):775–96.
[27] Son L, Fouladirad M, Barros A. Remaining useful lifetime estimation and noisy gamma deterioration process. Reliab Eng Syst Saf 2016;149:76–87.

[28] Peng W, Li Y, Yang Y, et al. Bayesian degradation analysis with inverse Gaussian process models under time-varying degradation rates. IEEE Trans Reliab 2017;66(1):84–96.
[29] Wang J, Ye L, Gao RX, et al. Digital Twin for rotating machinery fault diagnosis in smart manufacturing. Int J Prod Res 2019;57.
[30] Qiao Q, Wang J, Ye L, et al. Digital twin for machining tool condition prediction. Procedia CIRP 2019;81:1388–93.
[31] Booyse W, Wilke DN, Heyns S. Deep digital twins for detection, diagnostics and prognostics. Mech Syst Signal Process 2020;140(Jun). 106612.1-106612.25.
[32] Liu C, Mauricio A, Qi J, et al. Domain adaptation digital twin for rolling element bearing prognostics. Annual Conf PHM Soc 2020;12(1). 10-10.
[33] Astrium. A compact, Cost-effective, High Performance CMG Solution for Small Satellites [EB/OL]. http://www.astrium.eads.net/en/equipment/cmg -15-45s.html.
[34] Singleton RK, Strangas EG, Aviyente S. The use of bearing currents and vibrations in lifetime estimation of bearings. IEEE Trans Indus Inform 2016;13(3):1301–9.
[35] Khlaief A, Nguyen K, Medjaher K, Picot A, Maussion P, Tobon D, Chauchat B, Cheron R. Feature engineering for ball bearing combined-fault detection and diagnostic. In: 2019 IEEE 12th international symposium on diagnostics for electrical machines, power electronics and drives (SDEMPED). IEEE; 2019. p. 384–90.
[36] Ciszewski T, Swedrowski L. Comparison of induction motor bearing diagnostic test results through vibration and stator current measurement. Comput Appl Eng Educ 2012;10:165–70.
[37] Stoica P, Friedlander B, Söderström T. A high-order Yule-Walker method for estimation of the AR parameters of an ARMA model. Syst Control Lett 1988;11(2):99–105.
[38] Mei Y, Shujuan W, Shaopeng D, Zhuo P. Reconstruction of undersampled damage monitoring signal based on compressed sensing. In: Proceedings of 2014 IEEE Chinese guidance, navigation and control conference. IEEE; 2014. p. 2443–8.
[39] Hwang WL, Huang PT, Kung BC, Ho J, Jong TL. Frame-based sparse analysis and synthesis signal representations and parseval K-SVD. IEEE Trans Signal Process 2019;67(12):3330–43.
[40] Yin W, Osher S, Goldfarb D, Darbon J. Bregman iterative algorithms for l(1)-minimization with applications to compressed sensing. SIAM J Imag Sci 2008;1(1):143–68.
[41] Han D, Yu J, Tian L, Zhang L, Zhang Q. Online framework of prognostic and health management for CMG under multiphysics. In: Proceedings of the 30th European safety and reliability conference and the 15th probabilistic safety assessment and management conference; 2020. p. 11.
[42] Ren L, Sun Y, Wang H, Zhang L. Prediction of bearing remaining useful life with deep convolution neural network. IEEE Access 2018;6:13041–9.
[43] Chen Y, Peng G, Zhu Z, Li S. A novel deep learning method based on attention mechanism for bearing remaining useful life prediction. Appl Soft Comput 2020;86:105919.
[44] Wang B, Lei Y, Yan T, et al. Recurrent convolutional neural network: a new framework for remaining useful life prediction of machinery. Neurocomputing 2020;379:117–29.
[45] Schalkwijk W, Scrosati B. Advances in lithium-ion batteries. Springer Science & Business Media; 2007.
[46] Ure NK, Chowdhary G, Toksoz T, et al. An automated battery management system to enable persistent missions with multiple aerial vehicles. IEEE ASME Trans Mechatron 2014;20(1):275–86.
[47] Liu X, Wu J, Zhang C, et al. A method for state of energy estimation of lithium-ion batteries at dynamic currents and temperatures. J Power Sources 2014;270:151–7.

[48] Lin C, Mu H, Xiong R, et al. Multi-model probabilities based state fusion estimation method of lithium-ion battery for electric vehicles: state-of-energy. Appl Energy 2017;194(May 15):560–8.
[49] Stepniewski WZ. Rotary-wing aerodynamics. Volume 1: basic theories of rotor aerodynamics with application to helicopters. Washington, United States: National Aeronautics and Space Administration NASA, Scientific and Technical Information Office; 1979. p. 44–90.
[50] Holst GC. Electro-optical imaging system performance. SPIE-International Society for Optical Engineering; 2008.
[51] Ratches JA, Lawson WR, Obert LP, Bergemann RJ, Cassidy TW. Night vision laboratory static performance model for thermal viewing systems. ARMY ELECTRONICS COMMAND FORT MONMOUTHNJ; 1975.
[52] Prochnau M, Holzbrink M, Wang W, et al. Measurement methods to build up the digital optical twin. In: Components & packaging for laser systems IV; 2018.
[53] Jain P, Poon J, Singh JP, et al. A digital twin approach for fault diagnosis in distributed photovoltaic system. IEEE Trans Power Electron 2019. 1.
[54] Delvit JM, Leger D, Roques S, et al. Modulation transfer function estimation from nonspecific images. Opt Eng 2004;43(6):1355–65.
[55] Rabiei E, Droguett EL, Modarres M. Damage monitoring and prognostics in composites via dynamic Bayesian networks. In: Reliability & maintainability symposium. IEEE; 2017.
[56] Storvik G. Particle filters for state-space models with the presence of unknown static parameters. IEEE Trans Signal Process 2002;50(2):281–9.
[57] McAuliffe JD, Blei DM, Jordan MI. Nonparametric empirical Bayes for the Dirichlet process mixture model. Stat Comput 2006;16(1):5–14.
[58] Rasmussen CE. The infinite Gaussian mixture model. In: International conference on neural information processing systems, vol. 12; 1999. p. 554–60.

CHAPTER 9

Production process management for intelligent coal mining based on digital twin

Xuhui Zhang[1,2], Xinyuan Lv[1], Yan Wang[1,2] and Hongwei Fan[1,2]
[1]School of Mechanical Engineering, Xi'an University of Science and Technology, Xi'an, Shaanxi, China;
[2]Shaanxi Key Laboratory of Mine Electromechanical Equipment Intelligent Monitoring, Xi'an, Shaanxi, China

9.1 Introduction

Coal mine equipment (CME) is a kind of typical complex electromechanical system. The characteristics of intelligent coal mining are different from stable working conditions, controllable unmanned production mode, and digital mode. The key issues of intelligent coal mining are how to ensure that CMEs are extremely controllable, and production processes are highly safe in the complex and changeable working conditions. Fig. 9.1 shows the working conditions of the underground coal mine. Service performance degradation and high maintenance cost of CMEs are inevitable during the harsh environment of heavy load and low speed. It has become a challenging task to enhance the service performance, the reliability, and especially the security of CMEs. Due to the characteristics of safety requirements in the coal mining, the decision-making by engineering staff is necessary for the comprehensive intelligent coal mine system integrated with human, information system, and physical system.

a) Fully mechanized mining face

b) Mining working face

Figure 9.1 Working conditions of underground coal mine.

Digital Twin Driven Service
ISBN 978-0-323-91300-3
https://doi.org/10.1016/B978-0-323-91300-3.00008-5

Digital twin (DT) is the creation of a virtual model of a physical entity in a digital way, through virtual and real interactive feedback, data fusion analysis, decision-making iterative optimization, and other means to provide more real-time, efficient, and intelligent operations or operation services for a physical entity [1]. DT technology has been demonstrated as an effective technology for realizing the complex interaction process management of comprehensive coal mine system or intelligent CMEs. Although some potential applications have been explored, some outstanding common issues still exist in the implement of production process management for key intelligent CMEs, such as (1) building the high-fidelity digital model to thoroughly describe intelligent CMEs; (2) establishing the interaction between intelligent CMEs and their digital models to make them seamlessly support production process management; (3) converging the data from physical space and virtual space to generate accurate information for production process management.

In this chapter, by introducing real-time data stream of CMEs, a general interaction model for the convergence of physical entity and digital models of CMEs is applied to address the above three issues. Firstly, based on DT technology, a high-fidelity digital model for key CMEs is built in different levels of geometry, physics, behavior, and rule. It provides an access to the equipment even out of physical proximity. Secondly, the interaction mechanism of DT can detect disturbances from the complex environment in a coal mine, potential faults in the CMEs and defects in the models. It is a coupled optimization to make the equipment and digital model evolve continuously. Thirdly, the accurate production process information can be greatly enriched based on the condition data of CMEs, the digital model, and the fused data.

9.2 Production process management and control framework of coal mining face based on digital twin

9.2.1 Digital twin model of coal mine production process management and control

To realize the construction of smart coal mine, the core is to build a DT-driven *Coal Mine production control Model* (CMM) with data interaction, multisource fusion, holographic perception, process control. Based on the application objects of physical working face and physical entity characteristics of the coal mine, a virtual model is established based on these and then the realizations: the virtual and real information interaction and data synchronization mapping between the virtual space and physical space, and the convergence, fusion and analysis of twin data of the coal mine, and finally provide users with application services through the human–computer interaction. In order to promote the application of intelligent technology in coal mine, the basic system architecture of the DT-driven CMM must be established [2,3]. The basic system architecture of the DT-driven CMM consists of a five-layer structure, including coal production physical layer, model layer, data layer, service layer, and application layer, as shown in Fig. 9.2 [4].

(1) The coal mine production physical layer is the base layer of the DT-driven CMM. It provides system resources and physical support for each layer of the DT coal mine production control system structure. The interaction and integration of resources between each layer, the simulation of multi-dimensional virtual models, the verification of logical models, and the calculation of data models are all based on the physical entities [5,6].

(2) The coal mine production model layer is the core component that realizes various functions such as smart mine planning and design, production management, operation and maintenance, and failure prediction. The physical model, simulation model, logic model, and data model are mutually coupled, evolved, and integrated, and the object twin, process twin, and performance twin of the physical mining face are realized under the driving of coal mine data [2,3].

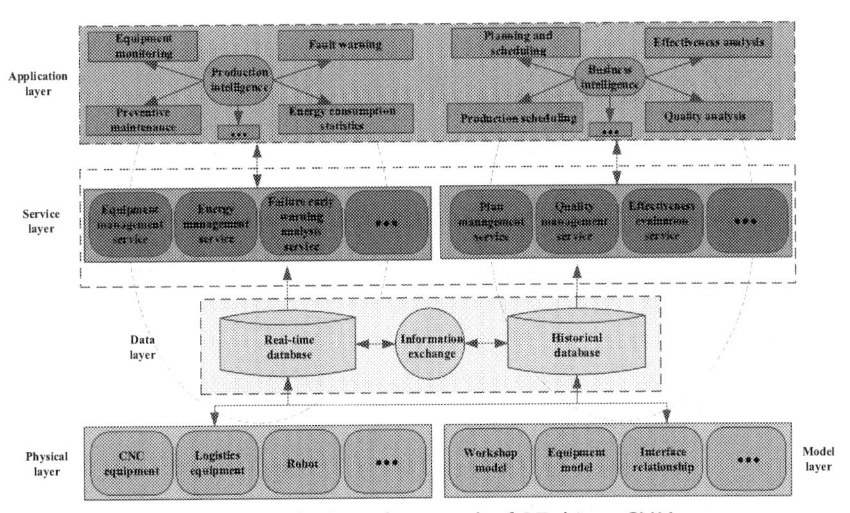

Figure 9.2 The logic framework of DT-driven CMM.

(3) The coal mine production data layer realizes the information interaction and synchronous feedback of the CMM through data driving. Coal mine twin data originate from physical entities, virtual models, and virtual twin application services. It connects the various layers into an organic entirety, so that information and data are coupled and interactive feedback between each part, realizing two-way communication and service interactive.

(4) The coal mine production service layer extracts real-time sensor data through the human–machine interface and uses the mine 5G gateway to realize the high-bandwidth and low-latency data transmission of CMM. After the integration of multi-source heterogeneous data, the data pass OPC UA, Web Service, and other communication interfaces. To realize real-time interaction and synchronous feedback between the physical mining face and the virtual DT model [7], through the VR/AR/MR human–machine interaction and intelligent monitoring terminal, the virtual operation and intelligent remote monitoring of the DT of the mining face are realized.

(5) The coal mine production application layer mainly realizes the application requirements for coal mine production control hardware resources on the industrial control network side, such as equipment monitoring, fault diagnosis, preventive maintenance, etc. On the classified network terminal, it mainly realizes the hierarchical implementation for operators and managers. Production command scheduling and centralized display information to improve production management decision-making level [8].

9.2.2 Intelligent interaction mode of HCPS of mining equipment

In the area of coal mining, the production and living system compose of engineering staff and simple physical tools/devices are called the *Coal-mining Human Physical System* (Cm-HPS). The essence of the Cm-HPS development is that humans continue to innovate and apply physical systems (power systems, mechanical structures, etc.), making mining equipment in physical systems replace and reduce human labor. With the emergence of modern information technology, the interaction between engineering staff and digital CMEs models has formed *Coal-mining Human Cyber System* (Cm-HCS). Now the scope of Cm-HCS has been extended to the augmented reality, mixed reality, social computing, and other fields. The essence of *Coal-mining Cyber Physical System* (Cm-CPS) is to build a set of closed-loop intelligent system based on automatic data flow between cyber space and physical space, it can control physical entities in real time, safely, and reliably to solve the problems of complexity and uncertainty in multidimensional complex systems.

A *Human-Cyber-Physical System* (HCPS) is constructed by exploring the interaction modes of Cm-HPS, Cm-HCS, and Cm-CPS. Obviously, the interaction in Cm-HPS is a kind of traditional mode in HCPS. The interactions between digital model and engineering staff, and the interactions between digital CMEs models and physical CMEs entities will become the key issue in the HCPS.

The information system is an information integration space that connects people and the mining face during the development of DT Technology. The system can help operators make analysis and decision-making and human-machine collaborative control, reducing the operator's mental work [9]. Through the integration of multisource data of people, information systems, and physical systems, a closed-loop intelligent interaction mode of CME based on HCPS is formed, as shown in Fig. 9.3. The virtual simulation and remote control technology of CME driven by DT realizes the collaborative control between people and equipment. VR/AR/MR technology is used to realize the interactive control between human and digital space [1].

HCPS can transcend the boundaries of time and space, realize the in-depth integration of human, physical system and information system, and

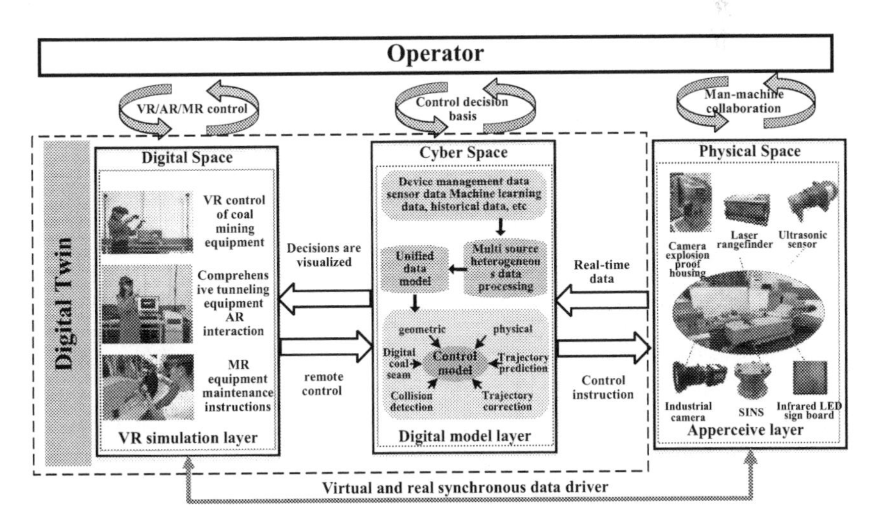

Figure 9.3 Coal mine equipment intelligent control system composition model based on HCPS.

enable the real-time interaction between the real mining face and the virtual working face. The physical system uses various sensor equipment to realize the perception of the status of CME and the environment and sends the perception information to the information system through the Industrial Ethernet. The information system analyzes and processes the perceived multisource information and heterogeneous information to form a unified data model and combines the coal mine GIS data to form an equipment group control model. Real-time data are combined with the control model to predict the potential behavior attributes in the control process. The digital space presents control information such as data, models, and prejudgment results through VR/AR/MR and other technical means to provide operators with a visual aid for decision-making. The operator issues control instructions through the observation digital space interactive platform, and feedbacks the control decision instructions to the information system through the virtual communication interface to match the predicted results, and finally uses the industrial Ethernet to send the control instructions to the end effector in the physical space to realize the entire system closed-loop control.

Based on the HCPS composition model of the intelligent control system of the cantilever roadheader, the research on the DT driving technology of the virtual equipment using underground production process data is studied, and the remote intelligent control system of the cantilever roadheader driven by multisource sensor data in real-time is constructed. The equipment status and environmental information changes of the coal mining face are mapped to the visualized three-dimensional virtual space in real time. On the basis of the construction of the virtual simulation scene of the heading face, the equipment state early warning analysis and historical data backtracking modules are added to provide reliable decision-making basis for remote control, ensure the safety of personnel and equipment, and realize automatic mining with few people or even unmanned coal mine.

The system is mainly composed of three parts: the physical space data perception fusion module of the cantilever roadheader, the DT module, and the HCPS-based human-machine collaborative control [10]. The geological survey department, the production department, and the electromechanical information department provide coal mine environment and equipment data to establish a basic database. During the production process, each equipment subsystem database is established according to the changes in the geometric characteristics of the roadway and the equipment operation states to realize the real-time collection, storage, and sharing of multisource heterogeneous data in industry and mining. The logical relationship between data is analyzed by accessing remote database, building a unified digital model and DT, providing twin-models, twin-control, and twin-services for visual decision-making.

Content of twin-control mainly includes remote control and man-machine collaboration control. With the help of virtual reality technology, a virtual coal mine underground multidimensional space with symbiotic interaction of "human, machine, and environment" is constructed. The data of coal mine face change and equipment pose are fed back to the virtual monitoring platform in real time. The operator of the control center can modify the running track of the equipment in real time by observing the interactive virtual interface to realize immersive remote control and online status monitoring. In the human-machine cooperative control, through mining and analysis of historical data, a prejudgment system of the equipment is established, and the delay in the data transmission process is eliminated in the loop as much as possible.

The virtual remote intelligent control system of the cantilever roadheader is based on the DT Technology and combines the immersive and interactive characteristics of virtual reality technology to realize the remote control and visual auxiliary cutting of the roadheader. As shown in Fig. 9.4, the system takes the cantilever roadheader as the research object for kinematic analysis and modeling, use image processing technology to obtain the position of the cutting head, and use multisensor fusion technology such as strapdown inertial navigation and ultrasonic sensors to obtain the position of the roadheader body. Through RS485 serial communication, a multisource heterogeneous data fusion matrix is established at the local control end to realize data acquisition. Using the TCP/IP data transmission protocol, a DT body of remote end was established for realizing the effective combination of virtual roadheader and real roadheader based on the

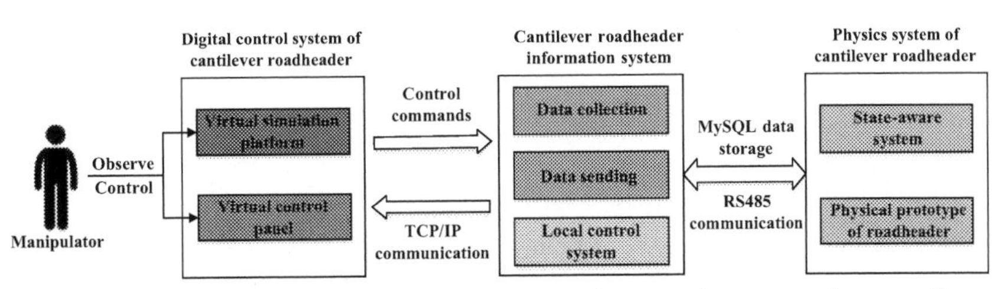

Figure 9.4 Human–machine interaction remote control system for cantilever tunneling machine.

functions of visual cutting, remote control, and virtual monitoring. A feasible solution was provided for the intelligent control of underground cantilever roadheaders.

The physical console cannot effectively integrate the information between people and physical environment in remote control process. The virtual interface research and development and hardware production costs of the hardware console are relatively high. This system adopts a control strategy built on the virtual workbench. The remote control system based on the virtual console uses the virtual reality GUI technology to program the controller virtual buttons in the virtual visualization window to form a virtual interactive platform. The virtual console can realize all the functions of the real cantilever roadheader, such as the start and stop of the pump station, the scraper, the star wheel and cutting, the lifting of the shovel, the various movements of the cantilever, and the expansion and contraction of the support. The operator can control the virtual roadheader through the corresponding buttons or levers, combine the control commands and real machine data in the visual virtual scene, and use the control strategy to realize the virtual and real synchronous movements of the roadheader to achieve the purpose of visualized remote control of the cantilever roadheader (Fig. 9.5).

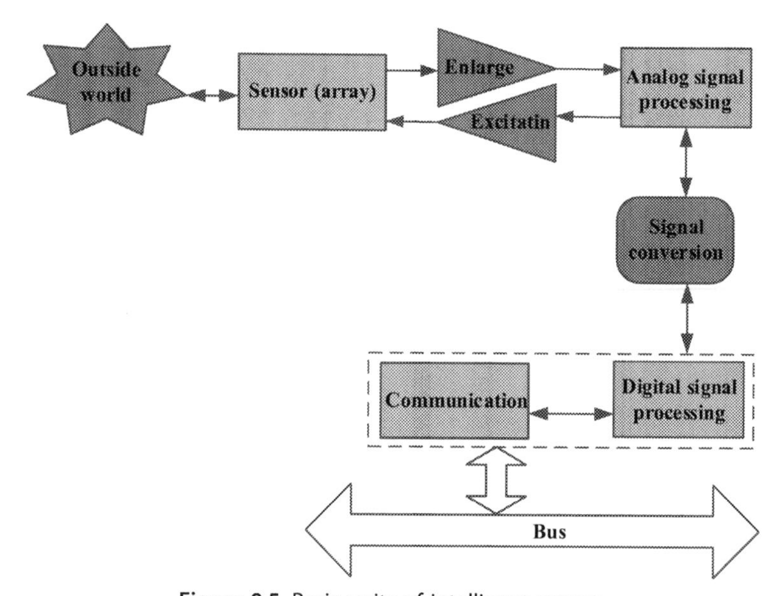

Figure 9.5 Basic units of intelligent sensor.

9.3 Key technologies

9.3.1 Intelligent sensing

Intelligent sensing technology is a building information platform established by integrating modern detection technology, computer technology, and information technology. In DT-driven health management for key CMEs, the intelligent sensing technology plays a primary role in fault identification, prediction, abnormal detection, and process control. It can collect all kinds of data in each stage of the life cycle of key CME and then realize the interactive ability of physical entities and these virtual models.

Intelligent sensor is a sensor system that can adjust the internal performance of the system to optimize the ability to obtain external data. It is more like a system with sensitive elements, data acquisition, processing, and transmission functions [1]. Its basic units include (1) sensitive components or array of sensitive components; (2) excitation control unit; (3) amplification; (4) analog filtering; (5) signal conversion; (6) compensation; (7) digital signal processing; (8) digital communication [1].

At present, the coal mine sensing equipment and transmission network cannot achieve the ubiquitous perception of the safety monitoring information of the whole mine, and there is no better solution for environmental monitoring in this kind of dangerous area. The tunnels, chamber, and equipment need to arrange a large number of sensing nodes, which not only requires the terminal to have the characteristics of low energy consumption, low cost, and high security, but also have higher requirements for the connection number density, transmission rate, and real-time performance of wireless network. The end-to-end delay of 5G technology is controlled within 10 ms, and the connection density reaches $106/km^2$, which meets the need for mine safety monitoring information collection. Underground environment monitoring, operating equipment monitoring, physical sign monitoring, instrumentation monitoring, ventilation and drainage monitoring, and other systems all need 5G networks to realize the intercommunication of large-scale sensing nodes. Use a network that accommodates a large number of sensors to realize the collection of safety monitoring information for the whole mine and access to the mobile edge computing server to analyze and process the collected information. Combined with advanced technologies such as cloud computing and artificial intelligence, the data processing results are returned (Fig. 9.6).

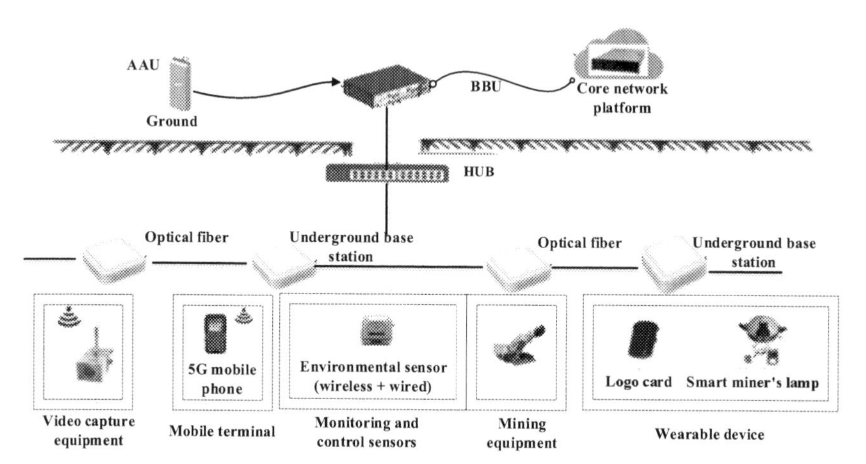

Figure 9.6 Architecture of mine 5G wireless communication system.

9.3.2 Position and attitude control

In order to ensure that CMEs advance along the planned path and direction, it is necessary to accurately measure the spatial location and operating trajectory of the equipment. However, the underground coal mine is a completely enclosed space and a complex electromagnetic environment, without the assistance of satellite navigation signal, it is very difficult to realize position and attitude control, positioning, and navigation [11].

At present, some experts and scholars have proposed various positioning and navigation technologies such as GIS geographic information system-based navigation, *Radio Frequency Identification*-based *Angle of Arrival*, and *Time of Arrival*, strapdown inertial navigation system, *Ultrawide Band* ultra-wideband positioning, etc., but there are still many issues.

The core technologies in urgent need of breakthrough include (1) low-cost, high-precision strapdown inertial navigation technology suitable for complex magnetic field environments; (2) core chip technology for local positioning and navigation application scenarios; (3) underground 5G high-speed wireless communication technology; (4) underground high-precision positioning system; (5) underground obstacle avoidance technology based on precise positioning and navigation; (6) precise guidance technology of roadheader; (7) unmanned driving system for auxiliary transportation vehicles.

In terms of intelligent mining technology, workers can observe the operation of the equipment on the working face in the automatic monitoring center of the *Fully Mechanized Mining Equipment* (FMME) and realize remote control. To realize the intelligent perception and control of fully mechanized mining equipment, it is necessary to carry out technological innovation on coal mining methods and processes, which mainly include: (1) accurate spatial positioning of the shearer. The position sensor is installed in the haulage part of the shearer. Through the data processing module of the controller, the displacement information of the shearer is calculated and input into the mathematical model to accurately realize the position of the mining equipment. (2) Real-time perception of device attitude. It is necessary to carry out real-time working condition detection, attitude detection and fault diagnosis of FMME equipment to ensure that the whole working face equipment runs in the best state. (3) Frequency conversion feedback of coal flow load, the distributed load of coal flow on the scraper conveyor is detected in real time, and the transmission speed is controlled using frequency conversion technology to realize the automatic and coordinated operation of mining and loading in the working face [12].

In DT, similar to intelligent sensing technology, pose detection and positioning technology are also supporting technologies. Combining data acquisition and transmission technology, the virtual model is predicted and optimized in the DT, and feedback and control the posture state of the physical world, so as to avoid various possible problems in all links of the CME operation.

9.3.3 Running condition identification and abnormal diagnosis

Running condition identification and abnormal diagnosis for CME, including quick faults capturing, accurate fault locating, service states evaluating, and performing predictive maintenance, is essential for maintaining the equipment in the extremely reliable state [13]. Based on real-time data stream, DT-driven fault prediction method was proposed to understand and master the operating status of equipment, identify abnormal performance of equipment, early discovery of potential equipment failure, and forecast failure development trends. A new health management mode for CMEs was formed using synchronous mapping, real-time interaction and precise service of physical and virtual equipment. Fig. 9.7 shows the execution logic of operation status recognition and exception diagnosis for remote monitoring [14—18].

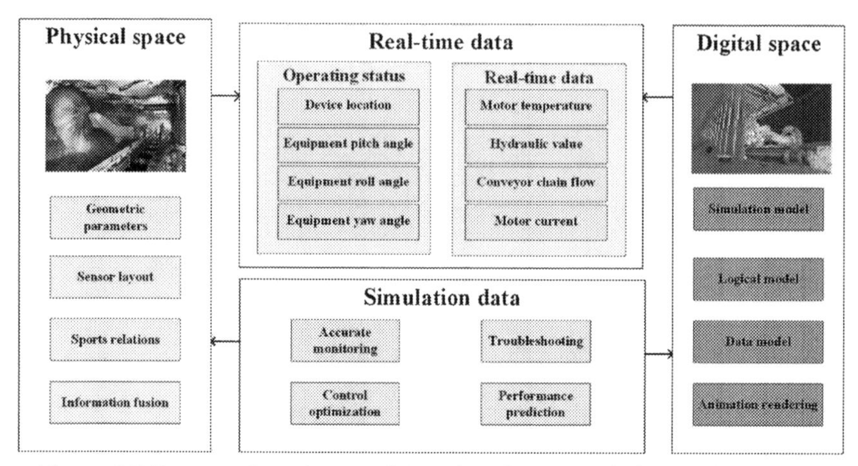

Figure 9.7 Process of running condition identification and abnormal diagnosis.

9.3.4 DT-driven prognostics and health management for key CMEs

Prognostics and Health Management (PHM) uses various sensors and data processing methods to evaluate the health of equipment, and to predict equipment failure and remaining life, thereby transforming traditional postevent maintenance into preexisting maintenance. DT-driven PHM is driven by twin data, based on the synchronous mapping and real-time interaction of physical and virtual devices, as well as accurate PHM services, forming a new device health management model to quickly capture failure phenomenon and accurately locate the cause of failure, design, and verify the maintenance strategy.

As shown in Fig. 9.8, in the PHM driven by the DT, the physical equipment perceives the operating status and environmental data in real time; the virtual device runs synchronously with the physical device driven by the twin data and generates data such as device evaluation, failure prediction and maintenance verification; integrating real-time data and existing twin data from physical and virtual devices, PHM services are accurately invoked and executed according to requirements to ensure the healthy operation of physical devices [19].

The PHM model driven by the DT brings the following new changes: (1) The fault observation mode changes from static index comparison to dynamic physical and virtual device real-time interaction and all-round state comparison; (2) the fault analysis method has changed from the analysis method based on physical device characteristics to the analysis method

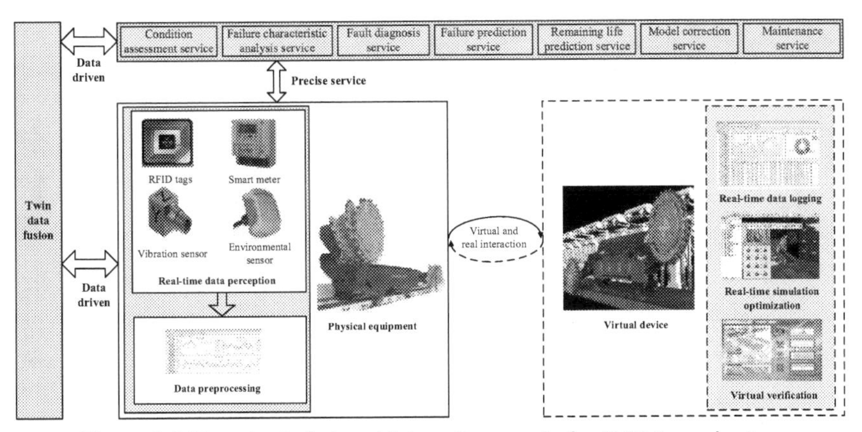

Figure 9.8 The physical-virtual interaction mode for PHM in coal mine.

based on the association and integration of physical and virtual device characteristics; (3) the maintenance decision-making mode changes from the decision-making based on optimization algorithm to the decision-making based on high fidelity virtual model verification; (4) the execution mode of PHM function has changed from passive allocation to autonomous precise service [20].

Research on PHM driven by DT requires breakthroughs in the following challenging issues:

(1) In the fault capture method, the autonomous interaction mechanism between the virtual and the real, and the inconsistency judgment rules between the virtual and the real are studied to capture and eliminate the interference that leads to the inconsistency between the virtual and the real.

(2) Research on fault mechanism, fault feature extraction, and fusion based on twin data, fault process modeling, and propagation mechanism, etc.

(3) Research on selforganization, selflearning, and selfoptimization mechanism of PHM service, requirement capture and accurate analysis, as well as accurate service execution based on virtual verification.

9.3.5 DT-driven adaptive control for intelligent mining

DT must rely on two key elements: data and virtual model. For intelligent CMEs, by constructing a DT, it can not only reflect the characteristics, operation and performance of the machine entity in a highly realistic manner, but also realize machine condition monitoring and evaluation and health management in a surreal form [21].

DT-driven adaptive control for intelligent mining is also one of the breakthrough directions of the key technologies of intelligent coal mining equipment. Mainly include the following aspects:

(1) High-efficiency adaptive cutting technology for coal and rock. In order for intelligent coal mining equipment to have adaptive and efficient coal and rock cutting functions, it is necessary to establish a mechanical-coal rock impact model and a new design theory and control method for identifying coal and rock fracture sensitivity [22,23].

(2) Adaptive control technology of hydraulic support. Hydraulic support of an FMME is a group of distributed support mechanisms. Its single support stability must be identified and controlled by the hydraulic support controller. While the support coordination and stability of the group hydraulic support are determined and regulated by the associated support controller. Therefore, the intelligent key elements of the hydraulic support electro-hydraulic control system are the selfadjustment, selforganization, and self-stabilization of the hydraulic supporting and moving movements. At present, the main research at home and abroad focuses on the optimization of the 4-bar linkage and parametric design of the hydraulic support. The electro-hydraulic control system realizes the automatic control of the hydraulic support. However, unmanned coal mining face requires hydraulic support to have strong adaptability of self-adjustment and selflearning for complex surrounding rock support. Solving the coupling effect of roof strata, supporting mechanism and hydraulic damping is the key problem of establishing hydraulic support.

(3) Adaptive control technology of power transmission. CME is a complicated electro-hydraulic system, which has long-standing problems of multiple overall failure, low reliability, and low power on rate, which cannot meet the requirements of a safe and reliable operation of intelligent CMEs. Therefore, it is necessary to study the gradual failure mechanism of power transmission components of intelligent CMEs under severe coal mining conditions, and propose dynamic and gradual reliability design methods for key power transmission components; the research on the principle of highly reliable and efficient power transmission under heavy load and damaging condition, and establish the adaptive control technology of high-power electro-hydraulic hybrid transmission.

(4) Coal-rock identification technology is a difficult problem to restrict the intelligent CME. Existing coal-rock identification methods, namely, ray detection method, radar detection method, infrared detection method, active power monitoring method, vibration detection method, sound detection method, dust detection method, and image recognition method, all have a certain dependence on geological conditions. There are still limitations and sensitivity problems when used as an airborne sensing environment technology. Therefore, it is necessary to study the new principles of coal—rock recognition and establish an accurate real-time recognition method of coal—rock interface.

(5) Intelligent diagnostic technology of CMEs. The reliability of equipment is essential for intelligent coal mining operations. Currently, it is necessary to establish a multisource information fusion model of intelligent CMEs and a real-time pickup method of three-dimensional spatial information to realize holographic perception and virtual scene reproduction of unmanned coal mining face. Study the operating conditions and fault identification models of intelligent coal mining equipment in harsh environments, and realize the online monitoring and prediction of intelligent CMEs faults.

9.4 Typical application case

9.4.1 DT-driven PHM for shearer

As complex electromechanical equipment is used for intelligent coal mining, the shearer has a harsh working environment, and strong noise has serious interference with fault response signals. In order to improve the working efficiency of the shearer and reduce the impact caused by a sudden failure, a DT-based shearer PHM method is proposed. Use various sensors and data processing methods to comprehensively consider equipment status monitoring, fault prediction, maintenance decision-making, etc., through the fusion of physical and virtual interaction, the failure mode of the shearer is accurately predicted, which effectively improves the service life and reliability of the shearer [24].

Therefore, the DT five-dimensional model is introduced into the PHM, and the PHM method based on the DT is used for fault prediction and health management of CMEs. This method first establishes and calibrates a five-dimensional DT model of the physical entity. Then, according to the model and interactive data, the consistency of physical entity parameters and virtual simulation parameters is judged; according to the consistency/inconsistency of the two, the gradual and sudden failures can be predicted and identified respectively; finally, the maintenance strategy is designed according to the failure causes and dynamic simulation verification. This method has been applied to the health management of the shearer, as shown in Fig. 9.9.

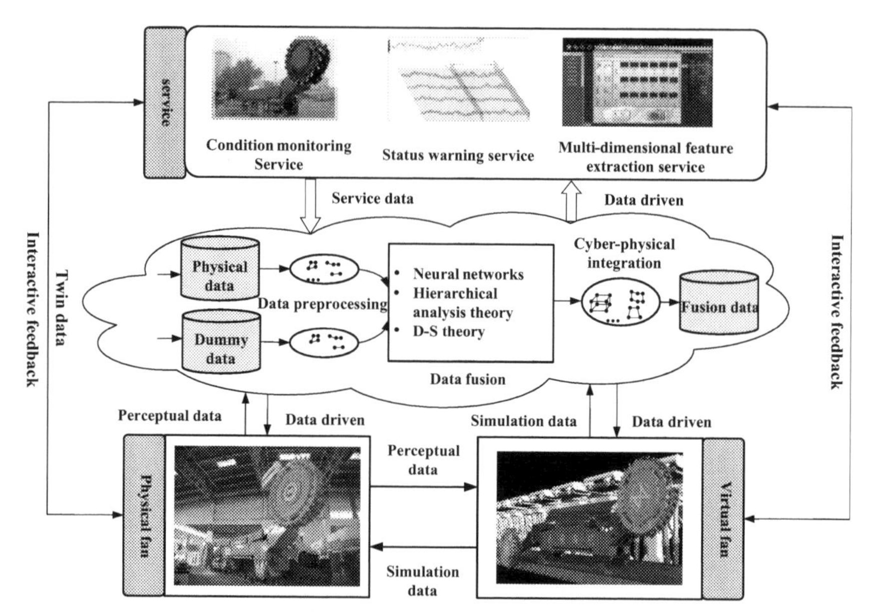

Figure 9.9 Fault prediction of the shearer based on DT.

First of all, the relevant sensors can be deployed on the gearbox, motor, spindle, bearing, and other key components of the physical shearer for real-time data acquisition and monitoring. Based on the collected real-time data, historical data and domain knowledge of the shearer, the multidimensional virtual model of the virtual shearer can be constructed to realize the virtual mapping of the physical shearer from the aspects of geometric, physical, behavior and rule. Based on the synchronous operation and interaction of physical shearer and virtual shearer, through the interaction and comparison of physical and simulation states, data fusion analysis and virtual model verification, the physical shearer state detection, fault prediction, and maintenance strategy design are realized. These functions can be packaged and provided to the users in the form of application software.

9.4.2 DT-driven interaction between virtual and real in remote monitoring

The remote control technology based on DT uses virtual visualization and other means, according to the equipment operation data, through the construction of virtual model, the equipment operation status can be reproduced in the virtual scene, so as to achieve stable, reliable, and intuitive remote monitoring. Currently remote control of CMEs has been researched [25].

Virtual remote control of CMEs driven by DT. The key is to digitize complex and multidimensional information that characterizes the status, interrelationship, and environmental changes of field equipment, and present multidimensional space with the help of VR/AR technology, which serves as the basis for remote control decision-making and overcomes the problems of incomplete, nonintuitive and difficult decision-making of field equipment information expressed by traditional remote control based on digital, simple graphics, and monitoring video [26].

DT-driven remote control of CMEs includes key technologies such as construction of scene and virtual device model, working face control model, dynamic modification of display and control model, anticollision warning of equipment, and remote control, etc.

Take the DT-driven roadheader remote control scheme as an example to introduce the processes of data acquisition, data transmission, and remote control, as shown in Fig. 9.10.

The sensing data in the remote control of the DT-driven roadheader are mainly the state data and environmental data of the roadheader, which constitute the data source of the DT-driven model [27]. The fuselage attitude measurement system is composed of parallel laser pointer and rear camera. Cutting head attitude measurement system composed of infrared LED target and front camera. The real-time data of roadheader status is obtained by using ultrasonic sensor. Gas sensor and dust sensor are used to obtain environmental data online.

Using the perception data to build a visual auxiliary cutting system on the local explosion-proof computer [28], real-time online monitoring of the tunneling process is used. At the same time, the sensing data are sent to the remote control terminal through the explosion-proof computer, and

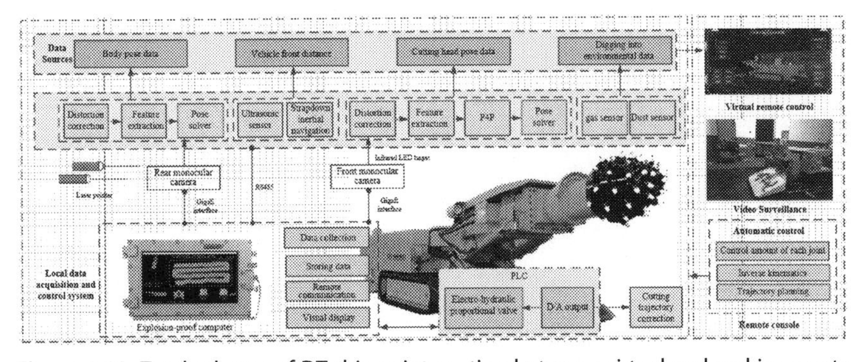

Figure 9.10 Total scheme of DT-driven interaction between virtual and real in remote monitoring.

the virtual equipment synchronization action is realized according to the DT model. The remote control terminal can realize virtual remote control and video monitoring functions.

In order to realize the real-time reproduction of the real scene change state in the virtual scene, in addition to obtaining the twin data of the real scene, the corresponding digital DT model needs to be established, including: (1) Establishing the coupling of virtual model and virtual scene of roadheader; (2) Establishing the kinematics model of roadheader and solving its forward and inverse solutions, the twin data are used to drive the virtual model action, and the running state of roadheader is restored in the virtual scene.

In the remote automatic cutting control of the roadheader, the data exchange medium is mainly the database. It realizes centralized management of sensor data, real-time acquisition of the current state of the sensor; through the analysis of the data group stored centrally, the prediction and perception of uncertain events can be realized. The DT data transmission process is shown in Fig. 9.11.

9.4.3 Verification of the DT-driven MR-aided maintenance method

In order to deal with the failure of underground mining equipment in a timely and effective manner, it is extremely important for the coal mine to resume normal production. With the Limitation of the technical level, how to maintain the failed CMEs depends on the guidance of professional people. When it is difficult to judge the maintenance point in the face of complex fault, technical experts are also required to provide remote guidance or arrive on-site for maintenance. It is difficult to interact with fault knowledge and maintenance skills, which lead to time-consuming and labor-intensive maintenance process.

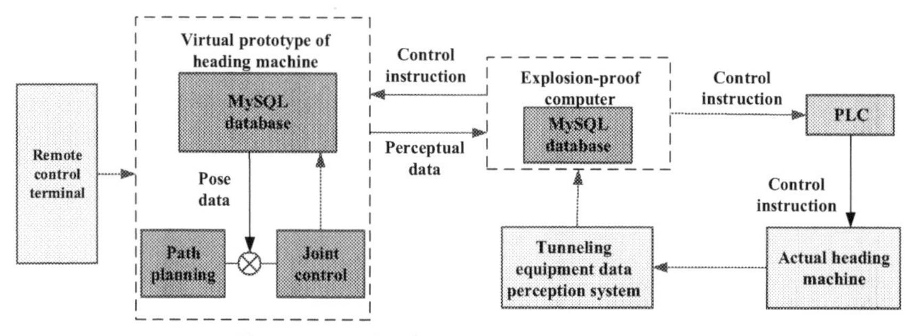

Figure 9.11 DT data transmission process.

The traditional fault maintenance guidance method of complex electromechanical equipment is not intuitive and timely. An integrated method of DT&MR (mixed reality) for maintenance guidance of electromechanical equipment is proposed, as shown in Fig. 9.12.

The overall model of the MR maintenance guidance system for equipment fault maintenance driven by the DT is developed in the platform of BIM-Unity3D-HoloLens. The functionalities of the system include 3D maintenance environment perception, interaction data, virtual-real registration, and fault maintenance guidance, which are also verified. To demonstrate the correctness of the proposed method, the maintenance guidance case of a transmission system on electric traction shearer rocker arm is examined. The results show that the integration method of DT&MR can realize real-time maintenance guidance and improve maintenance efficiency [29–32].

Equipment operating status, mechanical, electrical, hydraulic, key component failures, and other data from the physical maintenance environment are collected. After fusion processing, they are transmitted to the MR auxiliary maintenance guidance system of CMEs fault, providing the remote real-time monitoring data of CMEs for the service system. The

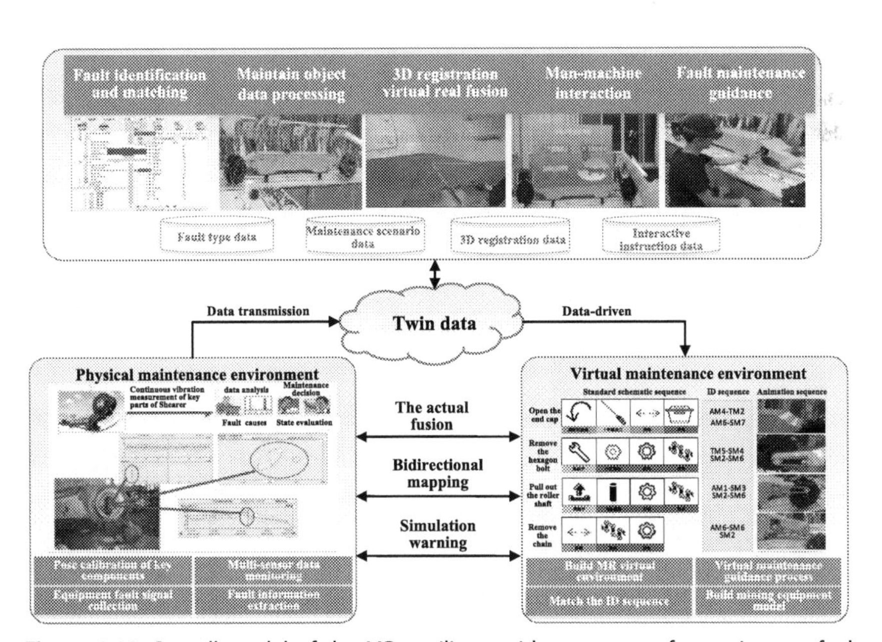

Figure 9.12 Overall model of the MR auxiliary guidance system for equipment fault maintenance driven by the DT.

virtual maintenance environment mainly completes the construction of the MR virtual maintenance development environment and the CMEs model. Through the virtual maintenance guidance process design, a fragment of the fault maintenance guidance process for key components is formed and matched with the fault ID.

The system uses twin data to fuse multisensor data from the physical maintenance environment and collects equipment fault data for equipment fault identification and matching, physical maintenance scene data are obtained through maintenance environment perception, and are integrated with the virtual guidance tools into the physical maintenance environment to obtain registration scene data. The human–computer interaction method is used to complete the data interaction between the human and virtual maintenance environment, and the formed interactive instruction data and maintenance process data can realize the data drive of the virtual entity of the virtual maintenance environment using the service system [33].

Aiming at the problem of insufficient auxiliary maintenance guidance for CMEs, the CMEs fault maintenance guidance method was adopted, and a DT-driven CMEs maintenance MR auxiliary guidance system overall model was proposed. Based on BIM-Unity3d-HoloLens system development platform, equipment fault diagnosis data are combined with mixed reality maintenance guidance, which provides an innovative data interaction method for the MR fault maintenance guidance field and solves the problem of lack of equipment data in equipment auxiliary maintenance technology.

The detailed functions of this platform are given as follows:

(1) A DT-driven MR auxiliary maintenance guidance system as shown in Fig. 9.13, which can effectively provide maintenance guidance

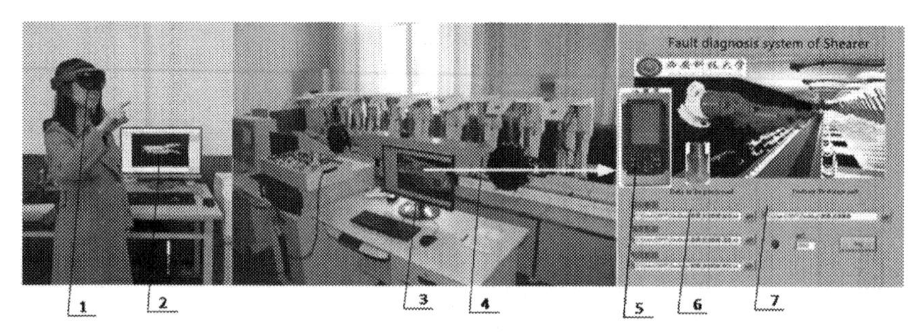

Figure 9.13 System experiment platform. *(1—HoloLens, 2—Remote PC, 3—Local PC, 4—5:1 experimental model of MG series electric traction shearer, 5—Vibration tester, 6—Piezoelectric sensor, 7—Fault diagnosis system of coal shearer.)*

solutions of virtual and real integration for maintenance personnel and provide corresponding maintenance guidance processes for different fault categories, maintenance personnel only need to wear one HoloLens glasses to complete the maintenance operation.

(2) By studying the maintenance behavior model of CMEs based on fault tree analysis, analyze and summarize the failure mechanism of CMEs, use the fault tree analysis method to analyze the key failure factors of CMEs, analyze the maintenance behavior and establish a maintenance behavior tree model according to the key failure factors. Through the establishment of general node modularization, the rapid development of CMES virtual maintenance guidance behavior simulation is realized, and the problems of large number of equipment parts and difficult system development are solved.

(3) Taking the virtual reality development platform Unity3D as the main platform, build an MR maintenance guidance development environment. The virtual reality fusion method of MR three-dimensional registration based on artificial identification and natural feature points is used to improve the effect of virtual reality fusion, design the standard schematic diagram suitable for the fault maintenance guidance of CMEs, which forms a matching maintenance guidance process for the gear failure of the rocker arm transmission system of the shearer.

(4) CMEs maintenance guidance system solves the problem of lack of information interaction in equipment auxiliary maintenance guidance. It mainly starts from the data interaction and feedback mechanism between the physical and the virtual maintenance space. From the perspective of MR auxiliary guidance system for CMEs driven by DT, the three key technologies of CMEs fault data interaction, human–computer interaction and remote expert online interaction are respectively realized.

(5) Build a DT-driven CMEs fault MR auxiliary maintenance guidance system platform to verify system functions. The test results show that when the equipment fails, the result of fault diagnosis can be sent to HoloLens glasses to invoke the corresponding maintenance guidance process. Maintenance personnel can complete the operations independently according to the system prompts. At the same time, for complex fault points, maintenance personnel can also use the remote expert online guidance function to further obtain guidance from experts, realize the virtual and real integration, two-way mapping and simulation warning between the physical and virtual maintenance environment, improve the equipment maintenance efficiency, and provide DT&MR auxiliary maintenance guidance method for CMEs.

9.4.4 Hands-on training

9.4.4.1 VR roaming training

The system allows the operation user to better understand the operating relationship between the equipment in the FMME, and to grasp the coal mining process as a whole. Users can not only roam in the virtual scene through the control handle, but also clearly see the various mechanical parts in the device through the GUI panel control. At the same time, they can choose their own route and change the viewing angle and position at will. As shown in Fig. 9.14, virtual models such as equipment parts, workbenches, and tool brackets are added to the virtual scene. The operator can use the interactive functions provided by the software development environment to drag the equipment parts in the scene for assembly operations.

The system is used for remote control of FMME equipment. It has a controllable virtual simulation three-machine (shearer, scraper conveyor, hydraulic support) model, as well as a three-dimensional display interface and a working condition data monitoring interface. Remote network communication is available. It realizes the remote control of the equipment in the FMME in a real sense. The control personnel can manually or automatically control the equipment at any remote end. The research of the system has laid a foundation for the underground "intelligent mining" and "semiunmanned working face."

9.4.4.2 MR operation training

From the perspectives of holographic simulation of CMEs mechanical structure and equipment working principles, as well as guidance of equipment assembly through MR experiment schemes, this system can fully

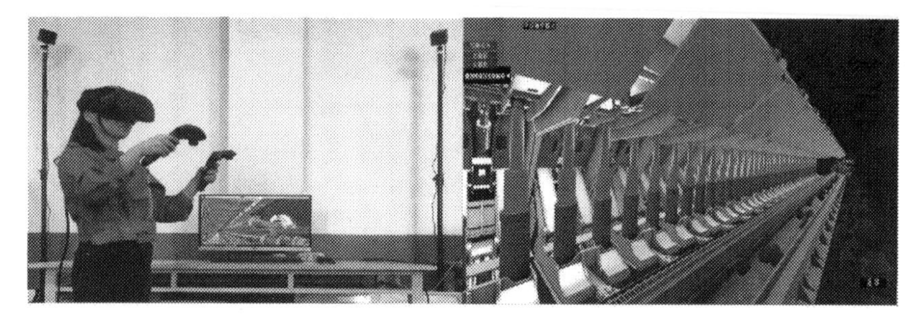

Figure 9.14 VR roaming scene diagram.

mobilize students' enthusiasm, initiative, and interest in participating in the experiment, and cultivate students' practical ability and innovation. At the same time, the experimental cost is reduced and the safety of the experimental process is enhanced. MR technology is a new digital holographic imaging technology based on further development of virtual reality technology and augmented reality technology. It can not only effectively combine computer logical thinking and human space perception ability, but also provide strong human–computer interaction. This technology is now used as an innovative training method for holographic training and teaching of CMEs. From the perspective of coal mine electromechanical equipment holographic training application, while studying MR holographic technology combined with CMEs virtual simulation experiments, the implementation plan and overall architecture of MR CMEs holographic training system are proposed, and the three-dimensional model of CMEs is built in 3Ds Max. The Unity3D-based MR development platform completes the basic functions such as environment perception and precise positioning, spatial topology, human–computer interaction, and remote multiperson data collaboration of virtual reality head-mounted display devices. From theoretical teaching, disassembly training, remote teacher online guidance and multiperson data collaborative teaching and other aspects have been studied to form a holographic training technology system for CMEs that integrates virtual and real. The effect diagram is shown in Fig. 9.15.

Combined with MR equipment holographic glass, the MR-based CMEs maintenance training system scheme and overall architecture are proposed. While receiving virtual information, trainers can operate on-site

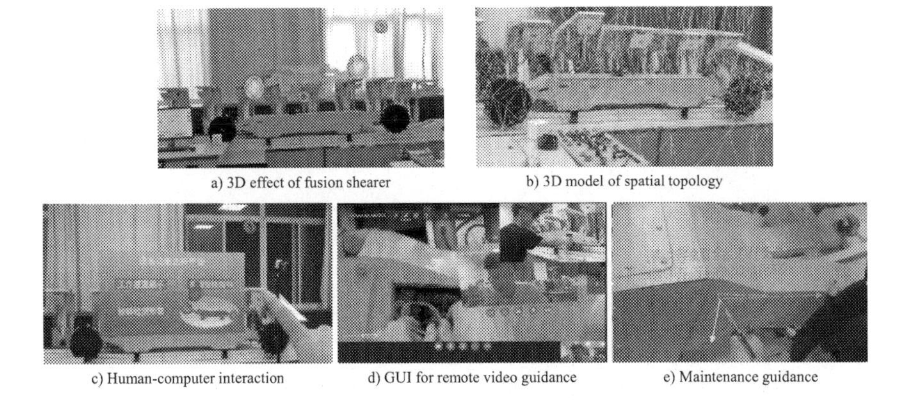

Figure 9.15 MR system operation effect diagram.

model equipment based on holographic annotations in real-time to form information interaction among humans, machines, and the environment. Starting from the basic requirements of professional training objectives for practical teaching, a system plan was developed for the unmanned, precise, and intelligent training and teaching of CMEs. In this program, the key technologies such as MR-based human—computer interaction, line-of-sight tracking, voice input, gesture recognition, equipment disassembly process guidance and multiperson collaborative remote expert collaboration are realized, and a complete set of virtual three-dimensional models and guiding animation of CMEs are built, and natural integration with physical reality space, forming a CMEs maintenance training system.

9.4.4.3 Robot remote control

This platform is used to simulate the tunneling operation of the roadheader in the actual coal production, as shown in Fig. 9.16. The main functions are the virtual simulation of the tunneling process of the roadheader and the intelligent control of the machine body and the cutting head. The virtual simulation part can simulate the environment and equipment of the underground tunneling face in the virtual environment; the visual cutting is mainly through multisensor information fusion such as condition sensors and vision sensors to intelligently control the cutting section of the Roadheader to perform intelligent cutting operations. The remote control console can also be used to synchronously control the virtual prototype and the real roadheader, and correct it through data feedback, so as to achieve the goal of minimum manpower in the tunneling face.

The system operation interface is shown in Fig. 9.16, which can realize the operation contents such as the start and stop of the roadheader, the

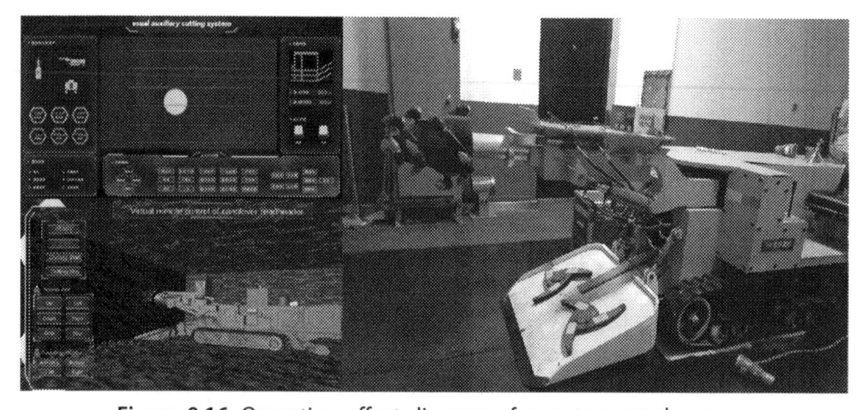

Figure 9.16 Operation effect diagram of remote control system.

cutting control, the movement of the cantilever up and down, left and right, and the perspective switching. The experimental project includes comprehensive mechanized tunneling technology, mechanical structure, electrical control, fault handling, etc. With the help of virtual simulation experiments, students can interactively learn coal mine machinery, electrical control principles, and roadheader technology and have a deep understanding of coal mining process, master the knowledge and skills of coal mine machinery design and application maintenance.

9.5 Summary

Starting from the DT framework and technical system of the mining face, this chapter analyzed the key enabling technologies for remote monitoring of the mining and production process in underground coal mine, constructed the DT application framework and technical system of mining face, and analyzed the key technologies to realize the DT.

The DT-driven production process management for key CME was proposed for the faulty maintenance guidance, using intelligent sensing technology, position and attitude control technology, data acquisition technology, and 5G-based transmission technology. Based on VR&MR, the typical application cases of DT-driven PHM for Shearer, DT-driven virtual-physical interaction, and DT-driven MR-aided Maintenance method have been realized. So that CMEs maintenance personnel who use the system can complete equipment maintenance operations according to the failure maintenance guidance solution given by the system and conduct equipment health management.

The new concept of fault prediction and health management for key CMEs provides an innovative data interaction method for the faulty maintenance guide field of the coal industry, promotes data interaction of the human-information-physical system. However, there are still some deficiencies in the authenticity of DT in the system. In order to realize the safe, efficient, green, and intelligent mining for FMME, it is necessary to further improve the data acquisition method, DT model, and life prediction model.

References

[1] Tao F, Liu W, Liu J, et al. Digital twin and its application exploration. Comput Integr Manuf Syst 2018;24(01):1−18.
[2] Ge S, Zhang F, Guan Z. Digital twin intelligent monitoring system for fully mechanized mining face: China, 201911388629. 4. December 30, 2019.
[3] Ge S, Zhang F. Digital twin intelligent monitoring system for fully mechanized excavation face: China, 20201000090126. January 6, 2020.

[4] Guo J, Hong H, Zhong K, et al. Production management and control method of aerospace manufacturing workshops based on digital twin. China Mech Eng 2020;31(07):808–14.
[5] Grieves M, Vickers J. Digital twin: mitigating unpredictable, undesirable emergent behavior in complex systems. Berlin, Germany: Springer-Verlag; 2017.
[6] Dou Y, Wang Q, Li J, et al. Data integration for aircraft digital assembly system. J Zhejiang Univ 2015;49(5):858–65.
[7] Qu T, Zhang K, Yan M, et al. Synchronized decision-making and control method for opti-state execution of dynamic production systems with internet of things. China Mech Eng 2018;54(16):24–33.
[8] Wei Y, Guo L, Chen L, et al. Research and implementation of digital twin workshop based on real-time data driven. Comput Integr Manuf Syst 2021-03-01;1–18.
[9] Zhang C, Liu J, Xiong H. Digital twin-based Smart Production management and control framework for the complex product assembly shop-floor. Int J Adv Manuf Technol 2018:1149–63.
[10] Yang W, Zhang X, Ma H, et al. Infrared LEDs-based pose estimation with underground camera model for boom-type roadheader in coal mining. IEEE Access 2019;7(7):33698–712.
[11] Pei A, Qi X, Liu Y, et al. Digital twin tree topology structure based on five-dimensional model. Comput Appl Res 2020;37(S1):240–3.
[12] Zhang X, Zhang Y, Wang M, et al. Mining equipment maintenance guidance system based on mixed reality. Ind Mine Autom 2019;045(06):27–31.
[13] Wang J, Huang L, Li S, et al. Development of intelligent technology and equipment in fully-mechanized coal mining face. J China Coal Soc 2014;42(8):1418–23.
[14] Zhang X, Zhang Y, Wang Y, et al. DT-driven Aided Guidance of Equipment Maintenance using MR. Comput Integ Manuf Syst 2021;27(08):2187–95.
[15] Zhou J, Zhou Y, Wang B, et al. Human–cyber–physical systems (HCPSs) in the context of new-generation intelligent manufacturing. Engineering 2019;5(4):624–36.
[16] Wang B, Yi B, Liu Z, et al. The development and research of intelligent manufacturing from the perspective of HCPS. Comput Integr Manuf Syst 2020:1–19.
[17] Wang B, Zang J, Qu X, et al. Research on a new generation of intelligent manufacturing based on human-information-physical system (HCPS). China Eng Sci 2018;20(04):29–34.
[18] Zhou J, Zhou Y, Wang B, et al. Human-information-physical system (HCPS) for a new generation of intelligent manufacturing. Engineering 2019;5(04):71–97.
[19] Tao F, Qi Q. Make more digital twins. Nature 2019;573:490–1.
[20] Tao F, Zhang M, et al. Digital twin driven prognostics and health management for complex equipment. CIRP Ann 2018;67:169–72. S0007850618300799-.
[21] Cao M, Hu G, Shen H, et al. Cyber-physical system architecture design for industrial equipment prognostics and health management system. Indus Technol Innov 2020;07(04):69–73.
[22] Tao F, Liu W, Zhang M, et al. Digital twin five-dimensional model and ten major applications. Comput Integr Manuf Syst 2019;25(01):1–18.
[23] Tao F, Zhang M, Cheng J, et al. Digital twin workshop——a new mode of future workshop operation. Comput Integr Manuf Syst 2017;23(01):1–9.
[24] Tao F, Cheng J, Qi Q, et al. DT-driven product design, manufacturing and service with big data. Int J Adv Manuf Technol 2018;94(9):3563–76.
[25] Nishimatsu Y. The mechanics of rock cutting. Int J Rock Mech Min Sci 1972;(9):261–70.
[26] Lang S. On the fault diagnosis and prediction of coal shearer. Contemp Chem Indus Res 2019;(14):16–7.

[27] Wang G, Ren huaiwei, Pang Y, et al. Research and engineering progress of intelligent coal-mine technical system in early stages. Coal Sci Technol 2020;48(7):1–27.
[28] Wang G, Liu F, Pang Y, et al. Coal-mine intellectualization: the core technology of high quality development. J China Coal Soc 2019;44(2):349–57.
[29] Zhang X, Zhang C, Yang W, et al. Research and development of visual auxiliary cutting system for cantilever roadheader. Coal Sci Technol 2018;46(12):21–6.
[30] Zhang X, Qiannan W, Wang M, et al. Research on remote virtual control system of cantilever roadheader. Coal Sci Technol 2020:1–9.
[31] Chakraborty S, Adhikari S. Machine learning based digital twin for dynamical systems with multiple time-scales. arXiv preprint arXiv:2005.05862 2020.
[32] Ong S. Beginning windows mixed reality programming10. Berkeley, CA: Apress; 2017.
[33] Zhang X., Zhang C., Wang M., et al. DT-driven virtual control technology of cantilever roadheader. Comput Integr Manuf Syst:1–18.

CHAPTER 10

Digital twin enhanced tribo-test service

Zhinan Zhang[1,2], Yufei Ma[2], Ke He[2], Ruiqi Hu[2] and Mingxuan Hu[2]
[1]State Key Laboratory of Mechanical System and Vibration, Shanghai Jiao Tong University, Shanghai, China; [2]School of Mechanical Engineering, Shanghai Jiao Tong University, Shanghai, China

10.1 Introduction

Tribo-test plays a crucial role in the evaluation of the material performance of tribo-pairs. Fundamental tribo-tests, for example, pin-on-disc, ball-on-disc, and four-ball, etc., are often used to perform the test of tribo-pairs. Traditionally, these tribo-tests are operated onsite and tightly depend on experienced human operators.

With the development of tribology, the concept and architecture of tribo-informatics were proposed by Zhang et al. [1]. Researchers can obtain tribology-related information from the database and use different models to implement behavior analysis, condition monitoring, and coefficient of friction prediction [2,3]. The relevant information includes data generated in tribo-tests, literature, theoretical models, and simulation data, etc. Tribo-informatics has laid the foundation for information monitoring and processing in tribo-tests to a certain extent, but it still needs the effort to investigate onsite tribo-tests further.

On the one hand, when some public health emergencies such as covid-19, they will substantially impact daily human life, including product and service systems. For tribo-tests, testers may not be able to conduct tribo-tests on the spot, thus resulting in the suspension of the testing processes and the results. On the other hand, some tribo-tests should be conducted under extreme working conditions, for example, space environments, extremely low temperature, high-temperature, high-humidity, corrosive surrounding. These working conditions are not suitable for the onsite participation of testers. In the above two cases, some auxiliary technologies are needed to complete the remote test process.

In recent years, with the development of the concept of everything as a service (XaaS), service has gradually permeated manufacturing and other fields. In terms of test services, some scientific research institutions or

Digital Twin Driven Service
ISBN 978-0-323-91300-3
https://doi.org/10.1016/B978-0-323-91300-3.00009-7

companies provide test services to clients. The current tribo-test service adopts the following mode: (1) The client submits an application for tribo-tests, which contains information such as test materials and test conditions; (2) The service provider conducts tribo-tests by developing new equipment or using the existing equipment and provides the client with the test report. In this mode, the service has the problem of insufficient information transparency. The client can only obtain a summary document without more information related to the test process. This results in clients being unable to judge the reliability of the test results when they have doubts about them. Therefore, it is necessary to develop new tribo-test service patterns that can expand the scope of application of tribo-informatics and the current tribo-test service. Remote tribo-test service, which takes the advantage of Internet of Things (IoT), Internet, and virtual reality (VR)/augmented reality (AR) can be a possible solution.

Digital twin is an emerging technology that can integrate and make use of the above tools. As a key technology to realize the mutual integration and coevolution of physical and virtual space, digital twin can reflect the attributes and real-time behavior of physical entities into virtual space, thereby realizing the monitoring of the status of physical entities [4]. On the other hand, digital twin can control remote physical entities by changing the status of their counterparts. Therefore, according to the functions of digital twin for real-time monitoring and physical control (see Fig. 10.1 [5]), it can be applied in tribo-test services to solve the problem of remote tribo-tests.

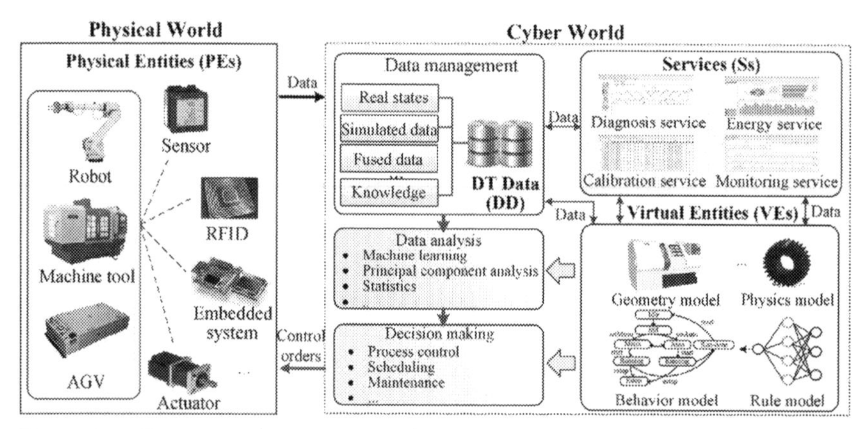

Figure 10.1 Two main characteristics of digital twin: process monitoring and operation control [5].

Currently, the digital twin concept has attracted much attention from both academia and industry because of its core function of realizing the convergence between physical and virtual space [6]. In recent studies, researchers have explored the functions and corresponding potential applications of the digital twin, including but not limited to process monitoring and operation control. For process monitoring, digital twin displays the operating status of the equipment in a visual interface in most cases, thereby realizing information interaction. As for operations control, digital twin serves as a controller to give commands to its corresponding physical entity, thus promoting the automation of processes. Table 10.1 lists some of the application cases related to digital twin services for process monitoring and operations control.

However, there is still a lack of research on applying the concept of a digital twin to empower tribo-test service. The digital twin concept was employed to enhance tribo-test service in this chapter, which can promote the process monitoring and operations control of tribo-tests.

Table 10.1 Application cases of digital twin for process monitoring and operations control.

Functions	Authors	Application cases
Process monitoring	Liu et al. [7]	Digital twin is applied in metal additive manufacturing systems in the form of cloud digital twin and edge digital twin, where cloud digital twin can manage data from edge digital twin to achieve process monitoring.
	Zhang et al. [8]	A digital twin and stacked auto encoder-based model is proposed to monitor the product quality.
Operations control	Brosinsky et al. [9]	Digital twin is applied to power system control centers to optimize performance and to rise OSA.
	Liu et al. [7]	Digital twin is applied in metal additive manufacturing systems in the form of cloud digital twin and edge digital twin, where cloud digital twin can send messages to edge digital twin to perform specific computing tasks to support its users.

The rest of this chapter is organized as follows. Section 10.2 presents related works about tribo-tests and digital twin-driven test service. Section 10.3 introduces the scope and characteristics of tribo-test service and reviews research progress on tribo-test service. Section 10.4 proposes the framework of the digital twin enhanced tribo-test service. Section 10.5 presents a case study of digital twin enhanced tribo-test service. Section 10.6 draws conclusions and outlines the development prospect of digital twin enhanced tribo-test service.

10.2 The related works

In this section, research progress on tribo-test is first reviewed. The result shows that deficiencies still exist in research on remote tribo-test services although some current research on tribology has begun to shift to the direction of improving research efficiency. Besides, little research has been conducted on digital twin-driven test service according to the review.

10.2.1 Research progress on tribo-test

The purpose of the tribo-test is to: (1) study the phenomenon and nature of friction; (2) evaluate the influence of various factors on friction energy; (3) optimize the structure of the tribo-pairs. In general, there are two types of tribo-tests; one is the standard test, the other is the test under simulated working conditions. Standard tests are usually carried out on general friction testing machines, including four-ball, ball-on-disc, pin-on-disc, etc. Tests under simulated working conditions are relatively complicated. Due to the wide range of friction materials, various forms of friction, complex friction phenomena, different experimental methods, and working conditions, the technology of development of tribo-test systems occupies an essential position in the research of tribological behavior of tribo-systems and tribo-pairs.

For the tribological behavior research methods, microscale tribo-tests have attracted a certain degree of attention. To measure the friction coefficient, using probes is widely accepted. Gee [10] has further improved this method by adding a lever counterweight in front of the probe, which reduces the system complexity and dramatically reduces the cost. To explore the tribological performance of the micromovement interface, molecular dynamics simulations were used to simulate the micromovement of different friction pairs [11]. The friction coefficient was calculated based on the simulation result. Finally, the accuracy of the simulation results is verified through experiments.

For the development of tribo-test instruments, Ford company [12,13] developed a tribological performance testing machine for the friction plate of automobile brake. However, its structure is complex, and the monitoring system is not perfect, which leads to low measurement accuracy. On this basis, He [12] improved the mechanical structure and overcame the shortcomings of large scale. Finally, the performance of the actuator friction plate was tested, and good results were achieved. However, due to its small volume, it is only suitable for the research of miniaturized brakes. To explore the tribological performance of the bushing in Variable Stator Vanes, Chen et al. [14] developed a new tribometer. It can operate under high temperatures, changing axial load and reciprocating rotation conditions.

Due to the diversity of friction materials and conditions, the tribo-test system has been developed in various fields. However, a particular test system requires a long development cycle, high manufacturing cost, and low versatility. In order to improve the genericity of the tribology test system, Wang [15] concentrated various friction properties on the same machine, which not only realized the reliable and accurate positioning of multisize workpiece installation but also tested through different friction pairs and transmission pairs and could be based on different test strips. It has the flexibility of modular upgrades. However, it is still not enough to study the change of friction form and lubrication condition, and it cannot be simulated and reproduced for more complex working conditions.

Nowadays, to improve the efficiency of tribological research, some researchers have not limited tribological research to physical field simulation, tests under simulated working conditions, and the development of friction testing machines. They have begun to explore cross-disciplines between tribology and other disciplines to improve the efficiency of tribological research. Taking the combination of tribology and information technology as an example, the concept and architecture of tribo-informatics have been proposed [1]. Researchers can obtain real-time data and information during the experiment and store it in the database, and such data can be processed further by the data science and machine learning approaches. In follow-up research, applications such as tribological and dynamics behavior analysis, lubrication and vibration condition monitoring, and friction and wear coefficient prediction can be realized by retrieving data from the database and calling algorithm models. Some researchers used a variety of machine learning algorithms to predict the friction coefficient and wear rate of materials based on the tribology-related data in the literature [2,3]. The effectiveness of the methods was verified compared with traditional methods.

Although researchers have begun to explore ways to improve the efficiency of tribological research, there is still no practical solution to the demand for remote tribo-test services. There is a great need for tribo-tests to be combined with other technologies, such as VR technology, computer-aided technology, digital twin technology to achieve multipurpose simulation, and even remote tests.

10.2.2 Digital twin-driven test service

The digital twin has proved to be applicable to multiple stages of the product life cycle, such as product design and manufacturing. A few researchers focus on digital twin-driven tests to make them more efficient. This kind of research can be reviewed from the following two aspects.

Some research focuses on using digital twin technology to synchronize the virtual test process with the real one. In this way, the purpose of process monitoring or even status prediction can be realized. To achieve this goal, the fidelity of the virtual equipment must be high enough to reflect the physical entity status accurately. In terms of test process monitoring or prediction, Qiao et al. [16] built a data-driven digital twin model for the machine tool. By collecting the vibration data of the tool in the test, the virtual tool model was updated in real time. The test verified the accuracy of the tool wear prediction algorithm based on the digital twin. Xie et al. [17] built a digital twin model for the hydraulic supports to monitor their attitudes. During the test, the attitudes of the virtual hydraulic supports will change with that of the physical supports. The error range was within 1%, which reflects the potential of the combination of digital twin and VR to achieve test process monitoring. Xu et al. [18] designed the electric cam structure and constructed its virtual simulation model based on the digital twin. Through real-time data interaction and virtual-real mapping, they realize the cam's real-time monitoring. To ensure the accuracy of the testing results, these studies have all attached much importance to improving the consistency of virtual entities and physical entities.

Also, some researchers combine digital twin technology and VR/AR to implement the virtual test process. This kind of research can provide support for users to control physical entities by driving their counterparts. Harvard et al. [19] proposed a real-time cosimulation system architecture based on digital twin and VR. Their research used digital twin to ensure the fidelity of the simulation system and VR to achieve visualization and human-machine interaction. Finally, to verify the feasibility of the architecture, they applied it to scenarios where an operator and a robotic arm work together. Pérez et al. [20] made use of VR to construct a visual interface for the digital twin of the manufacturing process. The system can provide a platform for visual testing before physical implementation. Similar to the previous case, the system was applied in the assembly process of human-robot collaboration, which verified its feasibility in applications such as operator training, production optimization. Sepasgozar [21] combined VR and AR technology with digital twin to build a virtual online teaching system of architecture, engineering, and construction. The system allows students to master the ability to operate excavators or plan the excavation process without onsite experience. At present, digital twin and VR/AR driven test services are still limited to manufacturing or construction scenarios, and there is a lack of exploration in the field of laboratory tests.

Currently, a digital twin is mainly used in product design and manufacturing processes. With the development of technologies such as IoT, the Internet, VR/AR, researchers have begun to study the application of digital twin in combination with different technologies in the testing process. At present, there is relatively little research on the application of digital twins in the testing process, especially in the field of tribo-tests. Therefore, there is still a need to combine different technologies with digital twins to achieve multipurpose simulation tests and remote tests.

10.3 Understanding of tribo-test service

Tribo-test service aims to provide specific forms of tribological motion testing for specific objects. In this section, the scope, characteristics, and research process of the tribo-test service are described. The content of this section mainly establishes the concept of basic tribo-test service for readers. It lays the foundation for the next section of the digital twin enhanced tribo-test service.

10.3.1 Scope of tribo-test service

The tribo-test investigates the characteristics and changes of a workpiece under actual working conditions and reveals the influence of various factors on friction and wear performance to reasonably determine the optimal design parameters per the service conditions. At present, the research of tribology depends on the test technology to a great extent. Therefore, the tribo-test technology is an essential part of tribology research, which covers physical field tests to laboratory simulation tests, and the development of friction testing machines. Tribo-test is still the primary tool.

Tribology testing service provides specific tribo-test technology to study specific tribological phenomena/behaviors. The essence of the tribo-test service is to provide a tribological test of samples under specific working conditions for the demander to analyze its tribological performance. Therefore, the scope of tribo-test service is generally determined by the working modes that the friction testing machine can provide. Taking the four-ball tester as an example, it can evaluate the performance of lubricating oil under point contact so that it can be analyzed according to the results of wear mark diameter and friction coefficient.

At present, the universal friction testing machine can provide the traditional contact motion forms of various workpieces, such as pin disk sliding and ring block rotation. It is mainly used to evaluate the performance of various materials or lubricants under different speed, load, and temperature conditions. It can also be used to study various wear mechanisms. Under specific conditions, many researchers have also designed many special friction testing machines, such as high-temperature rotating shaft friction testing machines. Thus, to some extent, the scope of tribology testing service is continuously developing and expanding.

Fig. 10.2 shows the contact and movement mode of the sample used in the universal friction testing machine. The movement mode of the sample includes rolling and sliding, which can be manifested as rotary motion and reciprocating motion; the contact forms include surface contact, line contact, and point contact. The form of the tribo-test should be determined according to the specific research objects. According to this form, researchers need to develop specific tribo-test technology to meet the tribo-test service.

There is a great need to design the sample and the friction testing machine for some specific conditions according to the determined

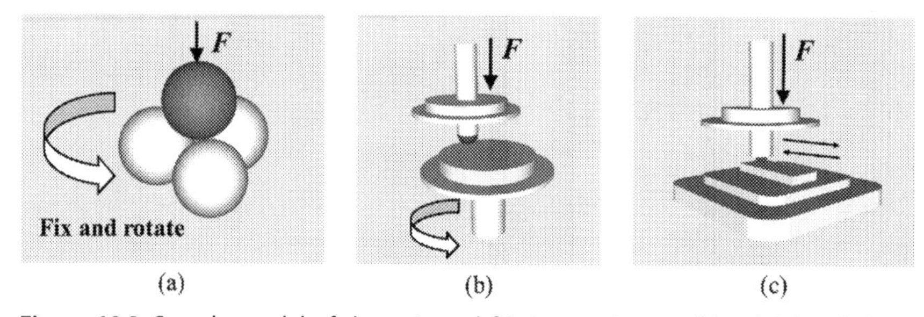

Figure 10.2 Sample model of the universal friction testing machine (A) four-ball (B) ball-on-disc (C) pin-on-disc.

parameters. The tests should be conducted under simulated working conditions. As the conditions are close to the actual working conditions, the reliability of the test results can be enhanced. Meanwhile, through the strict control of the working conditions, the system test data can be obtained in a relatively short time. Also, research on the influence of individual factors on the wear performance can be carried out on the special-purpose machine.

In general, the tribo-test service provides tribo-test technology for specific tribological forms, and these tribological forms constitute the scope of tribo-test services.

10.3.2 Characteristic of tribo-test service

Tribo-tests generally come from a specific tribological phenomenon or a tribological problem, which is of great interest to many researchers. Therefore, the tribology-test service is to provide specific tribo-test technology and physical tests for such researchers. However, at present, the domestic tribo-test service is not standardized and mature. Most tribo-test services only provide appropriate experimental equipment or experimental materials, so researchers need to design the corresponding tribo-test process. Fig. 10.3 shows the process of tribo-test service. First of all, a tribological problem or a tribological phenomenon has attracted the researchers' attention. Researchers have to determine the relative motion between friction pairs based on their analysis and design and process the corresponding test samples. Subsequently, in the case where the research institution they belong to cannot meet the test conditions, researchers need to contact a test service provider with corresponding qualifications and

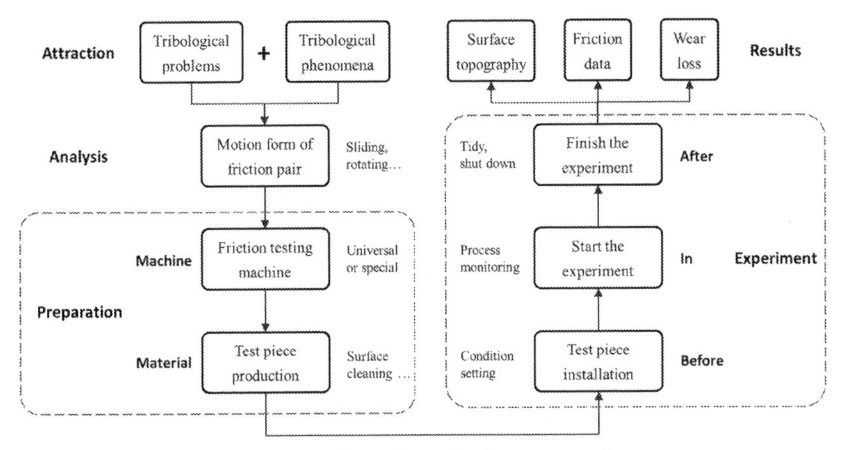

Figure 10.3 Procedure of tribo-test service.

determine the test machine and plan. The preparation before the test, the operation during the test, and the arrangement of the results after the test are all done by the service provider. Finally, the researchers can get a relatively detailed test report. The characteristics of tribo-test service are summarized as follows:

The first characteristic of tribo-test service is its wide variety. There are two main reasons for this feature: the variety of the workpiece motion forms and the complex operating conditions. At present, there are dozens of friction testing machines. Except for a few universal testing machines, most special testing machines are used for specific workpieces and specific working conditions. Therefore, the first point of the tribo-test service is to determine the motion form of friction pairs and to prepare friction testing machines and experimental materials.

The second characteristic of tribo-test service is the high degree of human participation. The main component of tribo-test service is a tribological test. In this stage, there is close interaction between the operator and equipment. Before starting the test, the operator should set the test conditions and fix the test piece. During the test, the operator also needs to monitor the test process to ensure that the test is carried out correctly. After the test, the operator should arrange the test bench and shut down the relevant experimental machines. Thus, deep interaction between operators and test machines is required in each stage of the test process.

The third characteristic of tribo-test service is time-consuming. The time consumption of the tribological test is mainly reflected in two aspects: The preparation period of test machines and test pieces; the test time required for each group. Considering the reliability and universality of tribological tests, many tests need to be repeated, which further strengthens the time-consuming characteristics of tribological tests. The preparation and experiment of Fig. 10.3 are the main time-consuming stage of tribo-test service.

10.4 The framework of digital twin enhanced tribo-test service

In this section, modeling of digital twin enhanced tribo-test service is explored, including modeling of the tribo-test machine and the service. Based on that, procedures of application of digital twin framework are described. The content of this section elaborates on the framework of digital twin enhanced tribo-test service based on the previous section. The framework aims to provide theoretical support for its practical application.

10.4.1 Modeling of digital twin enhanced tribo-test service

10.4.1.1 Modeling of virtual tribo-test machine

The virtual tribo-test machine contains models of four different dimensions: geometry model, physics model, behavior model, and rule model. The geometry model is constructed using three-dimensional modeling software, which reflects the structural characteristics and motion status of the tribo-test machine. The physics model is embodied in the form of mathematical formulas in the tribo-test services. It reflects the implicit mechanism of the test process and is used to mine the principles behind the test data. Physics models commonly used in tribo-test services include Coulomb's law, Archard model, etc. The behavior model establishes the interaction logic between the human and the virtual tribo-test environment during the test. When the operator performs operations, the virtual test environment will respond accordingly. For example, when the corresponding button is clicked, the tribo-test machine may be started or paused. The rule model reflects the phenomenon that occurs when the test conditions change. For instance, as the test temperature changes, the wear rate on the surface of the friction pairs will change to a certain extent. Such changes can be constructed in the form of mathematical formulas or neural network models. As the test progresses, the four models of the virtual tribo-test machine will be continuously updated to reflect the changes of the physical entity accurately.

10.4.1.2 Modeling of digital twin data

The digital twin data include data from the physical test environment, data from the virtual test space, and data from the tribo-test service. The physical test environment mainly generates some data obtained from the routine tribo-test process (see Fig. 10.4), such as the temperature data collected by

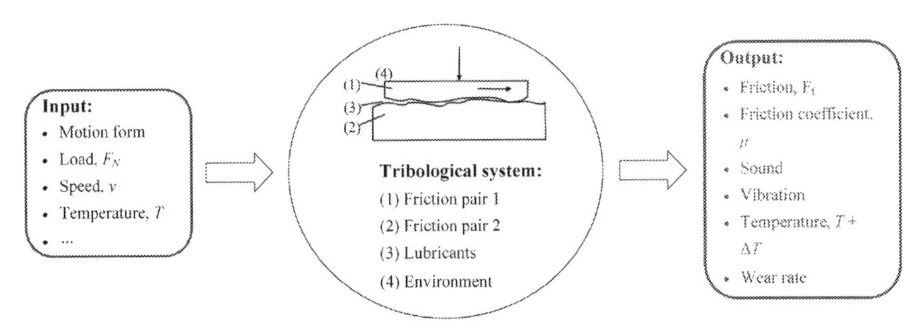

Figure 10.4 Digital twin data from the physical test environment.

thermal sensors, the friction force sensed by mechanical sensors, etc. This kind of data is what researchers want to monitor in real time. In addition, it also contains data that need to be further analyzed using the tribo-test service, such as sound and vibration data, etc.

As for data from the virtual test, before the physical test, the corresponding pretest simulation will be carried out on the virtual tribo-test machine and the virtual sample. The generated simulation results will be used for guidance of the physical test. Besides, during the test, as the physical test proceeds, the virtual test environment will change according to the data from the physical test environment. Similarly, data can be obtained from the virtual test environment to be compared with that from the physical test environment. In this way, the service provider can judge the consistency between these two environments, and the clients can make real-time decisions on the actual test.

Data from the tribo-test service is relatively complex as different functions contained in the service will produce different data. As mentioned above, data from the physical test environment will be processed by artificial intelligence algorithms, and the analysis results produced belong to this type of data. For another example, the function of viewing the structure and the status information of the tribo-test machine provided by the service are all stored in the service in the early stage by the service designer.

10.4.1.3 Modeling of tribo-test service

As shown in Fig. 10.5, tribo-test service is a data-and-model-driven function collection system. Generally, it provides clients with the functions of virtual model view, report view, process monitoring, and predictive tribological test simulation and provides the test operator with the functions of remote interface control. According to different users' needs, tribo-test service includes the following three modes.

Mode 1: The client may not have the ability to perform tribological tests but wants to obtain the test results for further analysis. In this case, the client can submit a test service application through the internet, indicating the friction pairs, friction form, and other information. The service provider will carry out the information review and follow-up tribo-test. Once the tribo-test machine is determined, the client can view the machine's structure and learn about the principle of how the machine works. During the test, the client can view the service provider's operations and real-time test data in the remote interface for relevant learning. Finally, the service will provide the client with a detailed report of this test. In this case, if the

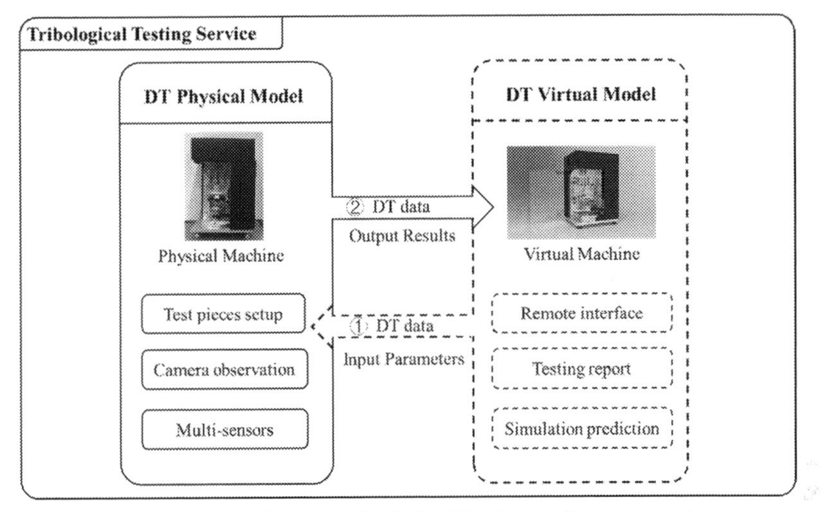

Figure 10.5 The framework of the DT-driven tribo-test service.

working condition is not suitable for onsite operation, the service provider will carry out the test remotely. This mode is designed mainly to provide monitoring functions for the clients.

Mode 2: The client has a certain tribo-test ability. In this case, the client can access the remote operating interface of the friction testing machine through the network to perform simple control. Similar to Mode 1, when the service application is submitted to the system, the service provider should provide the client with services such as the preparation of the specimen, the loading, and unloading of the specimen before and after the test. The client can set up and start the testing machine after the loading is completed. During the test, if the client finds out that the data deviates from what he/she expects, the test can be stopped in time to avoid the consumption of test time. After the test settings are adjusted, the test can be restarted. This mode provides users with certain control functions on the basis of Mode 1, and there is a certain degree of interactivity.

Mode 3: The client not only has a complete theoretical knowledge of tribology, but also sufficient tribo-test abilities to conduct a complete test process. In this case, the remote user can access the remote operating interface of the friction testing machine through the internet, submit personalized experiment requirements or remotely control the robot for experimentation. When the test requirements are given to the tribo-test service, the virtual friction testing machine can perform simulations in advance to predict the possible tribological and wear results according to different loads or other conditions; during the experiment, the physical friction testing machine will display the integrated data on the virtual model, and it has some degree of interactivity. When the test is over, the data obtained from the test can be passed to the remote user, or the test report generated by the software will be sent to the client directly.

10.4.2 Procedures of application of DT framework

As shown in Fig. 10.6, the process of tribo-test services based on a digital twin is divided into three parts.

1. Execution of control commands. Specified orders are entered in the control terminal and transmitted to the digital space. Functions of the virtual machines are started afterward. For example, dynamical analysis under the specified working condition is executed so that clients will have certain expectations about the possible results. If the analysis results do not meet the expectations, clients can check and change the test conditions in time before the test to avoid wasting test resources. After the experimental settings are clarified, the operation control of physical testing machines by virtual machines is started. Under the control of the digital system, the robot executes loading instructions. Then, physical machines begin to operate.

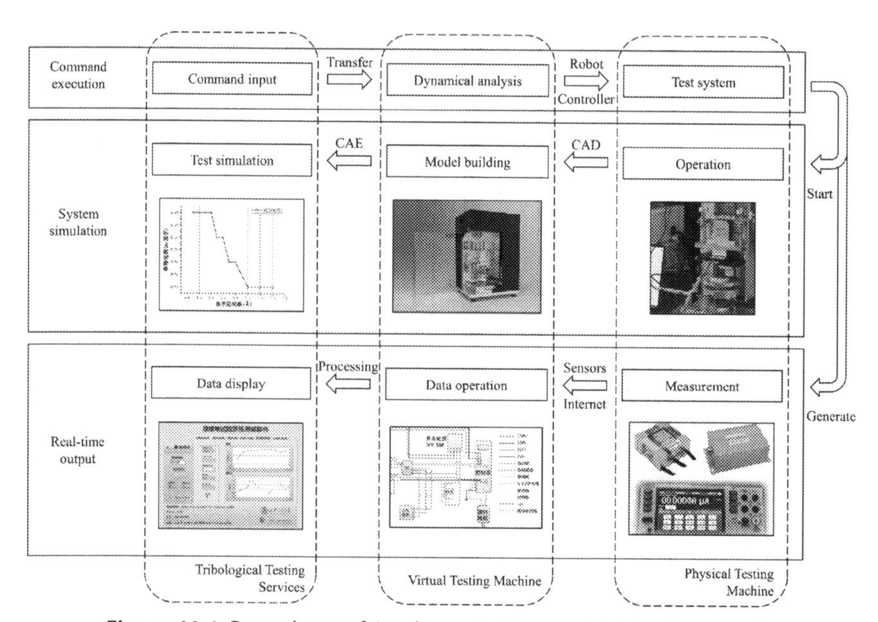

Figure 10.6 Procedures of DT-driven testing application framework.

2. Display of real-time output. When the test system operates, sensors automatically obtain the status information of the specimens and transmit the data to the digital space. Data from sensors are processed in the digital space, and the outcome is displayed on a real-time terminal. Meanwhile, the sensor data are also stored in the database for later viewing. Apart from data that can be directly analyzed by observing the trend, there are some data that need further analysis. This kind of data can be used for parameter prediction based on neural networks or other artificial intelligence methods. Finally, the prediction results will also be displayed on the interface in real time.
3. Test simulation. Before the test, the initial virtual test models are generated based on the system design. Virtual tests simulate through the digital models to give a prediction of test results. During the test, model simulation-related data (movement of the system structure, environmental temperature, etc.) is recorded and utilized by CAD/CAE software. In this process, the virtual test models change based on the data. Virtual tests proceed to show the system status.

10.5 Case study

Based on the framework of the digital twin enhanced tribo-test service, a case study on a triboelectric test system is discussed in this section, including its description and the development of the test service based on it. The specific functions of the test service are also described in this section.

10.5.1 Tribo-test machine

10.5.1.1 Basic description

To illustrate how digital twins enhance tribo-test service, the development of tribo-test machines and the test service based on the machine are given in this section. Considering the lack of standardized test support for triboelectrification research, a tribo-test machine for the investigation of the triboelectric performance of contacting surfaces was developed. The test system can be used to study the triboelectric properties of ball/pin-disk under rotating or reciprocating conditions and realize the real-time measurement of microcurrent, load, and friction. The test system was equipped with the microcurrent measuring device, and the LabVIEW measurement control software was developed. This test system can realize the smooth loading and accurate measurement under the condition of a small load and provide a reference for the standardized test of triboelectrification.

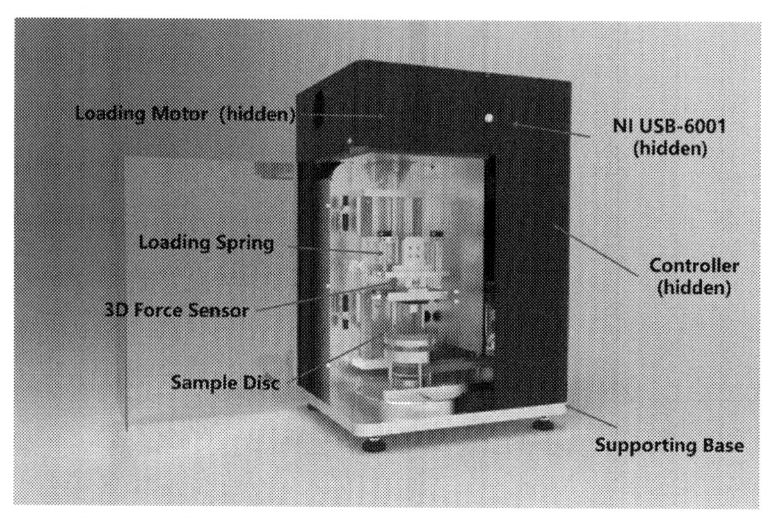

Figure 10.7 Main components of the triboelectric test system.

Fig. 10.7 shows the model of the triboelectric test system, which is mainly composed of the loading module, the rotating module, and the measurement control module. Table 10.2 shows the main technical parameters of the system. Besides, the test form is ball-disk rotation. The test system uses a USB + LabVIEW program for communication interaction and is equipped with a DMM6500 digital multimeter for measuring microcurrent signals.

10.5.1.2 Experimental procedure

Preparation

(1) Clean the aluminum disc with petroleum ether, spray alcohol, and let it dry naturally.

(2) Take the plastic measuring cup, pour the appropriate amount of isopropanol, and then put the copper ball sample into the measuring cup.

(3) Put the plastic measuring cup into the ultrasonic cleaning machine. After cleaning for 5 min, take out the copper sample and dry it naturally.

Table 10.2 Main technical parameters.

Parameter	Value
Maximum power	1000 W
Weight	<80 Kg
Microcurrent	0–1000 nA
Load	0–50 N
Rotation rate	0–600 r/min
Dimension	0.8 m × 0.6 m × 1.0 m

Experiment

(1) Clamp the ball sample and the disc sample in the specific position, wipe them with alcohol, and dry naturally.
(2) Connect the red and black probes of the DMM6500 to the linear slide and the rotary base, respectively, and test whether the circuit is connected.
(3) Open the test software, input the load, rotation rate, and time according to the experimental conditions, and then start the experiment.
(4) After the experiment, remove the pin sample and the disc sample, put them into the selfsealing bag, and copy the experimental data.
(5) Remove all equipment wiring and shut down the test system.

Operation specification

(1) Before the experiment, check whether the upper and lower samples are connected with other parts of the test system (connection indicates insulation failure).
(2) During the experiment, pay close attention to the fluctuation of microcurrent.
(3) Do not touch the upper and lower samples during the experiment.
(4) Generally, the experiment time is less than 1 min (wear debris interference).
(5) During the experiment, cut off the power at a time when the program is wrong.

10.5.2 Digital twin enhanced tribo-test service

10.5.2.1 Who needs to use

Clients: Individuals or organizations that need triboelectric experiments cannot be satisfied due to the lack of experimental facilities or conditions.

Providers: Qualified labs that are capable of performing triboelectric experiments.

10.5.2.2 Why need to use

1. Improve the level of information management in labs. For example, past records of experiments are stored in a structural system, and the repetition of experiments can be avoided.
2. Enrich auxiliary functions of the triboelectric test system. For example, researchers can execute ordered processing of experiments assisted by the software platform.
3. Make the best use of the triboelectric test resources to avoid the waste of idle resources and to provide convenience for researchers.

10.5.2.3 How to use

Information about orders and experiments is input by clients and saved in the database. The service provider reviews the order and gives corresponding feedback. Both of them can exchange information in the "remark" column. After the test requirements are clear, the provider performs experiments according to the information and upload the test data. Fig. 10.8 shows how clients and providers interact on the platform.

10.5.2.4 For clients

Clients have access to the status of their orders/tests and test data uploaded by providers. Clients can leave a message in the "remark" section of every order to communicate with providers about questions (such as unset materials).

To complete a delegation, the following steps should be followed:

1. Log in to the platform through a web browser. Anyone can get an account by registering.
2. Initiate a new order and set the deadline on a specified page.
3. Multiple experiments can be created under an order.
4. Set parameters (material, size, pairs, types of movement, operation time, frequency, etc.) separately for every experiment. Once a parameter is not determined, the system will prompt the client to fill in the blank to avoid misunderstanding.
5. Model view. After the parameter setting is completed, the system will automatically look for the tribo-test machine in the machine base and complete the machine matching. On this basis, clients can immediately view the corresponding three-dimensional model of the machine to achieve a basic understanding of the test equipment.

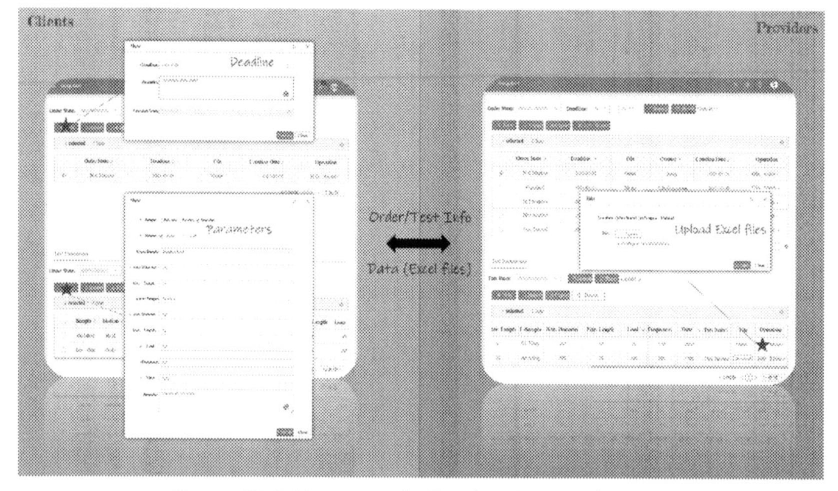

Figure 10.8 Use case of triboelectric test platform.

10.5.2.5 For providers

Providers should preset friction pairs and friction materials first. Different machines are designed for specified friction pairs (rotating between plates) and suitable for limited materials (iron, rubber, etc.), so providers are obliged to inform clients of optional friction pairs and available material types at the beginning of an order. Fig. 10.9 shows how to edit these settings on the online platform.

In the service request stage, the service provider should communicate the necessary information to the client as much as possible to avoid possible misunderstandings. For example, friction materials, types of movement should be confirmed over and over again.

After clarifying the needs, the service provider should perform tests in sequence, considering deadlines and create time for incomplete orders. After the test is completed, a person in charge should alter its status and upload the data, and report in time.

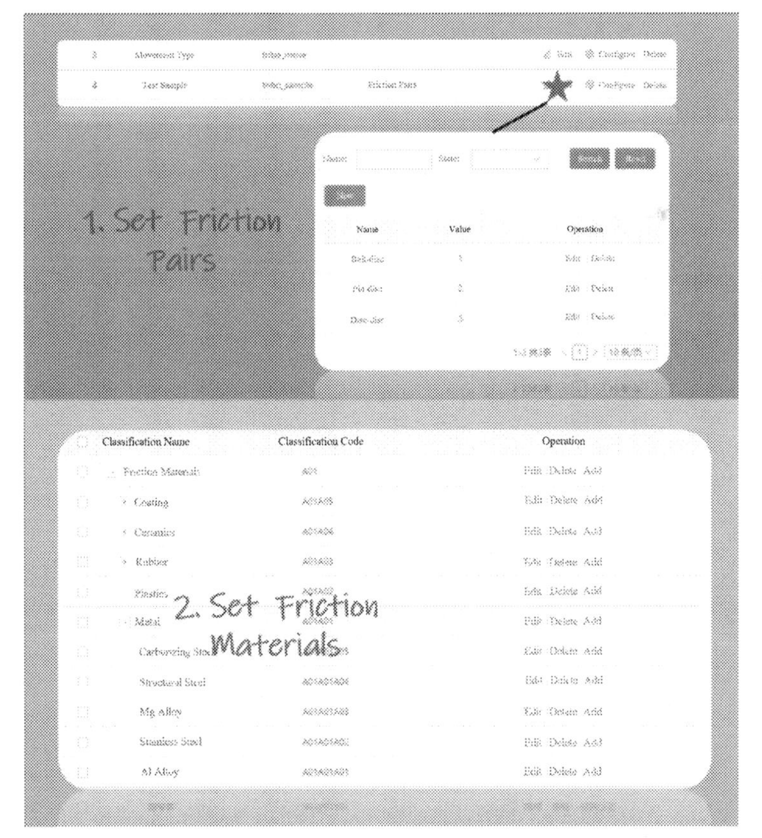

Figure 10.9 Settings of the triboelectric test platform.

10.5.2.6 Other functions

Tribo-test machine model view

This service provides the model view of the tribo-test machine, which is convenient for clients to view the testing machine structure and function (see Fig. 10.10). By moving the mouse to the corresponding module of the testing machine, its name and introduction will be displayed on the interface, which helps clients to learn about the equipment before the test. In addition, the material flow, information flow, and energy flow of the system will also be displayed in the form of a chart to provide support for clients to familiarize themselves with the equipment further. If the client does not want to know such detailed information, he/she can just skip or close the model view interface.

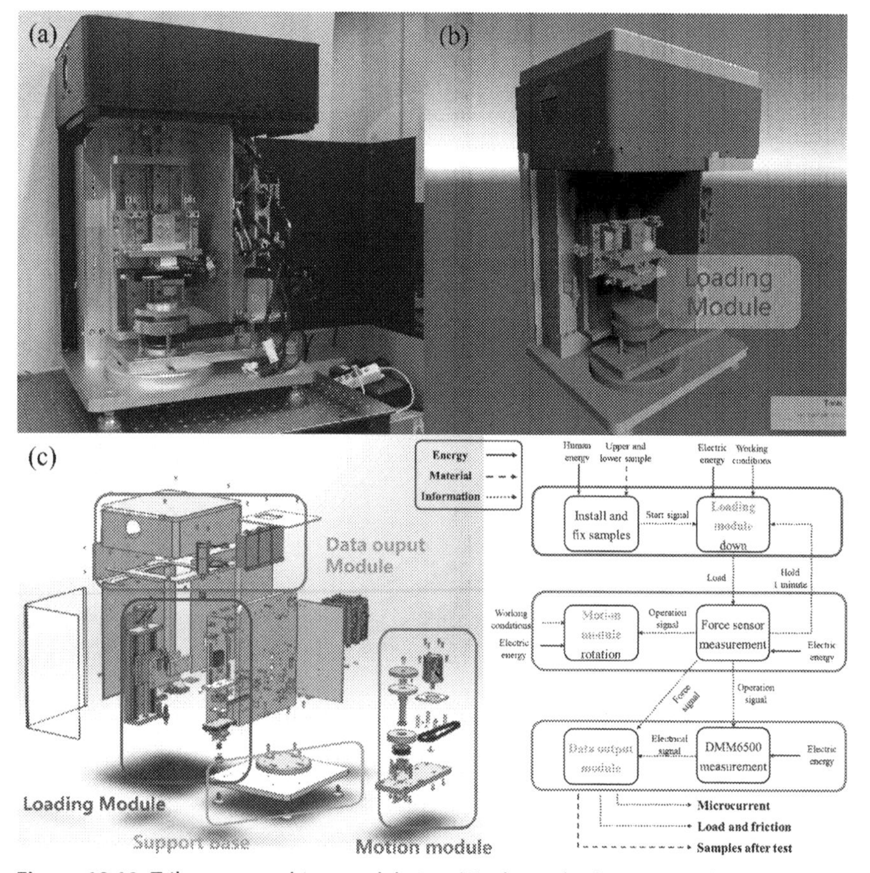

Figure 10.10 Tribo-test machine model view (A) physical tribo-test machine (B) virtual tribo-test machine (C) architecture of the tribo-test machine.

Also, the consistency of the physical machine and the virtual machine can be seen from Fig. 10.10. When the physical machine starts to run, the motion module will change its motion state in the physical space. Correspondingly, the virtual machine will present the same form of motion. This function provides visual support for the client to monitor the operation status of the machine. Different from directly using cameras to record, when the client wants to carefully check the status of each moving element, he/she can adjust the size or visualization settings of each part at will.

Real-time data monitoring

During the tribo-test, the sensor data that the client cares about will be presented on the interface in the form of a curve, including the temperature curve, load curve, and friction curve mentioned above. In addition, real-time predicted parameters obtained through analysis will also be presented on the interface, such as real-time wear rate. By monitoring and observing real-time data, when potential problems occur in the data, the client can adjust the test parameters or suspend the test in time to avoid damage to the tribo-test machine or waste of test resources. This function has dramatically improved the lack of transparency of information in the current tribo-test services.

Simulation analysis of tribological test

Before the test starts, the client can obtain the predicted result of the tribological test through the parameterized simulation provided by the service (see Fig. 10.11), which is convenient for them to understand the whole situation of the experiment and formulate a reasonable experimental strategy.

Taking the rotation tribo-test between a copper ball and a nylon disc as an example, the upper sphere of the constructed simulation model in Fig. 10.11B represents the copper ball. The lower disc represents the nylon disk. The cylindrical rod under the nylon disc represents the connection part between the lower sample and the testing machine. The distribution of three-dimensional spatial electric field lines and two-dimensional cross-sectional potential distribution map in Fig. 10.11C and D based on the simulation are consistent with those of the actual situation.

On the one hand, the simulation result can provide a simulation basis for the efficient use of the triboelectric test system and provide a reference for subsequent tests. On the other hand, test results from the physical machine can provide real-time data for the modification of the simulation model. Based on an improved model, users can further plan and perform a next step tribo-test.

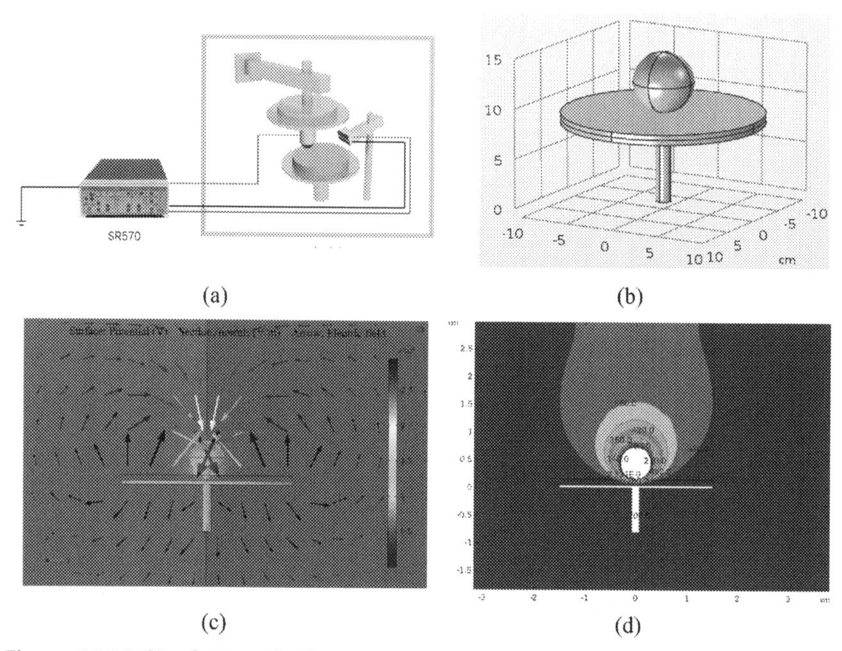

Figure 10.11 Simulation of tribo-test service (A) ball-on-disk device (B) simulation model (C) distribution of electric field lines in space (D) space potential distribution map.

10.6 Summary

Based on the current research status of digital twins and testing services, this chapter proposes a new digital twin enhanced tribo-test service framework, which maps the models and functions of the physical tribo-test machine to the virtual tribo-test service. The service designs the process corresponding to the testing service framework and establishes the tribo-test service method driven by the digital twin. According to the needs of tribo-test services, this research has developed a remote service webpage, which realizes remote control, report generation, simulation prediction, and other functions through the functional interface on the webpage. This research has guiding significance for the future development of digital twin-driven testing services. Future work will be devoted to extending the applications of the developed functions. For example, the tribo-test services can include educational function by combining digital twin with VR/AR. In this case, researchers can further learn how to conduct tribo-tests remotely even without a physical tribo-test machine.

Acknowledgments

This study is supported by the National Natural Science Foundation (Grant no. 12072191, 51875343, 51575340), State Key Laboratory of Mechanical System and Vibration Project (Grant no. MSVZD202108). Special thanks to Miss. Lu Xing, Mr. Hongxin Huang for discussion and insightful comments.

References

[1] Zhang Z, Yin N, Chen S, et al. Tribo-informatics: concept, architecture, and case study. Friction 2021;9(3):642–55.
[2] Hasan MS, Kordijazi A, Rohatgi PK, et al. Triboinformatic modeling of dry friction and wear of aluminum base alloys using machine learning algorithms. Tribol Int 2021;161:107065.
[3] Hasan MS, Kordijazi A, Rohatgi PK, et al. Triboinformatics approach for friction and wear prediction of Al-Graphite composites using machine learning methods. J Tribol 2021;144(1):011701.
[4] Tao F, Qi Q. Make more digital twins. Nature 2019;573:490–1.
[5] Tao F, Zhang M, Nee AYC. Digital twin driven smart manufacturing. London: Academic Press; 2019. p. 243–56.
[6] Li H, Wang H, Cheng Y, et al. Technology and application of data-driven intelligent services for complex products. China Mech Eng 2020;07:757–72.
[7] Liu C, Le Roux L, Körner C, et al. Digital Twin-enabled collaborative data management for metal additive manufacturing systems. J Manuf Syst 2020. https://doi.org/10.1016/j.jmsy.2020.05.010. In press.
[8] Zhang S, Kang C, Liu Z, et al. A product quality monitor model with the Digital Twin model and the Stacked Auto Encoder. IEEE Access 2020;8:113826–36.
[9] Brosinsky C, Westermann D, Krebs R. Recent and prospective developments in power system control centers: adapting the digital twin technology for application in power system control centers. In: Proceedings of 2018 IEEE international energy conference (ENERGYCON), Limassol; 2018. p. 1–6.
[10] Gee MG, Gee AD. A cost effective test system for micro-tribology experiments. Wear 2007;263(7):1484–91.
[11] Zhang Z, Pan S, Yin N, et al. Multiscale analysis of friction behavior at fretting interfaces. Friction 2021;9:119–31.
[12] He C. The design of the temperature test system of brake friction tester and data analysis. Lanzhou University of Technology; 2014.
[13] Daxin L, Lining W, Jun H. Research on standard of test machine for vehicle friction material. J Test Measure Technol 2016;030(003):215–20.
[14] Chen S, Yin N, Yu Q, et al. A novel tribometer for investigating bushing wear. Wear 2019;430–431:263–71.
[15] Wang J, Geng YX, Cheng ZT, et al. A multifunctional tribology and transmission machine and experimental investigation. J Mech Sci Technol 2016;35(11).
[16] Qiao Q, Wang J, Ye L, et al. Digital Twin for machining tool condition prediction. Proced CIRP 2019;81:1388–93.
[17] Xie J, Wang X, Yang Z, et al. Virtual monitoring method for hydraulic supports based on digital twin theory. Min Technol 2019;128(2):77–87.
[18] Xu J, Guo T. Application and research on digital twin in electronic cam servo motion control system. Int J Adv Manuf Technol 2021;112(3):1145–58.

[19] Havard V, Jeanne B, Lacomblez M, et al. Digital twin and virtual reality: a co-simulation environment for design and assessment of industrial workstations. Prod Manuf Res 2019;7(1):472–89.
[20] Pérez L, Rodríguez-Jiménez S, Rodríguez N, et al. Digital twin and virtual reality based methodology for multi-robot manufacturing cell commissioning. Appl Sci 2020;10(10):3633.
[21] Sepasgozar SME. Digital twin and Web-based virtual gaming technologies for online education: a case of construction management and engineering. Appl Sci 2020;10(13):4678.

Index

'*Note:* Page numbers followed by "t" indicate tables.'

E

F

G

H

I

R

S

Printed in the United States
by Baker & Taylor Publisher Services